EMERGING TECHNOLOGIES IN NON DESTRUCTIVE TESTING

PROCEEDINGS OF THE 3RD INTERNATIONAL CONFERENCE ON EMERGING TECHNOLOGIES IN NON DESTRUCTIVE TESTING, 26–28 MAY 2003, THESSALONIKI, GREECE

Emerging Technologies in Non Destructive Testing

Edited by

Danny Van Hemelrijck
Vrije Universiteit Brussel, Belgium

Athanasios Anastasopoulos
Envirocoustics S.A, Athens, Greece

Nikolaos E. Melanitis
Hellenic Naval Academy, Athens, Greece

A.A. BALKEMA PUBLISHERS   LISSE / ABINGDON / EXTON (PA) / TOKYO

Published by: A.A. Balkema, a member of Swets & Zeitlinger Publishers
       www.balkema.nl and www.szp.swets.nl

ISBN 90 5809 645 9 (volume)
ISBN 90 5809 646 7 (CD-Rom)

Printed in the Netherlands

Table of Contents

Acoustic emission

Infrared methods

Vibrational methods

Special techniques

Special applications

Preface

This is the third volume of a series of proceedings including papers presented in the respective International Conferences held in Greece since 1995. The initiative for the organization of the First Joint Belgian-Hellenic Conference on NDT, held in 1995 in Patras, was the long and fruitful co-operation between the Department of Mechanics of Structures and Materials of the Free University of Brussels and the department of Mechanical Engineering of the University of Patras. The enthusiasm of the two heads of department, Prof. W. P. De Wilde and Prof. S. A. Paipetis respectively, together with the efforts of the organizing committee developed this conference into a successful international event where representatives of Industry and Universities from over 15 countries attended presentations of forty scientific papers.

After the success of this first conference, the decision was made to organize the second international conference in 1999 in Athens. The Title "Emerging Technologies in NDT" was chosen better to represent the spirit of the conference and to highlight the state-of-the-art topics discussed. During the event, Innovation Relay Centre (IRC) Help-Forward organized a Technology Transfer and Business Partnership Event, where company/laboratory profiles offering or seeking technology know-how in the field of Emerging Technologies in NDT were distributed around the IRC network in order to encourage bilateral meetings in addition to the conference. Both events were well attended, leading to the establishment of an internationally recognized event, to be held on a four-yearly basis.

Within this framework and following the suggestions of the 1999 conference participants, the 3rd International Conference "Emerging Technologies in NDT", was held in May 26–28, 2003 in Sun Beach Hotel, one of the most picturesque resorts of Thessaloniki. The conference was jointly organized by the Free University of Brussels, The University of Patras, the Hellenic Naval Academy and the National Technical University of Athens.

The three-day event provided a forum for scientists, developers, engineers and practitioners as well as end users to review recent developments, identify outstanding needs and opportunities for further advances in the field of NDT, to exchange knowledge and experience with other NDT experts as well as broadening horizons for better utilization of the most modern technologies. Once again, the event provided the framework for establishment of links and partnerships throughout the Technology Transfer and Business Partnership meetings with the contribution of the Innovation Relay Centers Network and IRC Help-Forward. The conference also included an equipment exhibition, giving the opportunity for a close look at the latest products of the industry.

The new trend of mixed numerical–experimental analysis in NDT was the topic of the opening lecture of Prof. V. Papazoglou, highlighting both the multi-disciplinary nature of NDT and scientific developments as a result of recent advances in electronics and computer technologies. The state of the art in Emerging Technologies in NDT, presented in depth during the plenary lectures of Prof. B. Adams, Prof. G. Busse, Prof. B. Hosten, Prof. A. Mandelis, Prof. M. Ohtsu and Dr. T. Shiotani. The present volume, Proceedings of the 3rd Conference, includes more than 55 papers presented in the conference and is expected to form a valuable record of important contributions to the relevant literature.

The editorial board would like to express their great appreciation to the authors of this volume as well as the members of the scientific committee for their hard work, advice and constructive reviews in selecting the final papers.

The success of this conference was largely due to the enthusiasm of the Organizing Committee, who put in a great deal of effort, and managed to arrange such a successful conference by working far beyond his organizational remit. Special thanks are due to the conference secretary, Mrs. Myriam Bourlau, and to Mr. Frans Boulpaep for all his efforts and hard work. Special thanks are also due to the committed staff of IRC Help-Forward who organized the TT event, especially to Epaminondas Christophilopoulos, MSc.

We are particularly indebted to the conference chairmen, Prof. W. P. De Wilde, Prof. S. A. Paipetis and Prof. V. Papazoglou.

We look forward to seeing you all at the Emerging Technologies in NDT Conference of 2007.
The Editors
Prof. Danny Van Hemelrijck
Dr. Athanasios Anastasopoulos
Prof. Nikolaos E. Melanitis

Plenary lectures

Emerging Technologies in Non Destructive Testing, Van Hemelrijck, Anastasopoulos & Melanitis (eds)
© *2004 Swets & Zeitlinger, Lisse, ISBN 90 5809 645 9*

FE modeling of the S_0 Lamb mode diffraction by a notch in a viscoelastic material plate

B. Hosten, M. Castaings & C. Bacon
Laboratoire Mécanique Physique, UMR CNRS Université Bordeaux 1, Talence, France

ABSTRACT: This paper presents the use of a commercially available code FEMLAB for modeling the interaction of the symmetric Lamb mode S_0 with an opening notch in a plate made of a viscoelastic material. To speed up the computation, the code is used in a stationary mode for each frequency component of the temporal excitation. The displacement fields at the surface of the structure are given as a space/frequency representation that is transformed in a wave number/frequency domain with FFT procedures. This process permits to evaluate the proportion of the incident wave that is converted in antisymmetric A_0 mode, reflected and transmitted by the notch. The result is validated by a comparison with measurements made on a real system using an air coupled transducer for detecting Lamb waves in a Perspex plate. This receiver is scanned over the plate to analyze the propagation and the mode conversion due to the notch.

1 INTRODUCTION

Ultrasonic Lamb modes are usually used in NDE/NDT of structures. For metallic plate-like or pipe-like structures, no attenuation in the material has to be taken into account. However, in structures made of composite materials the attenuation due to their viscoelastic behavior strongly affects the propagation. Then it is important to include the viscoelastic properties in finite element models for predicting the interaction of Lamb waves with defects. This paper presents the use of a commercially available software FEMLAB for modeling the interaction of Lamb modes with notches in a plate made of a linear viscoelastic material. The viscoelastic behavior is taken into account by considering complex moduli as input data to the constitutive relations, which are solved in the frequency domain. A great advantage of this method is that the finite element code supplies stationary solutions for each frequency component of the temporal excitation. This approach considerably speeds up the computation by avoiding the numerical temporal differentiation and decreasing the number of variables used for each calculation step. Since a small number of frequencies are usually sufficient to achieve a correct representation, the computation time of the complex displacement and stress fields in the structure is reduced by a factor from 10 to 100. Numerical applications are made concerning the diffraction of the symmetric Lamb mode S_0 by an opening notch. These predictions are then compared to experimental data.

2 PROBLEM STATEMENT

The propagation of plane waves in a plate made of an orthotropic viscoelastic material of density ρ is considered. With the plane strain conditions ($\varepsilon_{xz} = \varepsilon_{yz} = \varepsilon_{zz} = 0$), the two-dimensional general equation is written as following in the frequency domain:

$$C_{11}^* \frac{\partial^2 \tilde{u}_x}{\partial x^2} + C_{66}^* \frac{\partial^2 \tilde{u}_x}{\partial y^2} + \left(C_{12}^* + C_{66}^*\right) \frac{\partial^2 \tilde{u}_y}{\partial x \partial y} = -\rho \omega^2 \tilde{u}_x$$

$$C_{22}^* \frac{\partial^2 \tilde{u}_y}{\partial y^2} + C_{66}^* \frac{\partial^2 \tilde{u}_y}{\partial x^2} + \left(C_{12}^* + C_{66}^*\right) \frac{\partial^2 \tilde{u}_x}{\partial x \partial y} = -\rho \omega^2 \tilde{u}_y \tag{1}$$

where $(\tilde{u}_x, \tilde{u}_y)$ is the Fourier transform of the displacement vector and x, y are the coordinates of the orthotropic directions. The above differential equations must be written in the following FEMLAB form [1]:

$$\nabla \cdot (c \nabla u) - a u = 0 \tag{2}$$

where c is a 2×2 matrix composed of four sub matrices such that:

$$c_{11} = \begin{pmatrix} C_{11}^* & 0 \\ 0 & C_{66}^* \end{pmatrix}, \quad c_{12} = \begin{pmatrix} 0 & C_{12}^* \\ C_{66}^* & 0 \end{pmatrix},$$

$$c_{21} = \begin{pmatrix} 0 & C_{66}^* \\ C_{12}^* & 0 \end{pmatrix}, \quad c_{22} = \begin{pmatrix} C_{66}^* & 0 \\ 0 & C_{22}^* \end{pmatrix} \tag{3}$$

Table 1. Complex viscoelasticity moduli measured for the Perspex plate.

Thickness (mm)	3.90 ± 0.01		
Density ρ(kg/m³)	1190 ± 10		
C_{11}^{*} (GPa)	8.5 ± 0.1	+I	0.3 ± 0.1
C_{22}^{*} (GPa)	8.3 ± 0.3	+I	0.4 ± 0.3
C_{66}^{*} (GPa)	2.3 ± 0.02	+I	0.06 ± 0.03
C_{12}^{*} (GPa)	4.4 ± 0.15	+I	0.2 ± 0.1

and a is a 2×2 matrix given by:

$$a = \begin{pmatrix} -\rho\omega^2 & 0 \\ 0 & -\rho\omega^2 \end{pmatrix} \qquad (4)$$

These equations are valid in any plane where x and y represent the orthotropic axes. In this paper, the plate is made of Perspex that is usually isotropic. In fact, the measurement of the viscoelastic properties C_{ij}^{*} made using a transmission method [3] reveals a slight anisotropy. The matrices in Equation (3) and the Perspex properties given in Table 1 are defined in the coordinate axis presented in Figure 1a where direction 1 is normal to the plate.

3 GEOMETRY OF THE PROBLEM

Figure 1a shows the geometry of the experimental set-up that is simulated using FEMLAB. A large rectangular (40×100 mm) piezoelectric transducer is placed in contact with the left edge of the plate for generating the symmetric mode S_0. The transducer is fed with a sinusoidal burst of 20 cycles at the center frequency of 150 kHz, multiplied by a Gaussian window. The frequency content of such temporal excitation is narrow and shown in Figure 1b. Around 12 values in the frequency domain are enough to represent the excitation.

4 MODELING THE PROPAGATION

The plate is considered as infinite in direction 3, which is normal to the plane of propagation formed by directions 1 and 2 (plane strain condition). It is modeled by a 3.9 mm high and 450 mm long strip region, these dimensions representing the plate thickness and length, respectively. The notch is located 150 mm away from the excited edge; its depth and width are 1.8 mm and 0.6 mm, respectively.

The spatial mesh of the plate is defined so that the smallest wavelength of any mode supposed to propagate in the frequency range of the excitation, i.e. ≈6 mm, is sampled by about 5 to 7 nodes. Triangular elements with linear, quadratic or third order behaviour are available in FEMLAB. Third-order elements of

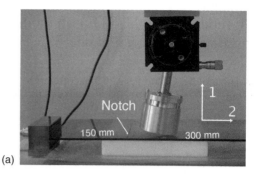

(a)

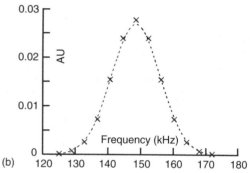

(b)

Figure 1. (a) Generation of the S_0 mode at the end of the plate by contact and reception with an air-coupled transducer. Angles of reception: 5° for S_0 mode, 20° for A_0 mode. (b) Frequency content of the temporal excitation.

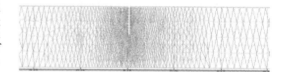

Figure 2. Elements repartition around the notch.

1 mm high and 0.5 mm wide were used to mesh parts of the plate where propagation is produced. However, to take into account the complex interaction between the Lamb mode and the notch, it was necessary to refine the mesh around this defect.

A procedure for optimizing the number of elements in the vicinity of this defect consists in computing the stress field for one frequency and in verifying whether the shear stress at the surfaces of the notch is null as it should be since the inner of the notch is vacuum. For the case considered in this paper, the total number of triangular elements was 4500 (Figure 2).

5 LAMB MODES IN A PERSPEX PLATE

The Rayleigh-Lamb equation (2) permits the phase velocities of Lamb modes in the Perspex plate to be

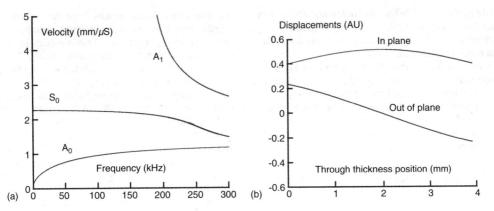

Figure 3. (a) Dispersion curves in the Perspex plate and (b) displacement field of the mode S_0 at 150 kHz.

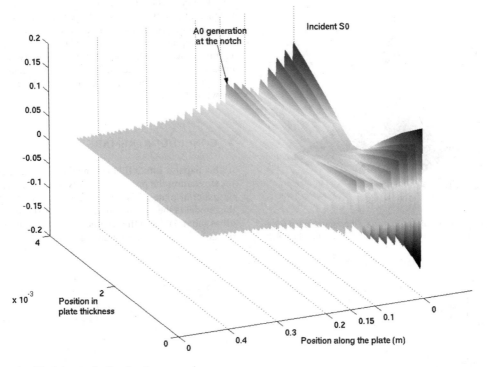

Figure 4. Displacement in direction 1.

computed. In the frequency region investigated in the experiment and the computation, i.e. below the frequency cut-off of the A_1 mode, only two fundamental modes A_0 and S_0 exist (Figure 3a).Since the symmetric S_0 mode is incident from one edge of the plate, its reflection by and transmission past the notch are expected, as well as a mode conversion from S_0 to the anti-symmetric A_0 mode, since the notch is not symmetric with respect to the middle plane of the plate. The through-thickness displacement field of the

incident S_0 mode, at the frequency 150 kHz, is computed (Figure 3b) and used as the initial displacement at the left edge of the plate in FEMLAB. Stress-free condition is imposed at any other boundary of the plate or of the notch.

The computation of the displacement and stress fields is done for all nodes of the plate, at the frequency 150 kHz.

Figure 4 presents the normal displacement at the top surface of the plate due to the propagation of the S_0

mode in the region before the notch and the conversion from mode S_0 to A_0 at the notch location. It is noticeable that the attenuation has a tremendous effect and cannot be ignored.

Once the computation for one frequency is done, a loop including all frequencies, 12 in this example, and the corresponding amplitudes of the frequency excitation (Figure 1b), gives the displacement fields for all nodes and for this set of frequencies. Then it is possible to reconstruct the temporal signal anywhere in the plate using a simple inverse Fourier transform.

6 MODE CONVERSION AT THE NOTCH

To analyze and compare the experimental and numerical results, a classical 2D FFT processing [4] is used

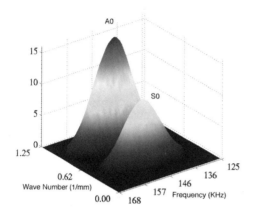

Figure 5. Wave number/frequency representation of the normal displacement at the surface of the plate after the notch.

that converts a set of signals from the time/position representation to the frequency/wavenumber representation. This permits the various modes to be well identified and their amplitudes to be quantified. A set of 100 numerical signals is predicted for nodes located past the notch at the surface of the plate, along a 100 mm long distance. Figure 5 shows that the amplitude of the A_0 mode after the notch is greater than that of S_0, although it was checked to be perfectly null before the notch. For each frequency, the maxima in amplitude of both modes are plotted in Figure 6a.

Two sets of experimental signals are obtained by scanning the air-coupled transducer (Figure 1a) above the surface of the plate.

It is scanned twice, with an angle of 5° for the reception of the S_0 mode and 20° for the reception of the A_0 mode. These angles are deduced from the Snell's law and the phase velocities of the modes in the plate and the phase velocity of the bulk mode in air [2].

The same procedure is applied on the two sets of signals captured in the same area after the notch. Figure 6b displays the relative proportion between the experimental amplitudes of A_0 and S_0. This result is in very good agreement with the numerical predictions shown in Figure 6a.

7 CONCLUDING REMARKS

In this paper a Finite Element method has been used for simulating the propagation of waves in viscoelastic material plates. The originalities and interests of this model is the use of complex viscoelasticity moduli as input data and the resolution of the equilibrium relations in the frequency domain. Other advantages

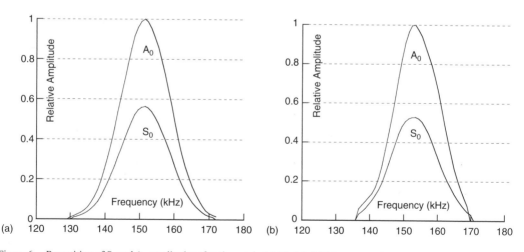

Figure 6. Repartition of S_0 and A_0 amplitudes after the notch: (a) Model, (b) Measurement.

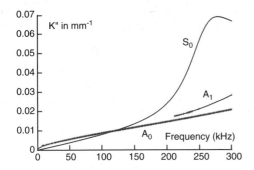

Figure 7. Imaginary part of the wave vector of three modes in the Perspex plate.

of this method are its simplicity and its fastness even if it is implemented on a small computer.

It was shown that for viscoelastic materials, it is mandatory to take into account the attenuation. Because of the presence of this attenuation, the definition of a crack transmission coefficient needs to be redefined. Figure 7 shows that the frequency content that was chosen in this paper corresponds to similar attenuations for modes A_0 and S_0. The study in another frequency region could give very different results due to strong variations of the modes attenuations.

An example was presented of interaction between the S_0 mode and an opening notch. More examples will be presented in further papers to appreciate the relation between the mode conversion and the geometry of the defect.

REFERENCES

1. FEMLAB – COMSOLAB, *User's Guide and Introduction.* Version 2.2 Page I-342.
2. B.A. Auld, *Acoustic fields and waves in solids.* Wiley-Interscience, Vol. 2, 1973.
3. M. Castaings and B. Hosten, *Air-coupled measurement of plane wave, ultrasonic plate transmission for characterising anisotropic, viscoelastic materials.* Ultrasonics 38, p. 781–786, 2000.
4. D. Alleyne and P. Cawley, *A 2-dimensional Fourier transform method for the measurement of propagating multimode signals.* J. Acoust. Soc. Am., 89 (3), p. 1159–1168, 1991.

Emerging Technologies in Non Destructive Testing, Van Hemelrijck, Anastasopoulos & Melanitis (eds)
© 2004 Swets & Zeitlinger, Lisse, ISBN 90 5809 645 9

Photo-carrier radiometry of semiconductors: a novel optoelectronic diffusion-wave technique for silicon process NDT

A. Mandelis

Center for Diffusion-Wave Technologies, Department of Mechanical and Industrial
Engineering, University of Toronto, Toronto, Ontario, Canada

ABSTRACT: Laser-induced infrared photo-carrier radiometry (PCR) is introduced as an emerging semiconductor NDT technology, both theoretically and experimentally through deep sub-surface scanning imaging and signal frequency dependencies from Si wafers. PCR completely obliterates the thermal infrared emission band (8–12 μm), unlike the known photothermal signal types, which invariably contain combinations of carrier-wave and thermal-wave infrared emissions due to the concurrent lattice absorption of the incident beam and non-radiative heating. The PCR theory is presented as infrared depth integrals of carrier-wave (CW) density profiles. Experimental aspects of this new methodology are given, including the determination of photo-carrier transport parameters (surface recombination velocities, carrier diffusion coefficients, recombination lifetimes and carrier mobilities) through modulation frequency scans. CW scanning imaging is also introduced. High-frequency, deep-defect PCR images thus obtained prove that very-near-surface (where optoelectronic device fabrication takes place) photo-carrier generation can be detrimentally affected not only by local electronic defects as is commonly assumed, but also by defects in remote wafer regions much deeper than the extent of the electronically active thin surface layer.

1 INTRODUCTION

In recent years the development of laser-induced infrared photothermal radiometry (PTR) of semiconductors in our laboratory as a quantitative methodology for the measurement of transport properties of semiconductors has led to several advances in the non-contact measurement of four transport parameters: bulk recombination lifetime, front and back surface recombination velocities and carrier diffusion coefficient in Si and GaAs. Reviews of the subject matter have been presented by Mandelis (1998) and Christofides et al. (2000). The major advantage of PTR over other photothermal techniques, such as photomodulated thermoreflectance (PMOR), has been found to be the higher sensitivity of PTR to the photo-excited free carrier-density-wave (the modulated-laser driven oscillating electronic diffusion wave (Mandelis (2001)) than PMOR (Wagner and Mandelis (1996); Salnick et al. (1997a)). This advantage exists due to domination of the free-carrier wave over the superposed thermal-wave (TW) contributions to the PTR signal. Even so, the ever-present thermal-wave contributions due to direct lattice absorption, followed by non-radiative energy conversion and blackbody (thermal infrared) emissions, have resulted in PTR signal interpretational and computational difficulties due to the large number of variables involved (Rodriguez et al. (2000)). Therefore, confidence in the measured values of the four electronic transport properties is always accompanied by the hurdle of having to assure uniqueness of the measured set of parameters in any given situation. Given the fundamental and practical importance of developing an all-optical, non-destructive and non-intrusive diagnostic methodology for monitoring the transport properties of semiconductors, and in view of the inability of photothermal semiconductor diagnostic methods (Christofides et al. (2000); Rosencwaig (1987)) to eliminate the thermal-wave contributions, we concluded that the search for a purely carrier-wave laser-based detection technique must move in the direction of isolating and filtering out the superposition of thermal-wave contributions to the infrared emission spectrum.

In a photo-excited semiconductor of bandgap energy E_G, an externally incident optical source such as a laser beam with super-bandgap energy photons $\Sigma\omega_{vis} > E_G$ will be absorbed and can generate free carriers which may subsequently follow several de-excitation pathways. Ultrafast decay to the respective bandedge (e.g. conduction band) through nonradiative transitions and emission of phonons, will raise the temperature of the semiconductor locally. The free carriers will further diffuse within their statistical lifetime and will recombine with carriers of the opposite sign across the bandgap or into impurity and/or defect states within the bandgap. The electron-hole recombination mechanism with or without phonon assistance will lead either to nonradiative energy conversion through phonon emissions (e.g. in indirect-gap semiconductors such as Si) which will further raise the temperature, or to radiative decay which will produce photons of near- or sub-bandgap energy. In actual semiconductor materials, there may be a distribution of impurity and defect states into which de-excitation may occur. Therefore, it is more relevant to consider the full spectral range of IR emissions from a photo-excited semiconductor crystal: $\Sigma\omega_{IR} = \Sigma\omega(\lambda_D)$. If the exciting super-bandgap radiation is intensity-modulated at frequency $f = \omega/2\pi$, then the photo-generated free carrier density constitutes a spatially damped carrier-density wave (CW) (or carrier-diffusion wave (Mandelis (2001))), which oscillates diffusively away from the generating source under its concentration gradient and recombines with a phase lag dependency on a delay time equal to its statistical lifetime, τ, a structure- and process-sensitive property (van Roosbroeck and Shockley (1954)).

Under conditions that apply to a number of semiconductors (Mandelis et al. (2003)), electronic transitions in these materials occur essentially adiabatically, with minimum thermal energy exchange interactions across well-defined electronic state densities, leading to validation of Kirchhoff's Law of Detailed Balance (Kirchhoff (1898)) through complete thermal decoupling of the CW oscillator ensemble. A by-product of adiabaticity is that the IR spectra of thermal and carrier recombination emissions are independent of each other, a feature which is central to the realization of Photo-Carrier Radiometry (PCR).

2 PRINCIPLES OF PCR

Figure 1 shows an elementary slice of thickness dz centered at depth z in a semiconductor slab. The crystal is supported by a backing, but is not necessarily in contact with the backing. A modulated laser beam at angular frequency $\omega = 2\pi f$ and wavelength λ_{vis} impinges on the front surface of the semiconductor. The super-bandgap radiation is absorbed within a (short) distance from the surface, typically, a few μm, given by $[\alpha(\lambda_{vis})]^{-1}$ where $\alpha(\lambda_{vis})$ is the visible-range absorption coefficient of the pump radiation. The ensuing de-excitation processes generally involve radiative and nonradiative energy release components, resulting in the generation of an IR photon field in the

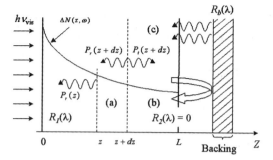

Figure 1. Cross-sectional view of contributions to front-surface radiative emissions of IR photons from a) a semiconductor strip of thickness dz at depth z; b) re-entrant photons from the back surface due to reflection from a backing support material; c) emissive IR photons from the backing at thermodynamic temperature T_b. The carrier-wave depth profile $\Delta N(z, \omega)$ results in a depth dependent IR absorption/emission coefficient due to free-carrier absorption of the infrared photon fields, both ac and dc.

semiconductor involving a relatively broad spectral bandwidth. At thermal and electronic equilibrium, assuming a one-dimensional geometry as a result of a large laser beam spotsize and/or thin sample, the emitted IR photons have equal probability of being directed toward the front or the back surface of the material. A detailed account of all IR emission, absorption, and reflection processes (Mandelis et al. (2003)) yields the expression for the total IR emissive power at the fundamental frequency across the front surface of the material in the presence of a backing support which acts both as reflector of semiconductor-generated IR radiation with spectrum centered at λ, and as emitter of backing-generated IR radiation centered at wavelength λ_b

$$P_T \approx \int_{\lambda_2}^{\lambda_1} d\lambda [1 - R_1(\lambda)] \left\{ \left(1 + R_b(\lambda)[1 + R_1(\lambda)]\right) \varepsilon_o(\lambda) \right.$$
$$\int_0^L \Delta W_P(z, \omega; \lambda) dz + \left[\left(1 + R_b(\lambda)[1 + R_1(\lambda)]\right) W_o(T_o; \lambda) \right.$$
$$\left. - W_P(T_b, \lambda)e(T_b, \lambda)[1 - R_1(\lambda)] \right] \int_0^L \varepsilon_{fc}(z, \omega, \lambda) dz \right\} \quad (1)$$

where R_1 is the front surface reflectivity, R_b is the backing support material reflectivity, $\varepsilon_o(\lambda)$ is the background IR emission coefficient of the material, $\varepsilon_{fc}(z, \omega; \lambda)$ is the IR emission coefficient due to the free photoexcited carrier wave, $e(T_b, \lambda)$ is the spectral emissivity of the backing material, $\Delta W_P(z, \omega, \lambda) dz$ is the harmonic IR emissive power due to the harmonically varying temperature of the sample, $W_o(T_o; \lambda)$ is the unmodulated emissive spectral power per unit wavelength due to both Planck-mediated $[W_{Po}(T_o, \lambda)]$ and direct radiative $[\eta_R W_{eR}(\lambda)]$ emissions, $W_P(T_b, \lambda)$ is the spectral emissive power per unit wavelength of the backing surface at temperature T_b, and $[\lambda_1, \lambda_2]$ is the spectral bandwidth of the detector. $W_{eR}(\lambda)$ is the spectral power per unit wavelength, the product of the recombination transition rate from band to band, or from bandedge to defect or impurity state, as the case may be, multiplied by the energy difference between initial and final states. η_R is the quantum yield for IR radiative emission upon carrier recombination into one of these states. During our experimental attempts to separate out carrier-wave and thermal-wave contributions which are always strongly mixed as in Eq. (1), we found that they can be separated out effectively only through spectral filtering and bandwidth matching at the IR detector, thus introducing the PCR technique.

Instrumental filtering of all thermal infrared emission contributions and bandwidth matching to the IR photodetector allows for all Planck-mediated terms to be eliminated from Eq. (1) yielding

$$P(\omega) \approx \int_{\lambda_2}^{\lambda_1} d\lambda [1 - R_1(\lambda)](1 + R_b(\lambda)) \eta_R W_{eR}(\lambda)$$
$$\int_0^L \varepsilon_{fc}(z, \omega; \lambda) dz \qquad (2)$$

The absorption coefficient (and, equivalently, assuming Kirchhoff's Law is valid, the emission coefficient) depends on the free-carrier density as (Smith (1978))

$$\varepsilon_{fc}(z, \omega; \lambda) = \alpha_{IRfc}(z, \omega; \lambda) = \frac{q\lambda^2}{4\pi^2 \varepsilon_{oD} c^3 n m^{*2} \mu}$$
$$\Delta N(z, \omega; \lambda) \qquad (3)$$

for relatively low CW densities. Here q is the elementary charge, ε_{oD} is the dielectric constant, c is the speed of light in the medium, n is the refractive index, m^* is the effective mass of the carrier (electron or hole) and μ is the mobility. This allows the PCR signal to be expressed in the form

$$P(\omega) \approx F(\lambda_1, \lambda_2) \int_0^L \Delta N(z, \omega) dz \qquad (4)$$

with

$$F(\lambda_1, \lambda_2) = \int_{\lambda_2}^{\lambda_1} [1 - R_1(\lambda)](1 + R_b(\lambda)[1 + R_1(\lambda)])$$
$$\eta_R W_{eR}(\lambda) C(\lambda) d\lambda \qquad (5)$$

The PCR signal is the integration of Eq. (4) over the image of the detector on the sample and thus is directly proportional to the depth integral of the carrier density in the sample. Consequently, the relative lateral concentration of any defects that affect the carrier density, either by enhancing recombination or altering diffusion coefficients, can be determined by scanning the surface of the wafer with the PCR probe.

In addition, frequency scan techniques can be used with the appropriate carrier diffusion model to obtain quantitative values for the four transport parameters (Rodriguez et al. (2000)). This quantitative technique can be combined with surface scans to provide quantitative imaging of the semiconductor sample.

3 PCR IMAGING NDT OF ELECTRONIC DEFECTS IN SI WAFERS

3.1 Instrumentation and signal characteristics

The experimental implementation of laser infrared photo-carrier radiometry is similar to the typical PTR set-up for semiconductors (Mandelis (1998); Rodriguez et al. (2000)), with the crucial difference being that the spectral window of the IR detector and optical filter, and the modulation frequency response of the preamplifier stage, must be tailored through spectral bandwidth matching to a combination of carrier recombination emissions and effective spectral filtering of the Planck-mediated thermal infrared emission band. Conventional PTR utilizes photoconductive liquid-nitrogen-cooled HgCdTe (MCT) detectors with spectral bandwidth in the 2–12 μm range. This includes the thermal infrared range, 7–12 μm, and only part of the electronic emission spectrum at shorter wavelengths. From experiments with several IR detectors and bandpass optical filters we concluded that emissive infrared radiation from electronic CW recombination in Si is centered mainly in the spectral region below 3 μm. Among near-IR photodetectors, the combination of variable-gain InGaAs detectors with integrated amplifiers, a near-infrared cut-on filter and a spectral response in the <1800-nm range, was found to be most suitable, exhibiting 100% efficient filtering of the thermal infrared emission spectrum from Si as well as maximum signal-to-noise ratio over InGaAs detectors with separate amplifiers and InAs detectors. PCR was implemented using an optimally spectrally matched room-temperature InGaAs photodetector (Thorlabs Model PDA400 "Switchable gain InGaAs photodetector", 10 MHz bandwidth, 1 mm diameter with NEP = 3×10^{-12} W//(Hz)) for our measurements. The incident Ar-ion laser beam size was 1.06 mm and the power was 20–24 mW. The detector was proven extremely effective in cutting off all thermal infrared radiation: Preliminary measurements using non-electronic materials (metals, thin foils and rubber) showed no responses whatsoever. Comparison with conventional PTR results was made by replacing the InGaAs detector with a Judson Technologies liquid-nitrogen-cooled Model J15D12 MCT detector covering the 2–12 μm range with peak detectivity 5×10^{10} cmHz$^{1/2}$W^{-1}.

Figure 2 shows two frequency responses from a test AlGaAs quantum well array on GaAs substrate

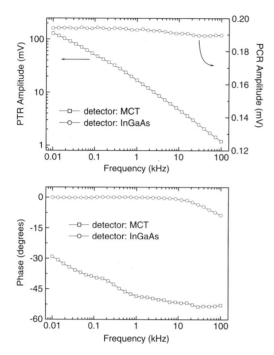

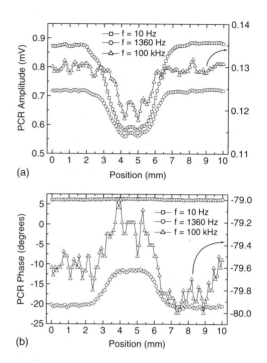

Figure 2. Comparison of normalized PTR (MCT detector) and PCR (InGaAs detector) signals from an AlGaAs quantum well array on a GaAs wafer. Incident laser power: 25 mW.

Figure 3. Line scans over an p-Si wafer region with back-surface mechanical damage. (a) PCR amplitude; (b) PCR phase. The wafer is resting on a mirror support. Laser power: 24 m.

using both the MCT and the InGaAs detectors. The MCT response is characteristic of thermal-wave domination of the PTR signal throughout the entire modulation frequency range of the lock-in amplifier. On the other hand, the PCR signal from the InGaAs detector/preamplifier exhibits very flat amplitude, characteristic of purely carrier-wave response and zero phase lag up to 10 kHz, as expected from the oscillation of free carriers in-phase with the optical flux which excites them (modulated pump laser). The apparent high-frequency phase lag is associated with electronic processes in the sample. The PTR signals were normalized for the instrumental transfer function with the thermal-wave response from a Zr alloy reference, whereas the PCR signals were normalized with the response of the InGaAs detector to a small fraction of the exciting modulated laser source radiation at 514 nm. To verify the effect of backing on the PCR signal obtained from a Si wafer, measurements were performed using a mirror and black rubber as backings, in the geometry of Fig. 1. From Eq. (2) and known values of the backing emissivity (Kreit and Black (1980)) it is expected that the ratio of PCR signals with mirror and black rubber backings should be approx. $[2 + R_l(\lambda)]/[1 + R_b(\lambda)[1 + R_l(\lambda)]] \approx 1.94$.

The measured ratio from the low-frequency end frequency scans was found to be ~1.8.

3.2 PCR imaging of deep sub-surface electronic defects

Figure 3 shows line scans with the excitation laser beam scanning the front (polished) surface of a 20 Ωcm p-type Si wafer and the IR detector on the same side. Based on the backing results, for maximum signal strength the sample was resting on a mirror. At all three selected modulation frequencies, the PCR amplitude decreases when the laser beam scans over the defect region, consistent with the expected CW density decrease as the back-surface defect efficiently traps carriers and removes them from further diffusion and potential radiative recombination. The PCR phase scan remains essentially constant at 10 Hz, Fig. 3b, as the diffusion-wave centroid is solely determined by the ac carrier-wave diffusion length (van Roosbroeack and Shockley (1954); Mandelis et al. (2003))

$$L_{ac}(\omega) = \sqrt{\frac{D * \tau}{1 + i\omega\tau}} \qquad (6)$$

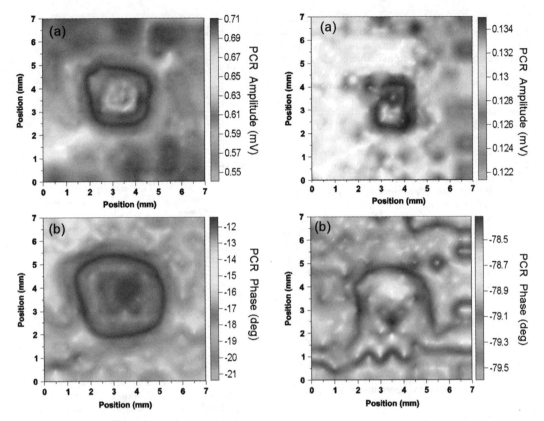

Figure 4. Scanning imaging of back-surface defect in the p-Si wafer using front-surface inspection. Laser beam radius: 518 μm. Frequency: 1360 Hz. (a) PCR amplitude; (b) PCR ph.

Figure 5. Scanning imaging of back-surface defect in the p-Si wafer using front-surface inspection. Laser beam radius: 518 μm. Frequency: 100 kHz. (a) PCR amplitude; (b) PCR phase.

where τ is the lifetime and D^* is the ambipolar carrier diffusion coefficient. This particular wafer was measured to have $\tau \cong 1$ ms and $D^* \cong 12$ cm^2/s, which yields an $|L_{ac}(10\,\mathrm{Hz})| \cong 1.1$ mm. Therefore, the CW centroid lies well beyond the thickness of the wafer (~ 630 μm) and no phase shift can be observed. At the intermediate frequency of 1360 Hz, $|L_{ac}| \cong 373$ μm, well within the bulk of the wafer. In this case, a phase lead appears within the defective region. This occurs because the CW spatial distribution across the body of the wafer in the defective region is weighed more heavily toward the front surface on account of the heavy depletion occurring at, and near, the back surface. As a result, the CW centroid is shifted toward the front surface, manifested by a phase lead. Finally, at 100 kHz, $|L_{ac}| \cong 44$ μm.

Nevertheless, Fig. 3a shows that there is still PCR amplitude contrast at that frequency, accompanied by a small phase lead, Fig. 3b.

To maximize PCR imaging contrast, differences in amplitudes and phases as a function of frequency

were obtained outside and inside the defective region. It is with the help of this type of analysis that the 1360 Hz frequency was chosen as one with the highest contrast in phase (but not in amplitude). Figure 4 shows images of the back-surface defect obtained through front-surface inspection at the optimum contrast frequency of 1360 Hz. Figure 5 shows the same scan at 100 kHz. At this frequency the PCR image clearly shows the highest spatial resolution of the back-surface defect possible. The PCR phase, Fig. 5b, shows details of the central defect as well as the radially diverging defect structures at the base of the central defect, like a "zoomed in" version of the 1360 Hz image, Fig. 4b. Both PCR images clearly reveal internal sub-structure of the central defect, which was invisible at 1360 Hz. In a manner reminiscent of conventional propagating wavefields, image resolution increases with decreasing carrier wavelength, $|L_{ac}|$.

Under front-surface inspection and precise depth profilometric control by virtue of the PCR modulation-frequency-adjustable carrier-wave diffusion length,

Eq. (6), Figs. 4 and 5 show for the first time that with today's high-quality, long-lifetime industrial Si wafers, one can observe full images of sharp carrier-wave density contrast due to underlying defects very deep inside the bulk of a Si wafer. Specifically, high frequency PCR imaging reveals so far unknown very long-range effects of carrier interactions with deep sub-surface defect structures and the detrimental ability of such structures to decrease the overall free photoexcited-carrier density far away from the defect sites at or near the front surface where device fabrication takes place. This phenomenon may be important toward device fabrication improvement through careful selection of substrate wafers with regard to deep bulk growth and manufacturing defects which were heretofore not associated with device performance. Further PCR imaging experiments with shorter lifetime Si wafers have shown that it may be beneficial to use lower quality starting substrates in order to avoid the full effects of deep sub-surface defects on the electronic quality of the upper (device-level) surface.

4   QUANTITATIVE PCR MEASUREMENTS OF ELECTRONIC TRANSPORT PROPERTIES

The PCR image contrast of Figs. 4 and 5 can, in principle, be quantified by use of the CW term in Eq. (4), appropriately modified to accommodate the defective region:

$$\Delta P(\omega) \approx F_2(\lambda_1,\lambda_2)\left[\int_0^t \Delta N(z,\omega)dz - \int_0^t \Delta N_d(z,\omega)dz\right] \quad (7)$$

where $\Delta P(\omega)$ is the difference in signal between the intact and defective regions. This is a complex quantity, so it can be separated out into amplitude and phase components. The apparent simplicity of this expression is due to the fact that the sub-surface defects considered here are on the back surface of the wafer and their presence mostly impacts the value of the back-surface recombination velocity S_2 (Mandelis (2001)), while the bulk transport parameters and the terms comprising the prefactor $F(\lambda_1,\lambda_2)$, remain essentially unaltered for a thin damage layer in an otherwise homogeneous semiconductor. If these conditions are not fulfilled, then a more complete expression of the carrier recombination related emissions must be used to quantify PCR contrast due to distributed sub-surface electronic defect structures.

The mild mechanical defect on the back surface of the p-type Si wafer that generated the images of Figs. 4 and 5 proved to be too severe for our sensitive InGaAs photodetector: upon scanning the affected surface the PCR signal vanished within the region of the defect, apparently due to the highly efficient trapping of the photogenerated free carriers by the high

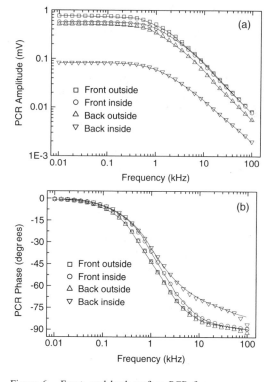

Figure 6.   Front- and back-surface PCR frequency scans inside and outside a defect area of a p-Si wafer on aluminum backing. Detector: InGaAs; beam size: 1.4 mm; Ar-ion laser power: 20 mW. (a) Amplitudes and (b) phases. Best fit parameters:
Front intact region: $\tau = 1$ ms; $D^* = 12$ cm^2/s, $S_1 = 10$ cm/s, $S_2 = 210$ cm/s.
Front inside the defect: $\tau = 1$ ms; $D^* = 14.9$ cm^2/s, $S_1 = 25$ cm/s, $S_2 = 300$ cm/s.
Back intact region: $\tau = 1$ ms; $D^* = 12$ cm^2/s, $S_1 = 10$ cm/s, $S_2 = 200$ cm/s.
Back inside the defect: $\tau = 1$ ms; $D^* = 5$ cm^2/s, $S_1 = 450$ cm/s, $S_2 = 130$ cm/s.

density of near-surface electronic defect states. Therefore, a different region of the same wafer was chosen to create a visually undetectable defect by simply touching the back surface of the wafer with paper. Then PCR frequency scans were performed on both sides of the material, outside and inside the defect region, Fig. 6. The inherent instrumental transfer function was removed by introducing an indirect normalization method based on the fact that in the high frequency regime both HgCdTe and InGaAs detectors monitor the same electronic processes (Mandelis et al. (2003)). All curves shown here were normalized by the same transfer function obtained by this method. The PCR theoretical model involved carrier-wave IR emissions using diffusion-wave field expressions

(Mandelis (2001), Chap. 9.12), with adjustable electronic transport coefficients (Rodriguez et al. (2000); Ikari et al. (1999)). The effect of the back-surface defect was modeled as a change in the recombination velocity S_2 (front-surface probing) only. When the wafer was turned over, the definitions of S_1 and S_2 were reversed. Regarding the D^* values, those outside the defect remained constant for both sides of the wafer, however, the D^* value from the back inside the defect region was relatively low. The high sensitivity of the InGaAs detector to the electronic state of the inspected surface is probably responsible for this discrepancy, as the theoretical phase fit is poor at high frequencies (>1 kHz) within that region, an indication of near-surface depth inhomogeneity of transport properties.

5   APPLICATION TO QUANTITATIVE DETERMINATION OF CARRIER MOBILITY

Carrier mobility (μ) is an important phenomenological parameter for describing the operation of semiconductor devices such as MOSFETs and solar cells. It is one of the basic input parameters for expressing electrical current in devices. In addition, the determination of doping level in wafers requires knowledge of carrier mobility. Several techniques have been used in determining carrier mobilities in semiconductors (Schroder (1998)). However, they all require samples specially prepared or, when this is not the case, they require electrical contacts for signal acquisition. PCR is the first all-optical and non-contacting NDT methodology for determination of carrier mobility. In the experiments the laser beam size at the sample surface was 1.34 mm. Temperature ramps were introduced by a heater/temperature controller. The entire process was controlled by a computer. The heating system was capable of varying and maintaining the sample temperature up to 675 K. One FZ n-type silicon wafer with resistivity $\rho = 10$–15 Ωcm, $N \sim 8 \times 10^{14}$cm^{-3}, and a 980Å thermally grown SiO$_2$ layer was studied. The wafer thickness (l) was 530 μm.

Figure 7 shows the PCR experimental amplitude and phase frequency responses at different temperatures. The measured temperature range was 300–575 K, given a fixed increment of 25 K. For the sake of clarity, however, only six of the twelve data sets are shown. In the same figure shown are the results of the best multi-parameter fitting using the theory based on Eq. (4). There is very good agreement between the theory and experimental data for all temperatures. With one frequency scan it is possible to obtain simultaneously the carrier lifetime, diffusion coefficient, and surface recombination velocity. As can be seen, the PCR amplitude increases with temperature. There are two phenomena contributing to this behavior. The

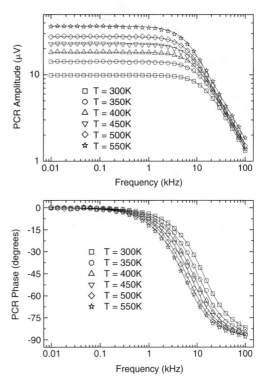

Figure 7. (a) PCR amplitudes and (b) phase frequency responses at different temperatures. Only six of the twelve data sets are shown. The symbols represent experimental data points, while the lines are the best fitting to data. Laser power: 15 mW; beam size: 1.34 mm.

first one follows the Shockley-Read-Hall theory (Shockely and Read (1952); Hall (1952)), which assumes that the thermally excited carrier density partially neutralizes existing ionized impurities by increasing the occupation of empty states. Therefore, the photo-injected density of free carriers increases, consequently increasing the plasma recombination emission. This is directly associated with an increase of free carrier lifetime, as shown in Fig. 8. The second contribution accounts for increasing of scattering mechanisms with temperature, which results in a decrease of carrier diffusivity, as shown in Fig. 9. The symbols represent experimental values and the line is the result of data fitting using a polynomial function of the form: $D(T) = a \times T^b$. The value obtained for the exponent $b(b = -1.49 \pm 0.01)$ is in excellent agreement with that reported in the literature (Salnick et al. (1997b)), where, using PTR detection, the diffusion coefficient is found to have the following dependence on temperature: $D(T) \sim T^{-1.5}$. Lower diffusion coefficients amount to a higher probability of a carrier to remain within the PCR detection area before the

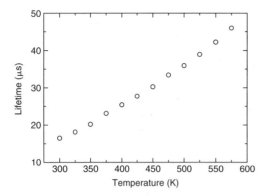

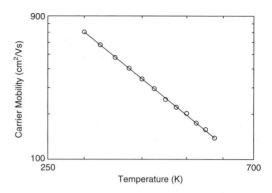

Figure 8. Lifetime temperature dependence obtained by simultaneously fitting amplitude and phase frequency responses of Fig. 7.

Figure 10. Temperature dependence of mobility. The symbols represent experimental values, while the line is the best fitting using the function $\mu(T) = a\%T^b$, where $a = (1.06 \pm 0.07)310^9$ and $b = -2.49 \pm 0.01$.

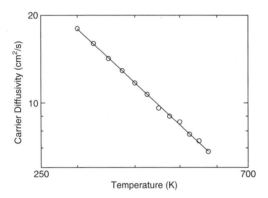

Figure 9. Temperature dependence of ambipolar diffusion coefficient obtained by simultaneously fitting amplitude and phase frequency responses of Fig. 7. The symbols represent experimental values, while the line is the best fitting using the function $D(T) = a \times T^b$, where $a = (8.75\ 6\ 0.73) \times 10^4$ and $b = -1.4960.01$.

de-excitation processes. A decreasing front surface recombination velocity with increasing temperature was observed: from 166 cm/s at 300 K to values less than 5 cm/s when the temperature approaches 575 K. This result was expected, since recombination velocity is inversely proportional to surface recombination lifetime. The PCR signal was not sensitive to changes in the back surface recombination velocity because the wafer is electronically thick ($L \ll l$) for all values of the carrier-wave diffusion length. The maximum value for L was about 177 μm at 10 Hz, much smaller than the wafer thickness (530 μm). The carrier diffusion coefficient is a key parameter for mobility determination. It is assumed that the carrier transport in wafers under intensity-modulated optical excitation is predominantly by diffusion. In the relatively low

injection level regime, the carrier transport can be described by Boltzmann statistics and hence the mobility is related to the carrier diffusion coefficient by the Einstein relation $D = (kT/q)\mu$ where k is the Boltzmann constant, T is the absolute temperature and q is the elementary charge. Therefore, if the diffusion coefficient is known at different temperatures, the temperature dependence of carrier mobility can be determined. Figure 10 shows the temperature dependence of the experimental carrier mobility (symbols) along with the best fit (full line) using the same polynomial function as used for carrier diffusivity. The values of mobility were appropriately normalized by their corresponding thermal voltages (kT/q). The temperature dependence of carrier mobility obtained could be expressed as

$$\mu(T) = (1.06 \pm 0.07) \times 10^9 \times T^{-2.49 \pm 0.01} \text{ cm}^2/\text{Vs} \quad (8)$$

A similar result ($\mu_n(T) = (2.160.2) \times 10^9 \times T^{-2.560.1}$ cm^2/Vs and $\mu_n(T) = (2.360.1) \times 10^9 \times T^{-2.760.1}$ cm^2/Vs) was reported previously (Ludwig and Watter (1956); Mette et al. (1960)) using the Hall effect for determining the mobility of electrons and holes in Si, respectively. In a different study (Mnatsakanov et al. (2001)), a semi-empirical model was proposed and applied to several sets of data from the literature, and values in between 2.2 and 2.4 were found for the constant b, the temperature power in Si samples.

The mobilities of carriers in non-polar semiconductors are determined by interactions with the acoustical vibrations of the lattice as well as by scattering by ionized impurities or other defects. The temperature dependence of mobility predicted by the deformation potential theory (Bardeen and Shockley (1950)) is $\sim T^{-3/2}$. However, as mentioned above, experimentally measured dependencies differ from this value of $(-3/2)$. Reasons for this discrepancy include: (a) contributions from other scattering mechanisms may be present (for example, above 100 K the contribution of optical phonon scattering becomes considerable, which lowers the value of the mobility); and (b) the non-parabolicity, distortion of equi-energy surfaces and the effect of split-off sub-band holes (Takeda et al. (1982)).

6 CONCLUSIONS

Laser infrared photo-carrier radiometry (PCR) has been introduced. The emerging technology is a novel all-optical method based on carrier diffusion-wave diagnostics. Based on the theoretical foundations and the first few experimental case studies using industrial-quality Si wafers, there are excellent prospects for PCR as an *in-situ* NDT quality control NDT technology in semiconductor processing.

Its local monitoring nature surpasses the currently available techniques for non-destructive, non-contact monitoring and imaging of deep electronic defects in Si wafers, for measuring free-carrier transport properties, and for measuring carrier mobilities without the need for auxiliary electric circuit fabrication and electrode application. PCR can become a valuable NDT technology as it can monitor local values of carrier mobilities and other transport properties at several intermediate stages of device fabrication. A noteworthy feature of high frequency PCR imaging is that it has revealed for the first time a very long-range effect of carrier interactions with deep sub-surface defect structures and the detrimental ability of such structures to decrease the overall free photoexcited-carrier density in locations far away from the defect sites at or near the front surface where device fabrication takes place. Therefore, PCR may become an important tool toward device fabrication improvement through careful selection of substrate wafers with regard to deep bulk growth and manufacturing defects which were heretofore not associated with device performance.

REFERENCES

Bardeen, J. & Shockley, W. 1950. *Phys. Rev.* 80: 72.

Christofides, C., Nestoros, M. & Othonos, A. 2000. in *Semiconductors and Electronic Materials*. Progress in Photo-acoustic and Photothermal Phenomena Vol. IV (A. Mandelis and P. Hess, eds, SPIE, Bellingham, WA, 2000), Chap. 4.

Hall, R.N. 1952. *Phys. Rev.* 87: 387.

Ikari, T., Salnick, A. & Mandelis, A. 1999. *Journal of Applied Physics*. 85: 7392.

Kirchhoff, G. 1998. Abhandlungen über Emission und Absorption. (M. Planck, ed., Verlag von Wilhelm Engelmann, Leipsig, 11–36.

Kreit, F. & Black, W.Z. 1980. *Basic Heat Transfer*. (Harper and Row, New York).

Ludwig, G.W. & Watter, R.L. 1956. Phys. Rev. 101: 1699.

Mandelis, A. 1998. *Solid-State Electron*. 42(1).

Mandelis, A. 2001. *Diffusion-Wave Fields: Mathematical Methods and Green Functions*, Springer-Verlag, NY, Chap. 9.

Mandelis, A., Batista, J. & Shaughnessy, D. 2003. Phys. Rev. B (In press).

Mette, H., Gartner, W.N. & Loscoe, C. 1960. Phys. Rev. 117, 1491.

Mnatsakanov, T.T., Pomortseva, L.I. & Yurkov, S.L. 2001. *Semiconductors* 35: 394.

Rodriguez, M.E., Mandelis, A., Pan, G., Nicolaides, L., Garcia, J.A. & Riopel, Y. 2000. *J. Electrochem. Soc.* 147: 687.

Rosencwaig, A. 1987. in *Photoacoustic and Thermal-Wave Phenomena in Semiconductors*. (A. Mandelis, ed., North-Holland, New York), Chap. 5.

Salnick, A., Jean, C. & Mandelis, A. 1997a. *Solid-State Electron*. 41: 591.

Salnick, A., Mandelis, A., Ruda, H. & Jean, C. 1997b. *Journal of Applied Physics*. 82: 1853.

Schroder, D.K. 1998. *Semiconductor Material and Device Characterization*. 2nd ed. (Wiley-Interscience, New York).

Shockley, W. & Read, W.T. 1952. *Phys. Rev.* 87: 835.

Smith, R.A. 1978. *Semiconductors*, 2nd ed. (Cambridge Univ. Press, Cambridge), 118–119.

Takeda, K., Sakudi, K. & Sakata, M. 1982. *Journal of Physics C: Solid State*. 15: 767.

van Roosbroeck, W. & Shockley, W. 1954. *Phys. Rev.* 94: 1558.

Wagner, R.E. & Mandelis, A. 1996. *Semicond. Sci. Technol.* 11: 300.

Emerging Technologies in Non Destructive Testing, Van Hemelrijck, Anastasopoulos & Melanitis (eds)
© 2004 Swets & Zeitlinger, Lisse, ISBN 90 5809 645 9

Theory and application of moment tensor analysis in AE

M. Ohtsu
Graduate School of Science and Technology, Kumamoto University, Kumamoto, Japan

M. Shigeishi
Department of Civil Engineering, Kumamoto University, Kumamoto, Japan

ABSTRACT: Nucleation of a crack in a solid emits elastic waves, which are readily detected as acoustic emission (AE). A powerful technique for quantitative AE analysis has been developed as SiGMA (Simplified Green's functions for Moment tensor Analysis). Thus, crack kinematics of locations, types and orientations can be determined. Because these kinematical data are obtained as three-dimensional (3-D) locations and vectors, visualization procedure is developed by using VRML (Virtual Reality Modeling Language). After describing theoretical background of the moment tensor analysis, applications to identify failure mechanisms of concrete structures are discussed.

1 INTRODUCTION

The generalized theory of acoustic emission (AE) was proposed on the basis of elastodynamics (Ohtsu & Ono 1984). It is clarified that AE waves are elastic waves due to dynamic dislocation in a solid. This is because theoretical treatment of elastic waves in a homogeneous medium is applicable to AE waves in concrete (Ohtsu 1982), whereas concrete is not homogeneous but heterogeneous.

Source characterization of AE waves leaded to the moment tensor analysis for identifying crack kinematics (Kim & Sachse 1984). Here only diagonal components of the moment tensor were assumed to characterize cracking mechanisms of glass due to indentation. Although they took only into account tensile cracks, mathematically the presence of tensor components is not actually associated with the type of the crack, but substantially related with the orientation of the coordinate system. Although the crack orientations are often assumed as parallel to the coordinate system (Saito et al. 1998), they are generally inclined to the coordinate system and the presence of all the tensor components is consequent.

In order to determine all the tensor components from AE wave, a powerful procedure has been developed, which is named SiGMA (Simplified Green's functions for Moment tensor Analysis) (Ohtsu 1991) and now commercially available. Applying the eigenvalue analysis to the moment tensor, crack kinematics of locations, types and orientations are quantitatively determined. Since these kinematical data are obtained as three-dimensional (3-D) locations and vectors, 3-D visualization is desirable. To this end, recently a visualization procedure is implemented by using VRML (Virtual Reality Modeling Language) (Ohtsu & Shigeishi 2002).

In the present paper, theoretical background of the moment tensor analysis is briefly presented. Then applications of visual SiGMA to identify failure mechanisms of concrete structures are described.

2 THEORETICAL BACKGROUND

2.1 *Integral representation*

Elastodynamic properties of material constituents are physically dependent on the relation between the wavelengths and the characteristic sizes of heterogeneity. In the case that the wavelengths are even larger than the sizes of heterogeneous inclusions, the effect of heterogeneity is inconsequent and the material can be referred to as homogeneous. This is the case of such massive solids as concrete and rock, if the sizes of specimens are large enough compared with the wavelengths. Thus, elastodynamics in the homogeneous material is available for analyzing AE waves in most of solids.

A solution of wave motions $\mathbf{u}(\mathbf{x},t)$ in an elastic solid is represented,

$$u_k(\mathbf{x},t) = {}_{\cdot S}[G_{ki}(\mathbf{x},\mathbf{y},t)*t_i(\mathbf{y},t) - T_{ki}(\mathbf{x},\mathbf{y},t)*u_i(\mathbf{y},t), \qquad (1)$$

where $\mathbf{u}(\mathbf{y},t)$ are displacements, and $\mathbf{t}(\mathbf{y},t)$ are tractions on boundary surface S. The asterisk * represents the convolution integral in time. $G_{ik}(\mathbf{x},\mathbf{y},t)$ are Green's functions and $T_{ik}(\mathbf{x},\mathbf{y},t)$ are the associated tractions with Green's functions,

$$T_{ik}(\mathbf{x},\mathbf{y},t) = G_{ip,q}(\mathbf{x},\mathbf{y},t)\, C_{pqjk}\, n_j \ . \qquad (2)$$

Here C_{pqjk} are the elastic constants, and $G_{ip,q}(\mathbf{x},\mathbf{y},t)$ are the spatial derivatives of Green's functions as they imply $G_{ip}(\mathbf{x},\mathbf{y},t)/x_q$. $\mathbf{n}$ is the unit normal vector to surface S.

Physical meaning of Green's function is readily derived, as applying a force, $\mathbf{t}(\mathbf{y},t) = \mathbf{f}(\mathbf{y},t)$, at only point $\mathbf{y}$ on surface S in Equation 1,

$$u_i(\mathbf{x},t) = G_{ij}(\mathbf{x},\mathbf{y},t) * f_j(\mathbf{y},t). \qquad (3)$$

Equation 3 implies that $G_{ij}(\mathbf{x},\mathbf{y},t)$ results in dynamic displacement $u_i(\mathbf{x},t)$ in the x_i direction at point $\mathbf{x}$ due to the force $f_j(\mathbf{y},t)$ in the x_j direction at point $\mathbf{y}$. Because Green's functions are dependent on not only material properties but also on configuration of the medium, generally they have to be computed numerically in a finite solid. Only for an infinite space, the analytical solution is known. Numerical solutions $G_{ij}(\mathbf{x},\mathbf{y},t)$ were already published in a half space (Ohtsu & Ono 1984) and in an infinite plate (Pao & Ceranoglu 1981). For a finite body, numerical solutions of Green's functions are reported by applying the finite element method (Hamstad et al. 1999).

A famous experiment of pencil-lead break (Hsu & Hardy 1978) is mathematically formulated by Equation 3. A pencil-lead break is known to generate Heaviside's step-function force, $H(t)e_j(\mathbf{y})$, where $e(\mathbf{y})$ is the unit direction vector at point $\mathbf{y}$. Since the convolution integral with the step function leads to the integration in time, Equation 3 becomes,

$$p_i(\mathbf{x},t) = \int G_{ij}(\mathbf{x},\mathbf{y},t)e_j(\mathbf{y})\, dt. \qquad (4)$$

Here $p_i(\mathbf{x},t)$ is the detected waveform by the displacement sensor due to pencil-lead break. Accordingly, Green's function of an arbitrary specimen, $G_{ij}(\mathbf{x},\mathbf{y})$, can be empirically obtained as,

$$G_{ij}(\mathbf{x},\mathbf{y},t) = dp_i(\mathbf{x},t)/dt. \qquad (5)$$

Where the orientation of AE sensor sensitivity is parallel to the x_i direction and the force due to pencil-lead break is released in the x_j direction.

Some attempts were made erroneously to apply these Green's functions in Equation 5 to source characterization. Equation 3 is rationally not applicable to AE waves due to cracking, because source representation is explicitly related with not Green's

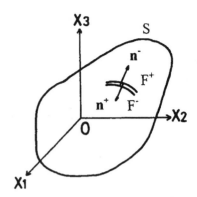

Figure 1. Dislocation (crack) surface F.

functions themselves but the spatial derivatives of Green's functions. This is the reason why seismic sources are often modeled by a dipole force and a couple force. Consequently, stress release due to the pencil-lead break has to be performed in two orientations either at one point or at two close points to physically model the derivatives and thus generate these equivalent forces.

In order to formulate a crack as an AE source, surface S in Equation 1 is replaced by internal surface F representing a crack surface. To introduce the discontinuity of displacements (dislocation), virtual two surfaces F^+ and F^- are considered as shown in Figure 1.

Before a crack is nucleated, these two surfaces make a coincident motion. Due to cracking, the discontinuity of displacement $\mathbf{b}(\mathbf{y},t)$ is nucleated between the two surfaces, using superscripts $+$ and $-$ on surface F^+ and F^-,

$$b_i(\mathbf{y},t) = u_i{}^+(\mathbf{y},t) - u_i{}^-(\mathbf{y},t). \qquad (6)$$

Vector $\mathbf{b}(\mathbf{y},t)$ is called the dislocation and is identical to Burgers vector in crystallography. Setting $\mathbf{t}(\mathbf{y},t) = 0$ on surface F, Equation 1 is rewritten,

$$\begin{aligned} u_k(\mathbf{x},t) &= \int_{F+}[-T_{ki}{}^+(\mathbf{x},\mathbf{y},t)*u_i{}^+(\mathbf{y},t)]dF, \\ &+ \int_{F-}[-T_{ki}{}^-(\mathbf{x},\mathbf{y},t)*u_i{}^-(\mathbf{y},t)]dF. \end{aligned} \qquad (7)$$

Here T_{ik}^+ and T_{ik}^- contain the normal vector $\mathbf{n}^+$ and $\mathbf{n}^-$, respectively. Assuming $\mathbf{n} = \mathbf{n}^- = -\mathbf{n}^+$, and $F = F^-$,

$$u_k(\mathbf{x},t) = \int_F T_{ki}(\mathbf{x},\mathbf{y},t)*b_i(\mathbf{y},t)dF. \qquad (8)$$

From Equation 2, Equation 8 is converted as,

$$\begin{aligned} u_k(\mathbf{x},t) &= \int_F T_{ki}(\mathbf{x},\mathbf{y},t)*b_i(\mathbf{y},t)\, dF \\ &= G_{kp,q}(\mathbf{x},\mathbf{y},t)*S(t)\, C_{pqij}\, n_j l_i \cdot_F\, b(\mathbf{y},t)dF \\ &= G_{kp,q}(\mathbf{x},\mathbf{y},t)*S(t)\, C_{pqij}\, n_j l_i \Delta V, \end{aligned} \qquad (9)$$

20

Figure 2.   A penny-shaped crack for debonding.

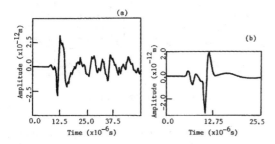

Figure 3.   (a) Detected AE wave and (b) simulated.

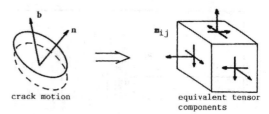

Figure 4.   Crack motion and moment tensor components.

where $\mathbf{l}$ is the unit direction vector and S(t) is the source-time function of crack motion. ΔV is the crack volume. It is noted that the amplitude of AE wave, $\mathbf{u}(\mathbf{x},t)$, is explicitly associated with the crack volume, integrating crack opening $b(\mathbf{y},t)$ over crack area F. This implies that some attempts to determine the crack area (Enoki et al. 1986) or the magnitude of shear slip (Dai et al. 2000) from AE waves might not be accurately made.

Equation 9 is the theoretical representation of AE wave due to cracking and is applicable to a simulation analysis (Ohtsu 1982). Modeling a penny-shaped crack for debonding as shown in Figure 2 (Ohtsu et al. 1987), AE waveform is successfully simulated in Figure 3 (b), as compared with AE waveform in Figure 3 (a). Here, the spatial derivatives of Green's functions in a half space are employed so that reflected waves after main motions are not recovered. Still, the shape and the amplitude in the detected wave are in remarkable agreement with those of simulated.

2.2   Moment tensor analysis

Since Equation 9 contains two vectors $\mathbf{l}$ and $\mathbf{n}$, a second-order tensor can be defined as,

$$M_{pq} = C_{pqkl}l_k n_l \, \Delta V. \tag{10}$$

Tensor, M_{pq}, is defined by the product of the elastic constants [N/m^2] and the crack volume [m^3], which leads to the moment of physical unit [Nm]. Consequently, it is called the moment tensor. In the case of an isotropic material,

$$M_{pq} = [\lambda l_k n_k \, \delta pq + \mu(l_p n_q + l_q n_p)]\Delta V, \tag{11}$$

where λ and μ are Lame's elastic constants. Equations 10 and 11 lead to the fact that the moment tensor is comparable to a stress due to crack nucleation, as the second-order tensor as shown in Figure 4. Then, Equation 9 is rewritten as,

$$u_k(\mathbf{x},t) = G_{kp,q}(\mathbf{x},\mathbf{y},t) \, M_{pq} * S(t). \tag{12}$$

In order to solve Equation 12 inversely, the spatial derivatives of Green's functions are necessary. As a result, numerical solutions are obtained by FDM (Enoki et al. 1986) and by FEM (Hamstad et al. 1999). These solutions, however, need a vector processor for computation, and are not readily applicable to processing a large amount of AE waves. Taking into account only the far-filed term of P wave, a simplified procedure is developed, which is suitable for a PC-based processor and robust in computation. The procedure is now implemented as a SiGMA code.

By picking up P wave motion in the far field (1/R term) from Green's function in an infinite space, displacement $U_i(\mathbf{x},t)$ of P wave motion is obtained from Equation 12,

$$U_i(\mathbf{x},t) = -1/(4\pi\rho v_p^3) \, r_i r_p r_q/R \, dS(t)/dt \, M_{pq}. \tag{13}$$

Here ρ is the density of the material and v_p is the velocity of P wave. R is the distance between the source $\mathbf{y}$ and the observation point $\mathbf{x}$, of which direction cosine is $\mathbf{r} = (r_1, r_2, r_3)$. Considering the effect of reflection at the surface and neglecting the source-time function, amplitude $A(\mathbf{x})$ of the first motion is represented,

$$A(\mathbf{x}) = Cs \, Ref(\mathbf{t},\mathbf{r})/R \, r_i \, M_{ij} \, r_j, \tag{14}$$

where Cs is the calibration coefficient including material constants in Equation 13. $\mathbf{t}$ is the direction of the sensor sensitivity. $Ref(\mathbf{t},\mathbf{r})$ is the reflection coefficient at the observation location $\mathbf{x}$. In the relative moment tensor analysis (Dahm 1996), this coefficient is not taken into consideration, because the effect of the sensor locations is compensated. Since the moment tensor is symmetric, the number of unknowns to be determined of M_{ij} is six. Thus, multi-channel observation

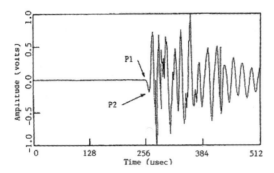

Figure 5. Detected AE waveform.

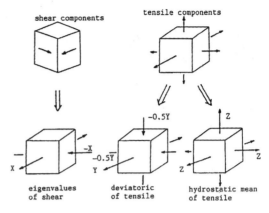

Figure 6. Decomposition of the eigenvalues.

of the first motions at more than six channels is required to determine all the moment tensor components.

2.3 *SiGMA analysis*

In the SiGMA code implemented, each AE waveform is displayed on CRT screen as shown in Figure 5. Then, two parameters of the arrival time (P1) and the amplitude of the first motion (P2) are determined. In the location procedure, source location **y** is determined from the arrival time differences. For 3-D location, 5-channel system is at least necessary. From a relation between the observation point **x** and the source location **y**, distance R and its direction vector **r** are determined. The amplitudes of the first motions at more than 6 channels are substituted into Equation 14, and then component M_{ij} are determined. Since the SiGMA code requires only relative values of the tensor components, the relative calibration of the sensors is sufficient enough.

Classification of a crack is made by the eigenvalue analysis of the moment tensor. Setting the ratio of the maximum shear contribution as X, three eigenvalues for the shear crack become X, 0, −X. Likewise, the ratio of the maximum deviatoric tensile component is set as Y and the isotropic tensile as Z. It is assumed that the principal axes of the shear crack is identical to those of the tensile crack. In a general case, the eigenvalues of the moment tensor are represented by combination of the shear crack and the tensile crack. Because relative values are taken into consideration, three eigenvalues are normalized and decomposed as,

$$1.0 = X + Y + Z,$$

the intermediate eigenvalue/the maximum
$$= 0 - Y/2 + Z,$$

$$(15)$$

the minimum eigenvalue/the maximum
$$= -X - Y/2 + Z,$$

where X, Y, and Z denote the shear ratio, the deviatoric tensile ratio, and the isotropic tensile ratio, respectively. These are illustrated in Figure 6.

In the present criterion, AE sources of which the shear ratios are less than 40% are classified into tensile cracks. The sources of X >60% are classified into shear cracks. In between 40% and 60%, cracks are referred to as mixed mode.

In the eigenvalue analysis, three eigenvectors **e1**, **e2**, and **e3**,

$$\begin{aligned}
\mathbf{e1} &= \mathbf{l} + \mathbf{n} \\
\mathbf{e2} &= \mathbf{l} \times \mathbf{n} \\
\mathbf{e3} &= \mathbf{l} - \mathbf{n},
\end{aligned} \qquad (16)$$

are also determined. Thus, vectors **l** and **n**, which are interchangeable, are readily recovered.

3 APPLICATIONS

3.1 *Bending test of a reinforced concrete beam*

In order to confirm an applicability to inspection of concrete structures (NDIS 2421 2000), a bending test of a reinforced concrete (RC) beam is carried out. An experimental set-up is given in Figure 7. Six AE sensors are arranged to cover the central zone of the beam. Applying two-point loading, bending cracks are visually observed after the test.

AE waves were recorded and analyzed by SiGMA code. Results are currently plotted in the two-dimensional (2-D) projection. They are displayed at their locations with symbols in Figure 8. A tensile crack is denoted by arrow symbol, of which direction is identical to that of crack opening. Mixed mode and shear cracks are denoted by cross symbol, of which two directions correspond to two vectors **l** and **n**. As can be seen, classification of cracks is readily made, although crack orientation is not easily recognized. This is because cracks are nucleated with 3-D orientations, which are not readily recognized in 2-D projection.

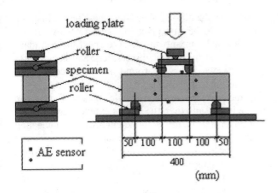

Figure 7. Bending test of a reinforced concrete beam.

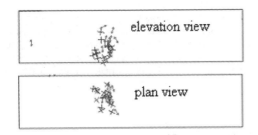

Figure 8. SiGMA analysis in a RC beam.

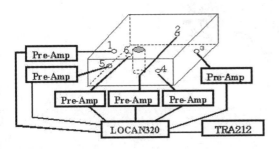

Figure 9. Test set-up for a crack-expansion test.

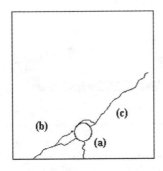

Figure 10. Crack patterns observed after the test.

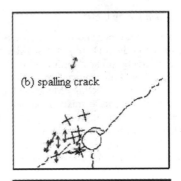

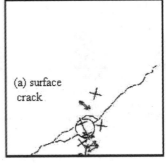

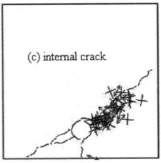

Figure 11. Results of SiGMA analysis in the expansion test.

3.2 Corrosion cracking in concrete

Another example is a crack-expansion test as shown in Figure 9. To identify cracking mechanisms in concrete due to corrosion of reinforcement, expansive agent was inserted in the hole and cracks patterns due to corrosion were experimentally simulated. After the test, three cracking patterns of (a) surface, (b) spalling and (c) internal cracks were observed as given in Figure 10. With respect to each crack pattern, results of SiGMA analysis are clustered and shown in Figure 11.

For the first two crack patterns of (a) surface and (b) spalling cracks, it is found that tensile cracks oriented perpendicular to the final crack plane dominate shear cracks. Concerning (c) internal crack, both of tensile and shear cracks are fairly mixed up.

23

3.3 *Visualization by VRML*

Because source locations and the two vectors **l** and **n** are determined in the 3-D space by SiGMA analysis, results are inherently suitable for 3-D visualization. In this respect, visualization procedure is developed by applying VRML (Virtual Reality Modeling Language). First, crack modes of tensile, mixed-mode and shear are visualized as shown in Figure 12 (Ohtsu & Shigeishi 2002). Here, an arrow vector indicate a crack motion vector **l**, and a circular plate corresponds to a crack surface, which is perpendicular to a crack normal vector **n**.

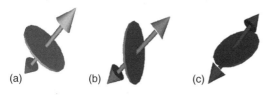

(a)   (b)   (c)

Figure 12.   Models for tensile, mixed, and shear cracks. (a) tensile crack (b) mixed-model crack (c) shear crack.

3-D visualization of SiGMA analysis on the bending test of the RC beam is given in Figure 13 as two views. Tensile cracks are mostly observed near the boundary of AE cluster. In contrast, mixed-mode and shear cracks are intensely observed around the reinforcement. In the bending test of an RC beam, it is known that bending cracks are nucleated as tensile cracks at the bottom of the beam, and then the cracks extend upward, generating tensile and mixed mode-cracks. After nucleation of tensile cracks (new stress-free surfaces), shear cracks are generated inside AE cluster. Thus, the orientations of shear cracks and mixed-mode cracks are not necessarily parallel to the final bending cracks. It is noted that shear cracks are intensely observed at the compressive zone. Besides AE cluster near the reinforcement, other cracks distribute widely, probably corresponding to nucleation of diagonal-shear (visible) cracks connecting between the loading point and the support.

The crack-expansion tests were conducted again, and results of SiGMA analysis are visualized by VRML in Figures 14 and 15. In the test shown in Figure 14, corresponding to two diagonal cracks similar to that in

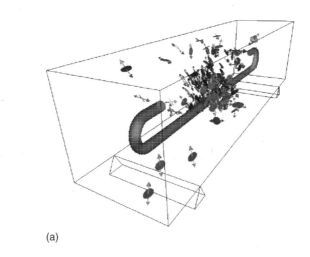

(a)

(b)

Figure 13.   VRML results of the RC beam. (a) Inclined view. (b) Elevation view.

Figure 10 (c), two AE clusters are observed. It is interesting that many AE sources are found in the extended zones including the final crack. This may imply nucleation of the fracture process zone, which is created ahead of the crack tip in advance to the final crack surface. Thus, AE clusters are found two diagonal directions from the hole.

In another specimen, one internal crack vertical to the bottom surface was observed. Results by SiGMA analysis are plotted in Figure 15. According to the previous research (Ohtsu & Yoshimura 1997), the internal crack is created if the surface crack is arrested by coarse aggregate. This is the case, because the concrete cover (25 mm) is shorter than the specimen (cover-thickness = 40 mm) in Figure 14.

The maximum size of coarse aggregate is 20 mm. Accordingly, it is likely that the cracks nucleated are arrested without extending in the zone of cover in the shorter cover-thickness comparable to the size

of aggregate. AE cluster in Figure 15 is so condense that mechanisms are not easily identified, but still orientations and crack-types are easily realized by VRML visualization.

4 CONCLUSIONS

Nucleation of cracks can be quantitatively analyzed from AE waveforms, applying the SiGMA code. Crack kinematics on locations, types and orientations are readily determined three-dimensionally.

Because visualization of results is desirable, 3-D visualization procedure for SiGMA analysis is developed by using VRML.

After clarifying theoretical background of the moment tensor analysis, applications to inspection of the reinforced concrete structure and to failure mechanisms of concrete due to corrosion of reinforcement are successfully reported.

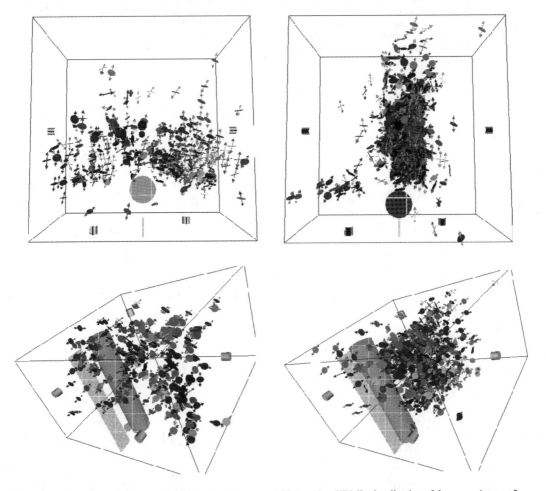

Figure 14.   VRML visualization of the expansion test 1.   Figure 15.   VRML visualization of the expansion test 2.

REFERENCES

Dai, S. T., Labuz J. F. and Carvalho, F. 2000, Softening Response of Rock Observed in Plane-Strain Compression, Trends in Rock Mechanics, Geo SP-102, ASCE, 152–163.

Dahm, T. 1996, Relative Moment Tensor Inversion based on Ray Theory: Theory and Synthetic Tests, *Geophys. J. Int.*, No. 124, 245–257.

Enoki, M., Kishi, T. and Kohara, S. 1986, Determination of Micro-cracking Moment Tensor of Quasi-cleavage Facet by AE Source Characterization, *Progress in Acoustic Emission III*, JSNDI, 763–770.

Hamstad, M. A., O'Gallagher, A. and Gary, J. 1999, Modeling of Buried Monopole and Dipole Source of Acoustic Emission with a Finite Element Technique, *Journal of AE*, 17(3–4), 97–110.

Hsu, N. N. and Hardy, S. C. 1978, Experiments in AE Wave form Analysis for Characterization of AE Sources, Sensors and Structures, *Elastic Waves and Nondestructive Testing of Materials*, AMD-Vol. 29, 85–106.

Kim, K. Y. and Sachse, W. 1984, Characterization of AE Signals from Indentation Cracks in Glass, *Progress in Acoustic Emission II*, JSNDI, 163–172.

NDIS 2421 2000, Recommendation Practice for In Situ Monitoring of Concrete Structures by AE, JSNDI, Japan.

Ohtsu, M. 1982, Source Mechanism and Waveform Analysis of Acoustic Emission in Concrete, *Journal of AE*, 2(1), 103–112.

Ohtsu, M. and Ono, K. 1984, A Generalized Theory of Acoustic Emission and Green's Functions in a Half Space, *Journal of AE*, 3(1), 124–133.

Ohtsu, M., Yuyama, S. and Imanaka, T. 1987, Theoretical Treatment of Acoustic Emission Sources in Microfracturing due to Disbonding, *J. Acoust. Sco. Am.*, 82(2), 506–512.

Ohtsu, M. 1991, Simplified Moment Tensor Analysis and Unified Decomposition of Acoustic Emission Source : Application to In Situ Hydrofracturing Test, *Journal of Geophysical Research*, 96(B4), 6211–6221.

Ohtsu, M. and Yoshimura, S. 1997, Analysis of Crack Propagation and Crack Indentation due to Corrosion of Reinforcement, Construction and Building Materials, Vol. 11, Nos. 7–8, 437–442.

Ohtsu, M. and Shigeishi, M. 2002, Three-Dimensional Visualization of Moment Tensor Analysis by SiGMA-AE, e-Journal of NDT, Vol. 9, No. 9.

Pao, Y. H. and Ceranoglu, A. N. 1981, Propagation of Elastic Pulse and Acoustic Emission in a Plate, *J. Appl. Mech.*, 48, 125–147.

Saito, N., Takemoto, M., Suzuki, H. and Ono, K. 1998, Advanced AE Signal Classification for Studying the Progression of Fracture Modes in Loaded UD-GFRP, *Progress in Acoustic Emission IX*, JSNDI, V-1–V-10.

Vibration measurements in nondestructive testing

R.D. Adams

Department of Mechanical Engineering, University of Bristol, Bristol, UK

ABSTRACT: Vibration and the associated acoustic response of components has been used for many years as a nondestructive means of assessing quality and integrity. With the advent of better testing techniques, especially instruments and computers, it has become possible to develop the method for new and challenging structures and materials. This paper describes both global and local methods of testing.

1   INTRODUCTION

The oldest method of nondestructive examination is almost certainly visual. But following closely would be aural inspection, using the ear to sense if a piece of pottery was defective. When tapped, a good pot will produce a clear note, while a cracked pot will sound quite different, depending on where and how big is the crack. Thus was born the concept of using vibration (although the ear senses the *acoustic* radiation from the vibrating surface of the pot) to check quality.

The ear detects the natural frequencies, and also the rate of decay of the vibration. The natural frequencies are a function of the size and shape of the pot, together with the density and elastic modulus of the material it was made from. The rate of decrease of the oscillation is a function of the damping (and frequency) and this is controlled partly by material quality and partly by the presence and position of cracks. Lead crystal glass has very low damping, whereas ordinary glass has a much higher level, and sounds relatively 'dead'.

Other vibration tests are tapping to detect disbonds in plasterwork or in honeycomb construction, the response of mine roof supports (acoustic emission), holographic measurement of vibration, and so on.

In effect, vibration methods can detect material quality, cracks and geometric or structural vibrations. The reader is referred to an extensive review concerning vibration methods for nondestructive testing (Adams & Cawley 1985): while now a little old, the basic principles are unaltered.

2   MATERIAL QUALITY

When examining the quality of materials, it is usual to use simple specimen shapes in well-conditioned

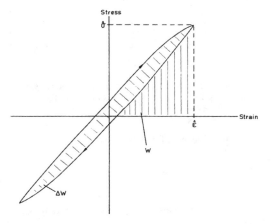

Figure 1.   Definition of specific damping capacity $\psi = \Delta W/W$, where $W = \hat{\sigma}^2/2E = E\hat{\varepsilon}^2/2$.

laboratory conditions. This enables the true material properties to be measured, providing data for later tests on real structures.

In metals, a given class of alloys such as steel, brass etc will show very little differences in their elastic modulus over wide ranges of heat treatment states. For instance, a quenched 1% carbon steel will have a Young's modulus (211 GPa) almost identical to that of an annealed, 0.1% carbon steel (209 GPa). However, their damping properties will vary enormously, as will be their response to cyclic stress amplitudes.

A convenient measure of damping is given in Figure 1. The specific damping capacity, ψ, is defined as the ratio of the energy dissipated per cycle, ΔW, to the maximum stored energy per cycle, W.

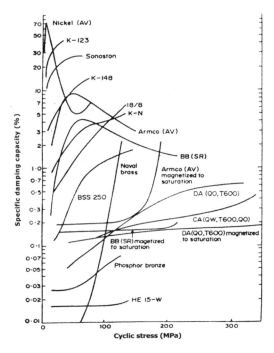

Figure 2. Variation of specific damping capacity, ψ, with cyclic stress for a wide range of engineering alloys (from Adams 1972a).

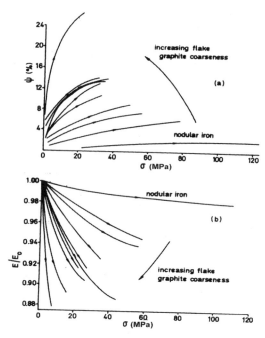

Figure 3. Non-linear variation of damping, ψ, and relative modulus change, E/E_0 with cyclic for a range of cast irons (From Adams & Fox 1973).

For small levels of damping, we have

$$\frac{\Delta W}{W} = \psi = 2\delta = 2\pi\eta$$

where δ is the logarithmic decrement, and η is the loss factor.

In steels, the damping is mainly due to magneto-elastic effects (Adams 1972b), whereas in non-ferrous metals, a wide range of mechanisms cause energy dissipation (Adams 1972a). Figure 2, from (Adams 1972b) shows the wide range of damping values in a series of structural alloys, as a function of the cyclic stress amplitude.

Cast iron presents an interesting variation to this pattern (Adams & Fox 1973, Fox & Adams 1972, 1973). Cast iron is more of a composite than an alloy. Its Young's moduli vary from below 100 GPa for coarse flake graphite irons, to 170 GPa for spheroidal graphite irons. Both the moduli and the damping are stress dependent, as shown in Figure 3. By correlating the damping at a cyclic stress level of 10% of the failure stress (other stresses can just as easily be used) with the tensile failure stress, a clear correlation can be found (Fox & Adams 1973, Adams 1987).

Advanced fibre reinforced composites have been extensively studied by the author. The Young's modulus

E_L, in the direction of the fibres in a unidirectional lamina is given by

$$E_L = E_f v + E_m (1-v)$$

where E_f is the fibre modulus; E_m is the matrix modulus; and v is the volume fraction.

This simple relationship holds well, provided the fibres are straight. It can therefore be used as a control on the fibre quality, provided the volume fraction can be measured. With modern composites and manufacturing methods, v varies little, but can be easily measured by burn-off or other tests. Since the fibre modulus relates directly to its strength, measurement of the composite properties can provide a nondestructive indication of strength. Composites are usually made with fibres at various angles, in different layers (laminae) so as to provide pre-determined strength and stiffness in certain directions. (They can also be woven). Once the composite has been made and the resin cured, it can be very difficult to determine if the correct fibre orientations have been created in the different layers. This aspect will be discussed later in the section on structures.

Polymers, whether used as the matrix material in a composite, as adhesives, or as semi-structural parts, have to be cured or processed correctly. One good

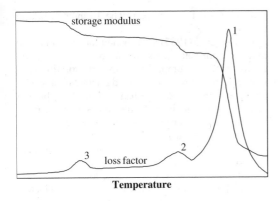

Figure 4. Variation of storage modulus and loss factor with temperature for a typical thermosetting polymer, showing the three identifiable transitions.

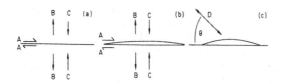

Figure 5. Cracks under static and dynamic loads: (a) zero volume crack, both sides touching; (b) normally open crack or void; (c) part open, part closed crack.

check for correct processing is to establish the various relaxation temperatures, often referred to as the glass-transition temperature (Tg). The matrix Tg is where the material changes from a glassy to a rubbery state.

It is measured dynamically from a rapid reduction in modulus, and a peak in the loss factor (η) as the temperature is increased, shown in Figure 4. Commercial DMTA (dynamic mechanical thermal analysers) apparatus can be used, as can differential scanning calorimetry (DSC), etc. Fully-cured epoxy resins will have a well-defined Tg. Unfortunately, the process of making the measurement means that the temperature has to be raised, thus often curing an under-cured material. The author has recently developed some new technology which enables a very rapid temperature scan to overcome this problem.

Cracks can also be detected in materials by vibration methods. Cracks may be open, closed, or part-open, part-closed. Shear loading of an open crack will produce no effect. If the crack is closed or partly so, there will be interfacial slipping and energy will be dissipated, resulting in an increase in damping. There will also be a reduction in stiffness. Tensile loading results in no energy dissipation, but a reduced stiffness will occur. Generally, cracks are part-open, part-closed, and will be loaded at same direction α to the plane of the crack as shown in Figure 5. Sliding friction will

occur at the edges, resulting in damping, while the open crack will produce a local reduction in stiffness.

There are, of course many forms of damage other than simple cracks. For instance, the material may have been subject to microplastic strain by creep or fatigue. Alternatively, some form of environmental attack may have taken place, such as by water or solvents in polymers, or by hydrogen embrittlement, in metals, or by nuclear radiation in any material. Under these circumstances, the defective zone may be local or general in the component.

Cracks in advanced fibre reinforced composites are complex because of its anisotropic layered structure. However, it has long been established (Adams & Cawley 1985) that cracks produce a reduction in stiffness and an increase in damping, whether they have been caused by static, fatigue, or even impact loading.

3 STRUCTURES – GENERAL

Global testing is where the overall response of a structure is assessed, as distinct from its local response. Obviously, if the 'defect' is local, it is best assessed locally, but this is not always possible. An example is the tapping of wheels in a railway train. By listening to the response, a skilled operator is supposed to be able to determine the presence of cracks. The wheel tapper hears the overall (or global) response and it does not matter where he strikes the wheel. However, coin-tapping, on such as on honeycomb structures, is measuring the local impedance, and the vibration response detected by the ear is due to the nature of the input force pulse from the tap.

4 GLOBAL VIBRATION MEASUREMENTS ON STRUCTURES AND COMPONENTS

Structures generally have many natural frequencies, some of which will predominate. Simple components such as uniform bars, beams and plates, have well-defined and calculable frequencies and mode shapes. More complex structures usually need such as finite element analysis for predicting natural frequencies and mode shapes. Even so, because of the difficulty of modelling, especially the boundary conditions, it is always necessary to check the theory by careful experimentation.

However, it is important to realise that the natural frequencies and the corresponding mode shapes are independent of the position of excitation (unless it is at a node, when no vibration will be caused).

If a structure is defective, there is a change in its stiffness and damping in the region of the defect. Usually, the stiffness decreases and the damping increases if the defect is in the form of a crack or series of cracks (micro or macro), or if the matrix of a fibrous composite has been attacked environmentally such as by steam or some other active agent. A reduction in stiffness implies a reduction in the structural natural frequencies and hence there is the possibility of using natural frequency measurements as a nondestructive test. Similarly, the damping of the different modes of vibration may be measured. The theoretical expressions for the flexural natural frequencies of a beam indicate that natural frequencies are very sensitive to dimensional changes and so natural frequency measurements may be used to check that a component is within the prescribed tolerances.

In fibrous composites, the omission or incorrect orientation of a ply layer may be difficult to ascertain without destroying the structure. However, such a defective structure will have changes in its natural frequencies compared with a perfect one and a vibration technique may therefore be used to obtain a quick indication of the structural condition. In structures such as filament wound tubes, incorrect fibre winding, the omission of a tape helix and incorrect cure also affect natural frequencies and so may be detected.

However, it is impossible to locate small cracks in large structures by global measurements of natural frequencies or damping.

Since natural frequencies are very sensitive to dimensions, their use for the detection of small cracks at the production stage is limited to components which are produced to strict dimensional tolerances. More generalised defects may be found in many components and structures. Frequency measurements can, of course be used to check whether the dimensions are within the specification.

Natural frequency measurements have been used to check crankshafts and other iron castings by General Motors, Fiat and British Leyland (Spain et al 1964, Magistrali 1975, Brown 1973). The test is used to detect the degree of spheroidisation in cast iron and the presence of internal defects. Magistrali (1975) says that the procedure permits a quick overall assessment of the integrity or otherwise of bulky parts where a thorough procedure of localised ultrasonic inspection would be too laborious and would be hampered by irregular coupling surfaces.

Adams and Vaughan (1981) developed a modulus parameter for assessing the grade of a cast iron from its stress nonlinearity, thus removing the constraint of close dimensional consistency which is demanded by the usual (sonic modulus) tests. The effective modulus at different amplitude levels was obtained from natural frequency measurements.

The quality of grinding wheels can readily be assessed by natural frequency measurements (Peters et al 1969, Shen 1982). The method is suitable for the determination of the hardness of both vitrified and resinoid wheels and gives a reliable indication of the behaviour of the grinding wheel under working conditions.

Cawley et al (1985) investigated the use of natural frequency measurements for the production quality control of fibre composites. Flexural vibration tests were carried out on a series of filament wound CFRP tubes, several of which had deliberately built-in defects. Some of the defective tubes had incorrect fibre volume fraction or winding angle while others had localised defects such as siliconised paper inserts and cut fibres. Figure 6 shows a graph of first mode natural frequency versus second mode frequency for all the tubes. The generalised defects, low-volume fraction and incorrect winding were readily detected. Until advances in production techniques improve the consistency of 'good' tubes, small localised defects will not be found since they cannot be discriminated from the normal distribution of frequencies for 'good' tubes. Lay-up errors in structures fabricated from pre-impregnated fibre sheet were also found.

Lloyd et al (1975) tested a batch of $152 \times 19 \times 7$-mm steel bars, some of which had saw cuts of varying depth at a single section. They showed that natural frequency changes could be used to detect the damaged bars. Adams et al (1978) showed that saw cuts and cracks have very similar effects on natural frequencies so these results were also applicable to cracks. Cawley (1984) has shown that natural frequency measurements provide a very quick and effective means of detecting cracks in mass produced components fabricated from powder metal.

Davis and Dunn (1974) have used natural frequency measurements to check the integrity of concrete piles. Lilley et al (1982) have extended this work to predict the location and severity of the defect using a technique similar to that developed by Cawley & Adams (1979 a, b, c). Both bulb and necking defects were located successfully, the maximum error in defect location being 3% while the maximum error in the length of the defect was 5%.

Natural frequency measurements may also be used as an in-service test, measurements after a period of service being compared with a baseline taken on the same component. The use of a baseline on the same component means that the effect of dimensional variations across a production batch is removed so greater sensitivity to small defects may be expected. One-off components may also be inspected which is impossible at the production stage in the absence of a standard for comparison.

Savage & Hewlett (1978) have used natural frequency measurements to monitor crack growth in concrete beams and also to check repairs made using epoxide resin. Chondros & Dimarogonas (1980) have shown that natural frequency measurements can be used to monitor crack growth in the weld at the root of a cantilever beam.

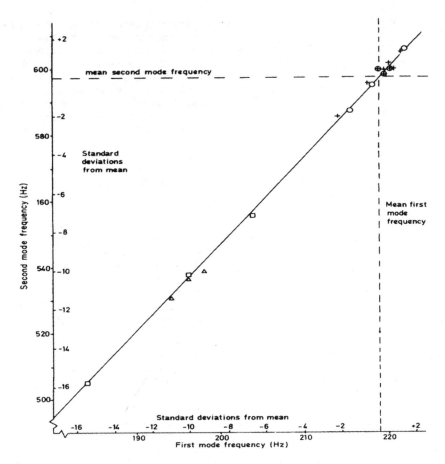

Figure 6. First and second mode flexural frequencies for good and faulty filament wound CFRP tubes. +, good tube (±45°); ○, tube with siliconised paper inserts (±45°); ⊕, tube with cut fibres (±45°); Δ, misaligned tube (±50°); □, tube with low fibre volume fraction (±45°) (From Cawley et al, 1985).

Natural frequency measurements may also be used to locate the possibility of damage within a structure. The stress distribution through a vibrating structure is non-uniform and is different for each natural frequency (mode). This means that any localised damage affects each mode differently, depending on the particular location of the damage. The damage may be modelled as a local decrease in the stiffness of the structure, and so, if it is situated at a point of zero stress in a given mode, it will have no effect on the natural frequency of that mode. On the other hand, if it is at a point of maximum stress, it will have the greatest effect. The location of the damage site requires the computation of the relative effect on several modes of damage at different sites within the structure. The experimentally measured changes may then be compared with the theoretically calculated changes for damage at different sites and the position of the damage deduced. The effect of damage may be determined by modelling the damage as a local decrease in the stiffness of the structure and carrying out a dynamic analysis of the system. Since the effect of damage is dependent on the stress distribution, which is in turn dependent on the mode shapes, the requirement of the dynamic analysis is that a reasonable approximation to the mode shapes be obtained. The dynamic analysis may be carried out by any convenient technique (e.g. finite element analysis).

The author and his co-workers carried out extensive tests on the method (Adams et al 1978, Cawley & Adams 1979a, b, c). Over forty tests were carried out on one- and two-dimensional structures with a variety of forms of damage including holes, saw cuts, fatigue cracks, crushing, impact damage and local heating. All these forms of damage were successfully detected and located by comparing the natural frequencies measured before and after damage. An indication of the severity of the damage was also obtained. One series

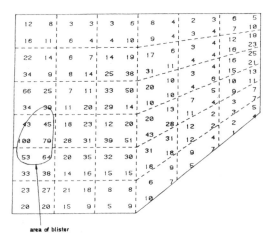

area of blister

Figure 7. Location chart for CFRP – skinned honeycomb sandwich panel damaged by local heating (From Cawley & Adams 1979c).

of tests was carried out on an automobile camshaft. Although this is a highly non-uniform structure, for the purposes of the damage location programme it was found to be sufficiently accurate to model the shaft as a uniform, straight bar. This test showed that it is unnecessary to expend large amounts of effort and computer time in producing an accurate dynamic analysis. The camshaft was initially damaged by a single, shallow saw cut of depth 1.7 mm. This corresponded to a 3.5% reduction in the cross-sectional area of that section of the bar. The damage was successfully detected and located by the vibration measurements.

Another structure tested was representative of the ways in which high-quality fibre reinforced materials are used in industry. This was a $610 \times 520 \times 10$-mm honeycomb panel with CFRP facings which is used for floor panelling in aircraft. The panel was made asymmetrical by removing one corner and was damaged by heating one face with a gas flame. This produced a large blister on this face but the damage was invisible from the remote side. Figure 7 shows that the damage was successfully located.

Pye & Adams (1982) used the vibration technique not only to locate a defect in a bar, but also to predict the mode I stress intensity factor (as used in fracture mechanics). This technique has great promise for monitoring fatigue cracks via the local reduction in specimen stiffness associated with the crack propagation. A similar technique was used by the Central Electricity Generating Board (Mayes & Davies 1976) to locate the position of transverse cracks in rotors. The location of the position of the crack was approached in a different manner to that described above. The effect of a transverse crack at various axial positions on the first four flexural natural frequencies of the rotor

was derived in terms of the stress intensity factor for a circumferentially notched bar and the experimentally determined mode shapes of the undamaged rotor.

Loland & MacKenzie (1974), Loland et al (1975a, b) and Loland & Dodds (1976) have applied the technique to off-shore oil rigs, using measurements of the natural frequencies of the framework obtained from the Fourier transform of the structural response to wave excitation. Their analysis did not attempt to locate the position of damage within a structural member but merely to predict whether a given member was fractured. There has been considerable interest in the technique in the oil industry and further discussion may be found in, for example, Coppolino & Rubin (1980) and Kenley & Dodds (1980). The main problem is that of excitation with these large structures, together with structural redundancy.

Global methods based on natural frequency and/or damping measurements have the major advantage that the whole component can be checked by measurements made at a single point. This means that very rapid testing is possible, particularly if a tap (impact) testing technique is employed. Not surprisingly, if a single point measurement is used to infer the quality of the whole component, the sensitivity of the technique tends to be lower than that obtained by local measurements such as ultrasonic inspection. Cawley & Adams (1979c) showed that damage equivalent to a crack through 1% of the cross-sectional area of a one-dimensional structure at a single section could be detected by natural frequency measurements. With two-dimensional structures, damage equivalent to the removal of about 0.1% of the area of the structure could be found. These values were obtained using base line measurements on the same component. The sensitivity to small defects is reduced by dimensional variations across a batch if the test is used at the production stage.

Damping is often more sensitive to damage than are natural frequencies and is insensitive to dimensional variations. However, it is much more difficult to measure accurately since care must be taken to minimise damping from the support system.

5 LOCAL VIBRATION MEASUREMENTS ON STRUCTURES AND COMPONENTS

Techniques in this category involve vibrating the test structure (usually at resonance) by applying an exciting force at a single point and measuring a local property of the structure in the particular mode of vibration at all the points of interest. If these measurements can be carried out by a scanning system, the test is potentially quick to carry out and, because local properties are being measured, it may be more sensitive than the global methods described earlier.

The presence of damage in fibre reinforced plastics results in the formation of cracks and crazes. When cyclic stresses are applied to a damaged composite material, relative motion takes place between the sides of the assorted cracks, resulting in the generation of heat (damping). The change in the overall level of damping in a structure is small for many form of serious but localised damage, while the damping change in this small local area may be large. Various authors have used this technique, usually referred to as vibrothermography, for nondestructive examining structures. Usually, it is necessary that the structure how low thermal diffusivity to prevent the rapid conduction of heat from the damaged area. It is therefore of least application to metals (unless the dissipation rate is large) and of greatest application to materials such as GFRP. Carbon-based composites have markedly higher thermal conductivity than glass-based ones, and are not so easy to test.

This technique was originally applied to fatigue tests on composites in which the stresses were applied by large, servo-hydraulic machines (Reifsnider & Williams 1974; Nevadunsky et al 1975). Henneke & Jones (1978) used ultrasonic excitation, while Pye & Adams (1981a, b) and Reifsnider & Stinchcomb (1976) used resonant vibration. This latter technique is better termed resonant vibrothermography. The advantage of using resonant vibration is that, for a given applied force, the stresses in the structure are greater than if the same force were applied at some other frequency, so that high-powered fatigue machines are not necessary.

Pye & Adams (1981b) used resonant vibration to detect shear cracks in unidirectional fibrous composites consisting of glass or carbon fibres in an epoxy resin matrix. They also considered the effect of the cyclic stress and frequency vibration on the temperature rises. They showed that, if other parameters remain constant, then the temperature difference ΔT between any two points of the structure is such that

$$\Delta T = K\sigma^n f$$

where σ is the peak cyclic stress, f the frequency of vibration, and the exponent n (usually greater than 2) a function of the damping-stress relationship in the region of the crack and K is a constant. Using resonant vibration, it is often possible to use higher cyclic frequencies than can be obtained using servo-hydraulic machinery. The above equation shows that this will increase the sensitivity of the test, thus allowing easier detection of the damage or, alternatively, a reduction in the cyclic stress used. The limit is how to couple the excitation to the structure without causing high energy loss in the coupling. Recent work by Busse & colleagues (Zweschper et al 2003) has shown that bursts of high-frequency high-amplitude vibration at about 20 kHz, together with Lock-in Thermography (Rantala et al 1996) can reveal damage in a CFRP plate.

Local damage in a structure tends to distort the vibration mode shapes. With the advent of laser holography systems, there has been an upsurge of interest in the exploitation of this effect in NDT. The technique involves vibrating the test structure at resonance and producing a time-averaged hologram of the motion. The method has been particularly successful for detecting skin-core disbonds in honeycomb panels. However, it is essential that the system be set up in a vibration-free environment. This means that it is only suitable for specialist applications.

The use of pulsed lasers make it possible to use the technique in the presence of background vibration. In this case, the best results are obtained by using impulse excitation and producing the hologram from views taken a few microseconds apart as the pulse propagates through the structure.

The coin-tap test is one of the oldest methods of nondestructive inspection. It has been used regularly in the inspection of laminates and honeycomb constructions. Indeed, Hagemaier & Fassbender (1978) found that it could detect more types of defect in honeycomb constructions than any other technique except neutron radiography. Until the early 1980s, however, the technique has remained largely subjective and there has been considerable uncertainty about the physical principles behind it. *It should be stressed again that this test is quite different from the wheel-tap test though the testing technique and subjective interpretation of the sound produced is similar in both cases. The wheel-tap test is a global test which investigates the whole test component from a tap applied at a single point, the difference between sound and defective components being detected from changes in the natural frequencies and damping. The coin-tap test will only find defects in the region of the tap, so it is necessary to tap each part of the structure under investigation.* The sound produced when a structure is tapped is mainly at the frequencies of the major structural modes of vibration. These modes are structural properties which are independent of the position of excitation. Therefore, if the same impulse is applied to a good area and to an adjacent defective area, the sound produced must be very similar. The difference in the sound produced when good and defective areas are tapped must therefore be due to a change in the force input. The impact on the damaged area has more energy at low frequencies but the energy content falls off rapidly with increasing frequency, while the impact on the sound area has a much lower rate of decrease of energy with frequency. This means that the impact on the defective area will not excite the higher structural modes as strongly as the impact on the good zone. The sound produced will therefore be at a lower frequency and the structure will sound 'dead'. The coin tap technique and its application have been described extensively by the author and Professor P Cawley (Cawley & Adams 1988, 1989).

6 CONCLUSIONS

The vibration techniques reviewed here make a very valuable contribution to NDT. The global methods involving measurements of natural frequencies and damping are extremely quick to carry out and give information about the integrity of the whole of the component from a single point test. They are not as sensitive to small localised defects as tests which involve point by point testing of the whole component, but their sensitivity is sufficient for many purposes and the speed and ease of testing which they offer means that their implementation is often worth investigating.

The local vibration techniques are particularly useful in circumstances where the use of the coupling fluids required for ultrasonic testing is undesirable. They are frequently used for the inspection of honeycomb constructions and composite materials.

The advent of microelectronics means that the cost and bulk of the equipment required to carry out vibration tests reliably and quickly is reducing rapidly. The time is therefore ripe for more attention to be paid to the opportunities which vibration techniques offer for improving inspection reliability and efficiency.

REFERENCES

Adams, R.D. 1972a. *Journal of Sound & Vibration* 23(2); 199–216

Adams, R.D. 1972b. *Journal of Physics* D5; 1877–1889.

Adams, R.D. 1987. Analysis of the damping properties of composites. *Engineered Materials Handbook, American Society for Materials*, 206–217.

Adams, R.D. & Cawley, P. 1985. Vibration techniques in nondestructive testing. *Research Techniques in Nondestructive Testing* R.S. Sharpe (ed.) 8; 303–360.

Adams, R.D., Cawley, P., Pye, C.J. & Stone, B.J. 1978. A vibration technique for nondestructively assessing the integrity of structures. *Journal of Mechanical Engineering Science* 20; 93–100.

Adams, R.D. & Fox, M.A.O. 1973. Principal mechanisms of damping in cast irons. *Journal of Iron Steel Institute* 211; 37–43.

Adams, R.D. & Vaughan, N.D. 1981. Resonance testing of cast iron. *Journal of Nondestructive Evaluation* 2; 65–74.

Brown, R.J. 1973. *Nondestructive Testing* 6; 81–85.

Cawley, P. 1984. *NDT International* 17; 59–65.

Cawley, P. & Adams, R.D. 1979a. The location of defects in structures from measurements of natural frequencies. *Journal of Strain Analysis* 14; 49–57.

Cawley, P. & Adams, R.D. 1979b. A vibration technique for nondestructive testing of fibre composite structures. *Journal of Composite Materials* 13; 161–175.

Cawley, P. & Adams, R.D. 1979c. Defect location in structures by a vibration technique. *ASME Paper* 79 DET–46.

Cawley, P. & Adams, R.D. 1988. The mechanics of the coin-tap method of nondestructive testing. *Journal of Sound & Vibration* 122; 299–316.

Cawley, P. & Adams, R.D. 1989. Sensitivity of the coin-tap method of nondestructive testing. *Materials Evaluation* 47; 448–563.

Cawley, P., Woolfrey, A.M. & Adams, R.D. 1985. *Composites* 16; 23–27.

Chondros, T.D. & Dimarogonas, A.D. 1980. *Journal of Sound & Vibration* 69; 531–538.

Coppolino, R.N. & Rubin, S. 1980. In *Proc. 12th Annual Offshore Technology Conference Paper OTC 3865.*

Davis, A.G. and Dunn, C.S. 1974. *Proc. Inst. Civil Engineers* 57(2); 571–593.

Fox, M.A.O. & Adams, R.D. 1972. Prediction of the damping capacity of cast iron from the variation of its dynamic modulus with strain amplitude. *Journal of Iron Steel Institute* 210; 527–530.

Fox, M.A.O. & Adams, R.D. 1973. Correlation of the damping capacity of cast iron with its mechanical properties and microstructure. *Journal of Mechanical Engineering Science* 15; 81–94.

Hagemaier, D. & Fassbender, R. 1978. *SAMPE Q* (July); 36–58.

Henneke, E.G. & Jones, T.S. 1978. *ASTM Cttee. D-30 Conference on NDE and Flaw Criticality for Composite Materials, Philadelphia.*

Kenley, R.M. & Dodds, C.J. 1980. In *Proc. 12th Annual Offshore Technology Conference Paper OTC 3866.*

Lilley, D.M., Adams, R.D. & Larnach, W.J. 1982. Location of defects within embedded model piles using a resonant vibration technique. *Proc. 10th World Conf. on Nondestructive Testing* 4; 77–84.

Lloyd, P.A., Joinson, A.B. & Curtis, G.J. 1975. *In Proc. Ultrasonics International*; 68–72.

Loland, O. & Dodds, C.J. 1976. *Offshore Technology Conference paper OTC 2551.*

Loland, O. & Mackenzie, A.C. 1974. *Mech. Res. Commun.* 1; 353–354.

Loland, O., Begg, R.D. and Mackenzie, A.C. 1975a. The dynamic response of a fixed steel offshore oil platform. BSSM/RINA Conf. Edinburgh.

Loland, O., Mackenzie, A.C. & Begg, R.D. 1975b. Integrity monitoring of fixed steel offshore oil platforms. BSSM/RINA Conf. Edinburgh.

Magistrali, G. 1975. Non-Destructive Testing 8; 32–37.

Mayes, I.W. & Davies, W.G.R. 1976. IMechE Conf. on Vibrations in Rotating Machinery, paper C168/76.

Nevadunsky, J.J., Lucas, J.J. & Salkind, M.J. 1975. *Journal of Composite Materials* 9; 394–408.

Peters, J., Snoeys, R. & Decneut, A. 1969. *In Advances in Machine Tool Design and Research*. S.A. Tobias & F. Koenigsberger (eds). *Proc. 9th Int. MTDR Conf. Part 2.*

Pye, C.J. & Adams, R.D. 1981a. Detection of damage in fibrereinforced plastics using thermal fields generated during resonant vibration. *NDT International* 14; 111–118.

Pye, C.J. & Adams, R.D. 1981b. Heat emission from damaged composite materials and its use in nondestructive testing. *Journal of Physics* 14; 927–941.

Pye, C.J. & Adams, R.D. 1982. A vibration method for the determination of stress intensity factors. *Eng. Fract. Mech.* 16; 433–445.

Ranatala, J., Wu, D. & Busse, G. 1996. Amplitude modulated lock-in vibrothermography for NDE of polymers and composites. In *Research in Nondestructive Evaluation* 7; 215–218.

Reifsnider, K.L. & Stinchcomb, W.W. 1976. In *Proc. Infrared Information Exchange.*

Reifsnider, K.L. & Williams, R.S. 1974. *Exp. Mech.* 14; 479–485.

Savage, R.J. & Hewlett, P.C. 1978. *NDT Int.* 11; 61–67.

Shen, C.H. 1982. Grinding wheel screening by the elastic modulus method. *Inspection & Quality Control in Manufacturing Systems PED* Vol. 6.

Spain, R.F., Schubring, N.W. & Diamond, M.J. 1964. *Materials Evaluation* 22; 113–117.

Zweschper, T., Riegert, G., Dillenz, A. & Busse, G. 2003. Ultrasound burst phase thermography (UBP) advances by ultrasound frequency modulation. *3rd Int. Conf. Emerging Technologies in Non-Destructive Testing, Greece.*

Emerging Technologies in Non Destructive Testing, Van Hemelrijck, Anastasopoulos & Melanitis (eds)
© 2004 Swets & Zeitlinger, Lisse, ISBN 90 5809 645 9

Evaluation of rock slope stability by means of acoustic emission

T. Shiotani

Research Institute of Technology, Tobishima Corporation, Chiba, Japan

ABSTRACT: Rock failure is one of the most frequent disasters as well as earthquakes in Japan. Evaluation of rock stability becomes thus an urgent and important issue to prevent eventual rock failure. Even for the transient failure processes as in rock slope, the deformation process in which micro-scale fracture develops macro-scale fracture might exist. Acoustic emission (AE) is a technique providing crucial information on a variety of fracture behavior as elastic waves generated due to the deformation. In the paper, essential aspects when performing AE monitoring in rock slope stability are methodically described, followed by a prospective method to install AE sensors into the rock slope (WEAD). Finally a successful application using WEAD is demonstrated and all of findings are briefly drawn.

1 INTRODUCTION

Geographically more than an 80% mountainous area covers Japanese national land, which resulted in forming numerous steep slopes in places in Japan. Development of the land into such steep slopes occasionally causes the tremendous disaster referred to as 'rock failure.' Since there are such a lot of inducements as earthquakes, typhoon, and torrential rains to trigger the rock failure, it could not be readily estimated that what a factor induced the final rock failure. The failure process of rock slope is thought to be a transient phenomenon comparing to those of soil slope, however, there should exist the fracture process in rock slope as well, irrespective to the duration of deformation. Acoustic emission (AE) is a powerful nondestructive technique, which detects elastic waves generated due to initiation, formation/growth and coalescence of cracks. Detected AE signals include a variety of information on AE-induced sources, and therefore fracture characteristics of AE sources can be evaluated when parametric/waveform analysis is properly carried out. A lot of research papers on AE application to rock monitoring have so far been reported (e.g., Goodman & Blake 1966; Yuda et al. 1984), while successful results that could evaluate and predict slope failure are quite few. This is, for example, because the type of AE sensors (resonant frequency of AE sensors) was chosen disregarding objective scale of fracture i.e., since micro-scale of fracture develops macro-scale of fracture even in the transient deformation of a vast rock slope, the resonant

frequency of AE sensors differs in objective scale of fracture. Thus, in the paper acoustic emission waves expected during rock deformation are described with reference to elastic waves due to the seismic activity. As an important issue for in-situ AE monitoring, AE signals due to precipitation are studied with model concrete specimens where the mechanical property (e.g., compression strength) develops with curing/aging processes. Then an effective method to detect AE waves for rock slope failure is introduced in comparison with conventional techniques using wave-guides in soil slope. Finally as for an example, long term in-situ AE monitoring in rock slope using a detecting technique proposed is reported and the verification of the proposed method is made in comparison to the slope behavior as borehole strains.

2 ACOUSTIC EMISSION WAVES EXPECTED DURING ROCK DEFORMATION

Is it possible to predict/evaluate rock failure only to set AE sensors onto/into rock slope? The answer would be 'No', because the AE technique is not a panacea to monitor a variety of fracture ranging from micro- to macro-failures. AE waves during rock deformation can be defined as ultrasonic waves in a narrow sense, whereas as seismic waves in a broad sense i.e., microscopic failure would be generated during the primary stage of failure process accompanied by an AE signal of high frequency (e.g., several hundreds kilo hertz), while during the tertiary stage, macroscopic failure would be occurred accompanied by an AE signal of low frequency, provided that fracture process is classified into three processes as primary, secondary and tertiary stage. Therefore, the resonant frequency of AE sensors should be determined corresponding to the scale of fracture to be focused. In other words, during the fracture process in rock slope, AE sensors with low resonant frequency is insensitive to early behavior of fracture stage, although they are very sensitive to the latter part of fracture (just before eventual failure). Thus, AE sensors with low resonant frequency would be not so effective as to know the early behavior of slope failure. With AE sensors of low resonant frequency, sometimes variation trends of both AE activity and displacement would be the same.

3 ENVIROMENTAL NOISES AFFECTING IN-SITU AE MONITORING

3.1 Environmental noises in in-situ AE monitoring

It can be readily expected that natural phenomena such as sunshine and precipitation would produce AE activity when performing in-situ AE monitoring. The AE activity due to the natural phenomena is referred to as 'environmental noises.' The environmental noises acquired only in short term monitoring could be manually excluded with waveform observation by AE experts. For long term rock AE monitoring, however, such a manual filter would be difficult to apply the raw data, since a large numbers of AE data would be collected. Therefore before the installation of AE sensors into the rock, studying factors influencing long term AE monitoring is a crucial thing. Among the environmental noises, especially for rock slope, precipitation becomes a municipal factor to induce noises for the long term monitoring. The following experiment was carried out to know the AE generating-area influenced by raindrops. Concrete specimens from early age are used to hit the raindrops. With development of concrete strength due to curing/aging, a variety of physical properties of concrete are reproduced and AE waves differed in materials properties are studied by the concrete.

3.2 Raindrops

Raindrops larger than 100 μm in diameter are defined as water droplets, whereas those of smaller than 100 μm are called cloud particles since the dropping velocity of those becomes smaller than that of up-current of air. A water droplet larger than 0.5 mm shows an ellipse due to both of friction of air and surface tension. Moreover, when the diameter grows larger than 3.5 mm, the raindrop is divided into two droplets namely there is no exist of rain drops of more

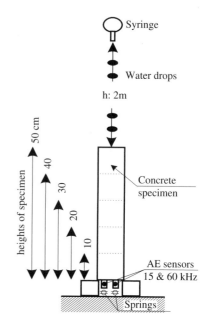

Figure 1. Experimental configuration for the raindrop test.

than 3.5 mm in diameter. In general for a representative of raindrops, it has a diameter of 1 mm and a terminal velocity ranging from 4 to 6 m/s. While in a larger raindrop, it has a diameter of 3 mm and a terminal velocity of 9 m/s.

3.3 Specimens and experimental condition

Figure 1 shows an experimental configuration of the specimen used for the raindrop test. Three prism specimens ($h\,150 \times w\,150 \times h\,500$ mm), consisting of Portland cement and aggregate were prepared. A sand-total aggregate ratio s/a was 60% and maximum size of aggregate G_{max} was 15 mm. The specimen was cured for one day in tap water and two specimens cut into several pieces with lengths of 10, 20, 30 and 40 cm. In this way, five lengths of the specimen as 10, 20, 30, 40 and 50 mm were prepared for the test. Raindrops were produced with a syringe at a height of 2 m from the top surface of each specimen. A diameter of one water drop was estimated 2.75 mm resulted from the average weight of total weight for ten water droplets. Thus 'large water droplets' were reproduced in this test. At ages of 1 day, 3 days and 15 days the test was carried out. Young's modulus corresponding to those ages are 1680 (12.9), 2249 (30.2) and 2761 N/mm² (46.5 N/mm²) respectively, where the value between parentheses shows the compression strength. Onto the bottom surface of the specimen, two types of AE sensors of 15 and 60 kHz resonance were placed with a constant pressure by two springs.

Table 1. Affecting range due to precipitation.

Age	σ_c^*	Z_a^{**}	AE detectable range (mm)			
			Threshold 40 dB		Threshold 50 dB	
			15 kHz	60 kHz	15 kHz	60 kHz
1	12.9	78256.2	2420	1500	1740	1000
3	30.2	88700.9	2030	1300	1500	950
15	46.5	95259.6	3100	2420	2230	1500

Age is shown in day, * N/mm^2, ** kN/m^{2}s.

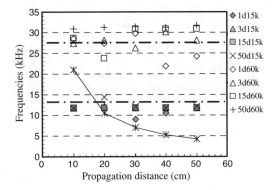

Figure 2. Apparent frequencies and peak frequencies as function of wave propagation distance. Legend shows an age of concrete and a type of sensor, respectively, e.g., X50d15k denotes a result of an age of 50 days concrete and using a 15 kHz resonant sensor.

Thus, AE waves generated by water droplets hitting the surface were propagated through the specimens and detected by the both AE sensors, and then the AE signals detected were amplified by 40 dB at preamplifiers and processed by a Mistras AE system (Physical Acoustics Corporation). Not only AE parameters but AE waveforms were recorded under the condition of 2 MHz sampling rate and 9 kwords wavelength. To avoid reflection of AE waves, a maximum lockout time of 65534 μs for an individual hit was used for the monitoring condition. Ten water droplets were successively produced with the syringe by five seconds, subsequently they were randomly generated for a few minutes.

3.4 Result

Affecting ranges by the raindrops is given in Table 1. Acoustic impedance Z_a ranges from 78256.2 to 95259.6 with the development of strength. Corresponding to the variation of acoustic impedance, the detectable range of raindrops changes from 950 mm to 3100 mm. For example, at an age of 15 days with a 50 dB threshold and by using a sensor of 60 kHz resonant frequency, 1500 mm would be a affecting area due to raindrops.

Correlations among AE parameters were studied, however, good relations among them could not be found without a correlation between ring-down counts and duration (Shiotani et al. 1999a). Figure 2 shows apparent frequencies exhibited by chain lines determined by both of ring-down counts and duration. Also in the figure, peak frequencies resulted from FFT calculations are drawn with several symbols with respect to the propagation distance. In addition, peak frequencies due to natural precipitation are presented by cross symbols as X and +, in which the specimen at an age of 50 days was subjected to the test. As shown in Figure 2, both the apparent frequency and peak frequency were distributed between 10 to 15 kHz in case of 15 kHz resonant sensors, and between 25 to 35 kHz in case of 60 kHz resonant sensors. These distributions of apparent/peak frequency of AE waves

appeared irrespective to the materials properties. Considering two aspects: 1) both frequencies distributed around 15 kHz in 15 kHz resonant sensors, suggesting that derived frequencies corresponded well to the resonant frequency of the sensor and 2) those distributed around 25 kHz in 60 kHz resonant sensors, AE waveforms generated by precipitation in rock materials have a resonant frequency around 30 kHz or less. This fact that there is a characteristic peak frequency around 30 kHz in rainy induced AE waves was further clarified by the follow up experiment using Davies's bar technique (Shigeishi et al. 2000).

Accordingly, a high pass filtering of 30 kHz can be thought to be effective to eliminate environmental noises due to precipitation, however, the high pass filtering of 30 kHz resulted in disregarding AE signals that would be useful information on macro scale of fracture, corresponding to a low frequency of AE signals. This leads to a conclusion that the filtering of environmental noises due to precipitation is in practical difficult to apply. Therefore, in order to eliminate those noises, it is essential to apply a spatial filter (or to place 'guard sensors'). In the WEAD monitoring which would be proposed, we perform the AE measurement in a deeper location namely deeper than 2.5 m from the rock surface, where rainy induced AE waves could not reach (see Table 1).

4 DETECTING TECHNIQUE FOR AE WAVES DUE TO SLOPE DEFORMATION

4.1 Conventional techniques

Generally in rock slope, effective detection of AE waves generated due to deformation of rock is difficult because AE waves are strongly attenuated when they

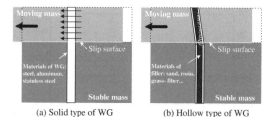

|(a) Solid type of WG|(b) Hollow type of WG|

Figure 3. Two types of conventional wave-guides and AE generation expected due to the movement of soil mass.

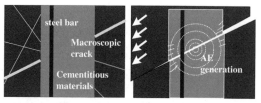

Figure 4. Schematic behavior of the WEAD during a deformation process of rock.

propagate across joints/cracks in rock slope. In geotechnical application of the AE technique, to avoid the energy attenuation of AE waves, wave-guides have so far been used to lead AE signals to AE sensors (Hardy & Taioli 1988; Nakajima et al. 1988). Two types of wave-guides have been employed. One is a low-attenuation type of wave-guide devised as to lead weak AE signals to AE sensors. Solid materials as metal are well adopted as this low-attenuation type of wave-guides (see Figure 3a). In the case using solid materials, AE waves due to friction between the wave-guide and a moving mass of soil would be generated. Another is deformation related wave-guides designed as to generate self-emissions of the wave-guide due to the deformation of the guides, along with the rock deformation as shown in Figure 3b. Pipes are used for this type of wave-guides and sand, rosin and fiberglass are filled in the pipes. For this hollow type, AE waves are expected to occur due to fracture of the filler. Since these are devised on relatively ductile materials such as soil materials, these wave-guides resulted in being difficult to apply the rock materials.

4.2 Rock slope (WEAD)

Following are desirable conditions of wave-guide for rock materials: 1) AE waves can propagate up to the AE sensor as to avoid strong attenuation influenced by rock conditions; 2) AE waves detected should reflect actual fracture mechanisms of rock; and 3) stability of the rock slope can be reasonably evaluated by AE signals acquired.

In order to meet the above conditions on the wave-guide for rock materials, WEAD (wave-guide for AE waves due to rock-deformation) has developed (Shiotani et al. 2001b). The WEAD consists of cement based materials and reinforcement with high resonant AE sensors of 60 kHz. In the WEAD application, borehole excavation is necessarily performed for both the installation of AE sensors and obtaining mechanical properties of the rock. A variety of physical tests are subsequently carried out for core samples, and a mixture proportion of cement-based materials is determined as to correspond to the mechanical properties of the core.

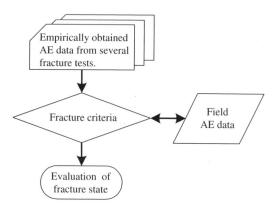

Figure 5. Evaluation flow of AE data acquired from in-situ monitoring with reference to empirically obtained fracture criteria.

Figure 4 shows the schematic behavior of the WEAD during the deformation process of rock. As in Figure 4, due to the deformation of rock, the WEAD simultaneously deforms along with the rock, followed by AE waves generating within the cement-based material. Fracture of the WEAD thus occurs similarly to that of the surrounding rock. Consequently in the WEAD application, characteristics of the AE waves detected would be compatible to those due to actual rock-deformation. Because an initial condition of the WEAD is intact, AE waves generated can efficiently propagate to the AE sensors without the influence of the cracks. The WEAD has another advantage. Because the filler of cement-based materials of the WEAD is mixed from known materials as cement, sand and chemical admixtures, AE characteristics of the filler can be experimentally obtained by several fracture tests. Thus, comparing field data with empirically obtained data, a fracture state of the rock can be evaluated in accordance with Figure 5. In the following stages, it becomes difficult to detect AE signals due to rock deformation: the stage where the WEAD has already contained a large number of cracks, namely cracking condition of the WEAD is similar to that of surrounding rock; and that where macroscopic fracture are dominantly generated namely macroscopic fracture occurred, accompanied by AE signals of low frequency components less than frequency response of used AE sensor. To solve this problem, reinforcement of steel bar is installed into the borehole with AE sensors. In this stage, directly fracture-related AE activity

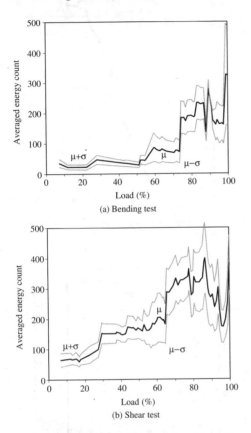

(a) Bending test

(b) Shear test

Figure 6. Average energy for bending (a) and shear (b) test as function of load applied.

cannot reach the AE sensors, however, it can be possible by the installation of reinforcement as mentioned above that large magnitude of AE signals are generated resulted from the friction between the reinforcement and surrounding cement-based materials. Consequently they can travel to the AE sensors through the reinforcement. It is noted that these AE waves may have nothing to do with the actual fracture mechanism of rock, however, AE counting and AE parameters could provide reasonable information on the fracture state. The reinforcement thus plays an important role in the WEAD. More details on the WEAD can be found in the reference (Shiotani et al. 2001b).

4.3 Fracture criteria in the WEAD

Since the WEAD consists of known materials, characteristic AE parameters can be evaluated by using parameters variations resulted from several fracture experiments in the laboratory (see Figure 5). Figure 6 shows the average AE energy for a bending failure test (a) and a shear failure test (b) with respect to applied load. The averages in the charts are drawn by means of 'moving average' on the basis of 100 data of AE hit. From the bending chart (see Figure 6a), the energy of AE is distributed around 50 in the primary fracture stage of bending. In the secondary fracture stage of bending, the energy appeared between 50 and 100, followed by the value ranging from 100 to 200 in the tertiary stage of bending. The value during the tertiary stage equivalents to those in the first stage of shear fracture (see Figure 6b). In this way, both of fracture types and fracture levels are potentially classified as shown in Table 2. Based on the detected AE waves from the monitoring site, the stability of the rock slope is evaluated with reference to Table 2. In the table, Ib-value stands for improved b-value and obtained from a negative gradient of cumulative type of amplitude distributions. The calculation method of

Table 2. Fracture criteria based on experimental results.

AE parameters					Fracture conditions &
C	E	Ib	G	FL	patterns expected
<40	<50	Increase up to 0.15	10	I	Early stage in b
	50–100	Decrease down to 0.04	–	II	Intermediate stage in b
		Repetition between I & II	5–10	III	Final stage in b
>40	100–200	Variation between 0.02 & 0.06	<5	IV	Final stage in b and early stage in s
	200–300	Variation between 0.02 & 0.06	<5	V	Intermediate stage in s
	>300	Variation between 0.02 & 0.06	<5	VI	Final stage in s

C: ringdown count, E: energy count, Ib: Improved b-value, G: grade, FL: fracture levels estimated, b: bending and s: shear.

Ib-value can be seen in the reference (Shiotani & Ohtsu 1999b; Shiotani et al., 2001a). During the increasing process of Ib-values, small scale of fractures predominantly appear in comparison with large scale of fractures whereas the Ib-value shows decrease in case of intensive generation of large scale of fractures. A great advantage of Ib-value analysis is to know the onset of fracture, however, it is noted that Ib-value cannot distinguish fracture types among tensile, shear and mixed mode between them. Dividing a peak amplitude (dB) by a rise time (μs) in an AE waveform, an apparent initial gradient can be obtained. Grade (dB/μs) is a parameter obtained from this kind of calculation (Shiotani et al. 2001b). The same ideas have been applied to concrete materials (see Iwanami et al. 1997; Uchida et al. 1999). A small value of grade corresponds to a rapid deformation of fracture as tensile type of fracture, while a larger value corresponds to a slow deformation of fracture as shear type of fracture.

5 IN-SITU AE MONITORING

5.1 Conditions of monitoring site and measurements

Geologically the monitoring site is made of hornfels, and the slope dips at an 80° angle. In the slope, diagonal joints are observed perpendicular to the slope surface. AE monitoring was started to perform with the WEAD after removal of unstable rocks to avoid disaster. Making reference to mechanical properties of rock experimentally determined, a mixture of cement based material for the WEAD was designed. Figure 7 shows the sensor arrangement and joint condition. Five AE sensors of 60 kHz resonance were installed into the slope at even space of 1.5 m. Also one AE sensor (60 kHz) was placed in the control room as well to monitor electric noises. To eliminate undesirable emissions in general induced on the slope surface by rain, wind and so forth (see chapter 3: environmental noises affecting in-situ AE monitoring), the space between slope surface and Ch-5 (located closest to the surface) was set in 2.55 m, moreover the surface portion of 1.0 m was filled with fine sand. AE signals generated from the slope were amplified by 40 dB at sensor-integrated preamplifiers, and both of AE parameters and waveforms over the threshold 40 dB were acquired by a remote Mistras AE system (Physical Acoustics Corporation). A sampling time and a wavelength of the waveforms are 1 MHz and 1 kwords respectively. To collect data and to change the initial condition remotely, the system, which can be controlled by a modem communication, is developed.

Besides AE monitoring, a seismometer, crack gauges in three dimensions and borehole strain gauges were set into/onto the slope. A long term monitoring

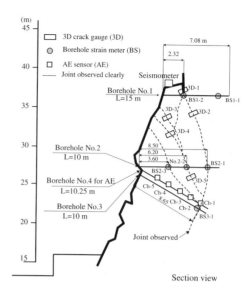

Figure 7. Monitoring rock slope and arrangement of a variety of sensors.

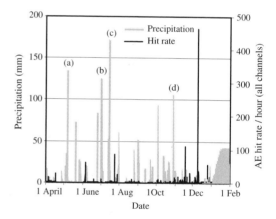

Figure 8. AE hits activity between April 1999 and February 2000 together with precipitation.

for three years was carried out, however, consequently a large scale of rock failure could not be observed. Therefore in the following results, the microscopic behavior of the slope would be only focused.

5.2 Result and discussions

5.2.1 Influence of precipitation in AE monitoring
Figure 8 shows AE hits activity between April 1999 and February 2000 together with precipitation. It is found that there is no correlation between precipitation and AE activity even in cases where precipitation over 100 mm, shown by (a) to (d), are recorded. This

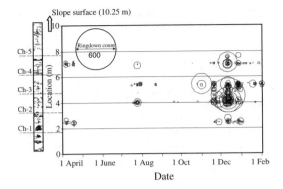

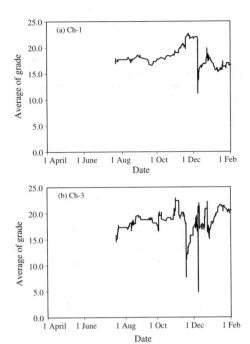

Figure 9. AE sources along the borehole with respect to elapsed time.

suggests that AE data acquired were not directly induced by rainfall, and thus the method of installation for AE sensor avoiding rainy induced AE activity resulted in successful.

5.3 AE source locations and AE parameters

Because AE sensors are linearly placed along the borehole in the WEAD, one-dimensional sources can be identified as shown in Figure 9. The diameter of circles reflects the magnitude of averaged ringdown counts from a set of AE hits in an event. At the beginning of December, AE sources are intensively observed between 4 m and 5 m (around the location of Ch-3) along the borehole. Figure 10 shows the behavior of grade for Ch-1 (a) and Ch-3 (b) during the period. A sudden drop is found in the beginning of December in Ch-1, and bending fracture is estimated from the Table 2 since the grade value appears larger than 10. While in Ch-3, twice decreases including a drop down to 5 are observed between November and December. Because the grade around 5 implies the generation of shear type of fracture (see Table 2), shear fractures were potentially generated around the location of Ch-3 in the beginning of December. Again, December is the month when intensive AE generation occurred around the location of Ch-3 (see Figure 9).

Figure 11 shows borehole strains measured in borehole No. 2 during the period. In the borehole No.2, three borehole strain-meters denoted as BS2-1, BS2-2 and BS2-3 from the bottom are installed. In BS2-1 which was set in the deepest location, a remarkable variation was not observed, whereas in BS2-2 and BS2-3, long periodic variations accompanied by rapid changes were observed. The rapid changes were thought to be induced by the local deformation of the rock, whereas the long periodic variation was influenced by the volumetric expansion/shrinkage of rock masses along with the change of outside temperature. To discriminate the rapid changes from the variation

Figure 10. Average grade of 100 data as function of time. (top: Ch-1, bottom: Ch-3).

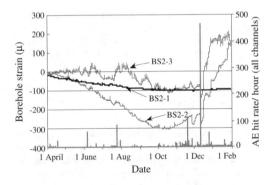

Figure 11. Borehole strains and AE hit rate as function of time.

of the strains, a strain increment per one hour are drawn in Figure 12. In BS2-1, the long periodic variation disappeared, and good agreement with the AE activity was not observed. In BS2-3 however, the strain increments changed rapidly when higher AE activities were obtained. From the detail observations of the rock conditions by borehole investigation with a CCD camera, the AE monitoring well No. 4 is actually associated with a well of borehole strain monitoring No. 2 with several closed cracks. Hence, strong relationship between borehole strains BS2-3, and AE activity around the location of Ch-3 is thought to be

43

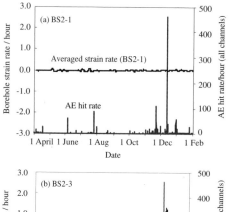

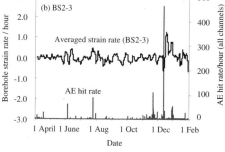

Figure 12. Strain increment and AE hit rate as function of time (top: BS2-1, bottom: BS2-3).

reasonable, although induced phenomena is still unsolved. Consequently, from all of the results obtained, it becomes clear that the slope had transiently experienced the local fracture in the beginning of December.

6 CONCLUSION

In the paper, essential aspects when performing AE monitoring in rock slope stability were methodically described, including a prospective method to install AE sensors called WEAD, followed by an in-situ application. All of the findings of this paper can be summarized as followings:

1. Environmental noises should be studied in advance of AE monitoring. AE signals due to precipitation, for example, are difficult to exclude from acquired AE data, and therefore AE sensors should be properly installed in the rock slope considering affecting areas by precipitation.
2. AE detecting techniques for soil materials are difficult to apply for rock materials from several reasons. The WEAD introduced had a prominent performance for evaluating rock stability.
3. With the WEAD application, in-situ AE data showed a cross correlation with increments of borehole strains. For the eventual failure of rock slope, further

studies should be needed, however, the WEAD has potential to follow and evaluate the final failure.

ACKNOWLEDGEMENT

The experiments on water droplets were made as the joint research project with Public Works Research Institute, and also the in-situ AE monitoring was carried out as the joint research with Civil Engineering Research Institute of Hokkaido. All those who involved in the research are greatly acknowledged for their contributions.

REFERENCES

Goodman, R.E. & Blake, W. 1966. Rock noise in landslides and slope failure, Highway Research Record 119: 50–60.
Hardy, H.R. & Taioli, F. 1988. Mechanical waveguides for use in AE/MS geotechnical applications. In K. Yamaguchi, I. Kimpara & Y. Higo (eds), Progress in Acoustic Emission IV: 293–301. Tokyo: JSNDI.
Iwanami, M., Kamada T. & Nagataki, S. 1997. Application of AE technique for crack monitoring in RC beams, Proceedings of Cement & Concrete 51.
Nakajima, I., Sato, J., Taira, N. & Kubota, N. 1988. The observation of landslide by the acoustic emission monitoring rod. In K. Yamaguchi, I. Kimpara & Y. Higo (eds), Progress in Acoustic Emission IV: 273–281. Tokyo: JSNDI.
Shigeishi, M., Shiotani, T. & Ohtsu, M. 2000. A consideration about the rainy influence in field AE measurement. In T. Kishi, Y. Higo, M. Ohtsu & S. Yuyama (eds), Progress in Acoustic Emission X: 177–182. Tokyo: JSNDI.
Shiotani, T., Miwa, S. & Monma, K. 1999a. Characteristic of elastic waves induced by raindrops, Proceedings of the 6th Domestic Conference on Subsurface and Civil Engineering Acoustic Emission: 7–12. Tokyo: MMIJ (in Japanese).
Shiotani, T. & Ohtsu, M. 1999b. Prediction of slope failure based on AE activity. In S.J. Vahaviolos (ed.), Acoustic Emission: Standards and Technology Update, ASTM STP 1353: 156–172. PA USA: American Society for Testing Materials.
Shiotani, T., Li, Z., Yuyama, S. & Ohtsu, M. 2001a. Application of the AE improved b-value to quantitative evaluation of fracture process in concrete-materials. In K. Ono (ed.), Journal of Acoustic Emission 19: 118–133. Los Angels CA USA: Acoustic Emission Group.
Shiotani, T., Ohtsu, M. & Ikeda, K. 2001b. Detection and evaluation of AE waves due to rock deformation. In M. Ohtsu (ed.), Construction and Building Materials 15(5–6): 235–246. London: Elsevier Science.
Uchida, M., Okamoto, T., Tsuji, N. & Ohtsu, M. 1999. NDT of concrete members by ultrasonic, Proceedings of Structural Faults & Repair-99 (CD-ROM).
Yuda, S., Hashimoto, Y., Takahashi, K. & Kumagai, M. 1984. Prediction of slope failure by acoustic emission technique, Progress in Acoustic Emission II: 660–667. Tokyo: JSNDI.

Visual and optical inspection

Emerging Technologies in Non Destructive Testing, Van Hemelrijck, Anastasopoulos & Melanitis (eds)
© 2004 Swets & Zeitlinger, Lisse, ISBN 90 5809 645 9

Non-destructive testing (NDT) and vibration analysis of defects in components and structures using laser diode shearography

W. Steinchen, Y. Gan, G. Kupfer & P. Mäckel
Institut für Maschinenelemente und Konstruktionstechnik (IMK),
Labor für Spannungsoptik, Holografie und Shearografie (Lab SHS),
Universität Kassel, Kassel

ABSTRACT: The demands for greater quality and product reliability has created a need for better techniques of non-destructive testing (NDT/NDI), particularly, technique for on-line inspection. Optical interferometric methods, such as speckle pattern and speckle pattern shearing interferometry, enjoy the advantages of being full field and non-contacting. Therefore, they have already been accepted by industry as a tool of NDT/NDI evaluation in the recent years and they are gaining more and more applications for non-destructive testing, strain measurement and vibration analysis. However, a relatively large coherence length for the used laser are usually applied in these measuring devices. It is obviously that these lasers are expensive in price and large in volume. Thus the development of a simple, mobile and inexpensive industrial testing tool for these measuring methods is desirable.

1 CLASSIFICATION OF THE MEASURING JOBS

1.1 Shearographic material testing and quality assurance

The first step towards recognising defects in components using the principle of shearographic speckle interferometry is to determine a suitable measuring process from the point of view of the recording technique and both the loading and the excitation methods.

The scope of the test, which is determined by means of shearographic laser speckle interferometry on the surface of the object, comprises the relative deformations or the deformation gradients or strains depicted in the selected part of the image.

1.2 Allocation of the weighting and recording techniques

Basically, according to the nature of the loading we distinguish three deformation processes. These may include the static, non-stationary (e.g. thermal, impact), and dynamic deformation processes.

Each of these deformation processes requires a different recording technique, in conjunction with the conventional temporal phase shifting technique.

2 PHASE SHIFTING TECHNIQUES

2.1 Conventional temporal phase shifting method

Deformations as produced by means of a constant static differential load (contact stress, uniformly distributed compression load internal pressure) can be measured and evaluated using the conventional temporal phase shift technique beside the spatial phase shifting method. This involves plotting at last three, as a rule four speckle intensity distributions both in their unstressed state (first deformation stage) and in their stressed state (second deformation stage). The phase distribution ϕ is calculated from the speckle pattern of the first deformation stage, and the phase distribution ϕ' from the speckle pattern of the second deformation stage. Through the digital subtraction of ϕ' and ϕ one arrives at the relative phase change Δ, which with a vertical sensitivity vector k_s is proportional to the relative out-of-plane slope of deformation gradient of the tested structure (Yang et al. 1995) in the boundary value observation.

For detecting the defect positions of the specimens they must be tested under similar conditions of operation.

2.2 Loading principles and defect detection

The specimen is loaded by different kinds of loading like tension, compression, bending, and torsion respectively. However not only the type of loading is important but the direction of the introduced force acts as an aid for detecting the defect. The complex interrelation between exposure technique, type and direction of the applied loading makes it difficult to implement this excellent testing method especially since the appearance of the defect in the phase map is unknown. Thus pretests are required.

2.3 Temporal phase shifting with large deformation changes

A variation of the simple recording technique of the temporal phase shifting method becomes evident when a very high load has to be applied to the tested object. If a very high load is induced, causing a correspondingly large deformation, it can be that the appropriate measuring range is slightly exceeded. The test range is restricted, on the one hand, by the decorrelation between the speckles and, on the other hand, by the limited resolution of the fringes connected with the pixel array. The dividing-up of the load to be applied into smaller loading steps (the so called cascade measurement) provides a solution for the detection and evaluation of large deformations. This gauging option is especially important for quantitative measurements. As long as linear-elastic problems are targeted, this should therefore be irrelevant for the qualitative measurements needed for quality assurance (Hung 1986, Hung 1991).

In the following the separation of the total load to be applied into smaller loading steps is shown. Shock absorber disks are applied in steering columns of passenger cars for damping. Such disks are manufactured from steel, fibers and rubber as high stressed compound elements subjected by shocks and vibrations. Twenty textile fibres are combined in a fiber package. Interferometric methods like ESPI and digital shearography are used for quality assurance and quality control. The following different design variants were investigated by ESPI resp. shearographic measurements in series:

- Shock absorber disks with and without defects (missing of all fiber packages (Fig. 1c) resp. regular reinforced disk)
- One until four fiber packages between the holes resp. around the bushs on the outer diameter of the disk (Fig. 1a right)
- A varying number of fibres in the fiber packages (Fig. 1b)
- A disk with two opposite fiber packages
- A disk manufactured from fibers covered by oil
- A disk existing of insufficiently coated fibers.

Figure 1a. Left: Fiber reinforced steering shock absorber; right: the shearogram shows the unequal interference fringes due to non-uniformly covered textile fibers (top and bottom).

Figure 1b. Shock absorber disk reinforced; left: by 10 fibers per package; right: by 20 fibers per package; same torque clockwise.

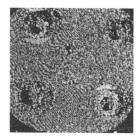

Figure 1c. Shock absorber disk without reinforcement, just rubber material.

The loading device contains an over-mounted shaft where a loading disk is fixed at the end. The pins of the loading disk fit in the opposite built-in steel bushs in the shock absorber disk. A lever with two pins introduces the external forces in the tested object so that a static torque is created in the shock absorber disk which is subjected by a preload of 5.3 Nm due to the dead load of the lever. In this state of loading the first shearogram and after the additional loading the second shearogram are taken. The first shearogram is subtracted from the second one by the program ShearwinNT and the final interferogram represents the fringe pattern describing the deformed state of the shock absorber disk. Then the load is increased. Out of the numerous design variants only a few rubber disks were selected to show the results of the investigations (Fig. 1).

2.4 *Temporal phase shifting with alternate loading and unloading levels*

In the case of periodic levels of loading, it is necessary to modify or adjust the recording technique of temporal phase shifting to suit the corresponding deformation behaviour.

To do this, four intensities are recorded, during the constantly loaded maximum loading stage and four intensities in the minimum loading stage, and the phase positions ϕ and ϕ' are calculated. The relative phase distribution is calculated, in turn, by means of digital subtraction of the two phase distributions.

The alternate loading and unloading of the tested object is necessary if the testing range is very large and can only be measured one step at a time. Step by step, the individual segments of the tested area are loaded and observed. The essential factor is that each loading and unloading is mapped singly for the examination of each consecutive segment. The frequency basically depends on the time factor in which the loading and unloading can be carried out, or in the time it takes to shift the testing range. The time needed to record and store the four intensities amounts to approximately one second according to (Yang et al. 1995). If this testing method is to be adopted by industry and the measuring process is automated, the loading and the recording of the intensity distributions have to be synchronised.

A cylinder of the kind actually used in conveying engineering for the transportation of paper and similar materials is examined. This roller basically consists of a metallic cylinder glued with an outer rubber layer. During production, impurities on the metal surface can lead to partial delamination defects between the metal and the rubber.

In this application, the cylinder is evacuated in a vacuum chamber. The environmental pressure between the metal and the rubber in the area of the delaminations causes an out-of-plane deformation of the rubber at those points where there is no bonding between the metal and the rubber. These deformations are visible in the phase image (Fig. 2). In order to completely scan the cylinder, divided into separate test areas, for possible disbonds, a test cycle, according to the alternate loading and unloading, is called for. From a segment of the cylinder a phase image recording ϕ_1 is made under loading (p max) and subsequently a phase image recording ϕ_1' in its unstressed state (p min).

Then the cylinder and the camera are moved nearer to each other without altering the state of stress and a new testing area is set up. Without changing the load, a phase image recording ϕ_2 is made of the new test area and, following reloading, a further phase image

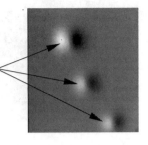

Buckling in the rubber layer around delaminations

Figure 2. Demodulated phase image a section of the of conveying cylinder under vacuum.

recording ϕ_2'. It should be noted that this procedure saves one load change at every gauging.

2.5 *Temporal phase shifting during a noncontrolled deformation process*

Another type of deformation that is relatively difficult to control and which allows the deformation of the test specimen under investigation, is thermal loading. The warming-up and cooling-down process can also be mapped by means of the recording technique of temporal phase shifting. However, the state of deformation is rarely long enough or steady enough to allow the recording of all four intensities, since the linear, time-dependent warming-up or cooling-down process can only be controlled very roughly.

The phase changes $\Delta\phi_1$ and $\Delta\phi_2$ occurring within this test period, measured in observation of the uncontrollable deformation of the tested object, is relatively small compared with the average phase change $\Delta\phi$ actually required, which is limited by the average measuring time and normally takes the maximum time. When calculating the phase change $\Delta\phi$ that is actually required, a relative phase change is measured, due to the uncontrolled deformation. However, the resulting inaccuracy is negligibly small for quality assurance testing purposes. It is comparable with a gauging error that occurs when quoting an incorrect wavelength from a laser source. When calculating the phase position, the error amounts to approximately 1% (Owner-Petersen 1991) with a wavelength difference of, for argument's sake, 532 nm instead of 685 nm.

The task is to detect the delaminations and to make them visible using digital shearography. In the laboratory experiment the rubber layer is heated using a hot-air drier. The rubber is deformed relatively homogeneously with steady thermal loading and there is no areal expansion of the rubber where the lamination is intact.

The stationary radial expansion becomes noticeable in the phase image through a constant fringe or greyish distribution (Fig. 3a). Only where the lamination is missing does the rubber layer show signs of

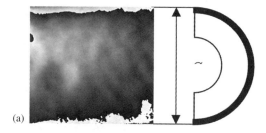

(a)

Bulging resulting from different thermal expansion
coefficients of the rubber layer compared with steel core

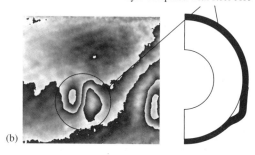

(b)

Figure 3.　(a) Cylinder segment without flaws. (b) Cylinder segment with flaws.

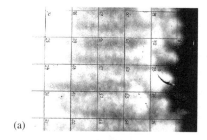

(a)

(b)

Figure 4.　NDT by means of vibration analysis showing a prepared default by improper glueing of a honeycomb structure. (a) Honeycomb panel (size: $1000 \times 1200 \times 12 \, mm^3$), size of the grid: $100 \times 100 \, mm^2$. (b) Time averaged digital shearogram with natural frequency f = 514 Hz.

bulging, due to the areal expansion of the rubber layer in view of the greater thermal expansion coefficient of rubber, as compared with steel, to balance the different change of length with respect to the steel body. The bulging is interpreted as an area of disbond in the phase image (Fig. 3b).

The first illustration (Fig. 3a) shows the longitudinal length of cylinder without any signs of delamination. Figure 3b shows the longitudinal length of cylinder displaying disbonding, visible in the form of buckling of the rubber layer in the phase map as a result of the thermal expansion.

2.6　Recording of periodic deformation processes using real-time subtraction method

If a body is repeatedly deformed by a certain force under recurrent periodic operations, the recording of just one intensity distribution of an almost stationary process, from a certain frequency level compared with the video rate (25 pictures/s) onwards, can prove to be fairly difficult. This periodic time-dependent deformation occurs when the tested object is excited, for instance, by a piezoelectric crystal as a very small shaker. In order to map the deformations, a recording technique allowing the real-time observation of the dynamic deformation of a tested specimen using the loading method is required. The recording technique that permits a real-time observation synchronised

with the videos is the 'real-time subtraction' described in (Steinchen et al. 1996, Steinchen et al. 2002).

When the tested object vibrates in a stationary state at a frequency f, which is as an order higher than the video rate, the (n−1) th image that is plotted is a time-averaging shearogram that was recorded during a single video rate. The following n th image is now digitally subtracted from the previous (n−1) th image i.e. the continuously renewed reference image technique. This makes the recording technique relatively insensitive to outside influences such as environmental disturbances using the continuous refreshment of the reference image. The result is a simple time-averaged digital shearogram showing a fringe pattern depending on the vibrating object. By using real subtraction with the continuously refreshed reference frame, the resonant frequencies of the vibrating object (Fig. 4.1) as well as the natural frequency of the oscillating defect (Fig. 4.2) can be determined simply and rapidly.

3　EXPERIMENTAL SET-UP

The optical apparatus used is a compact shearing set-up with phase shift unit, CCD camera and laser beam source. This measuring head allows a tested object, which might measure several square metres, to be gauged systematically in separate stages. Either the tested object stands still and the camera is moved, or the camera is mounted on a tripod and the specimen is

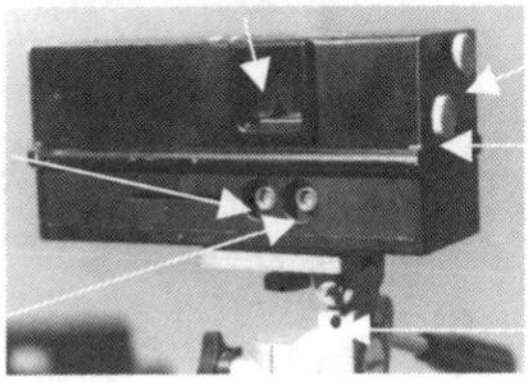

Figure 5. The compact measuring device of digital shearography for Non-destructive testing (NDT) with a CCD camera, the shearing unit and two 50 mW laser diodes (wavelength: 688 nm) without temperature stabilization and without expanding lens.

shifted. For the measuring campaign, the different specimens were clamped in a vice as a rule, and the vice was placed somewhere on the table. This arrangement allows for the tested object or the camera to be shifted on the measuring table, enabling either the specimen to be examined as a whole or a special feature of the object to be observed more closely.

The device can be mounted on a tripod for moving (Fig. 5). When applying one laser diode (wavelength $\lambda = 688$ nm and output power $P_{out} = 50$ mW) it can illuminate an area $A = 300 \times 300$ mm^2 in a distance $L = 800$ mm. In case of a larger area one laser diode with higher output power or a few e.g. two or three laser diodes with the same nominal wavelength are required; the coherence length is 0.5 mm or less. The illumination by a few laser diodes of the same nominal wavelength is possible due to the self-referencing property of the shearographic arrangement. Therefore the experimental investigation of a large measuring area can be performed by a few, simple and inexpensive laser diodes.

4   SUMMARY OF THE PRACTICAL ASPECTS IN SHEAROGRAPHIC NDT APPLICATIONS

Applying digital-shearography a few number of practical aspects and adjustable parameters have to be taken into account. The major part of this know-how especially of the image acquisition process such as recording and evaluation of the shearogram is controlled via the computer by the used program ShearwinNT. So the complete automation of the NDT process including a final automatic detection of object defects is not limited by the shearographic measurement technique or principle. Nevertheless the practical tests of this measurement campaign have shown, that the key position of a successful application and possible automation is found in the application of different loading principles in combination with the defect of interest and the following interpretation of the measurement results. The latest development is the stroboscopic illumination by laser diode array driven by an electronic driver in conjunction with the phase shifting technique for quantitative evaluation (amplitude and phase of the vibration mode) of the digital shearogram.

REFERENCES

Hung, Y.Y. 1986. Shearography versus Holography in Nondestructive Evaluation. *Proceedings of the International Society for Optical Engineering* 604: 18–29, SPIE, Bellingham, USA.

Hung, Y.Y. 1991. Recent Development in Practical Optical Nondestructive Testing. *Proceedings of the International Society for Optical Engineering* 1554B: 29–45, SPIE, Bellingham, USA.

Owner-Petersen, M. 1991. Decorrelation and Fringe Visibility: on the Limiting Behavior of Various Electronic Speckle-Pattern Correlation Interferometers. *Journal of the Optical Society of America* 8: 1082–1089.

Steinchen, W., Yang, L.X. & Kupfer, G. 1996. Vibration analysis by digital shearography. *SPIE*, Bellingham, USA 2868: 426–437.

Steinchen, W. & Yang, L.X. 2002. Digital shearography-theory and application of digital speckle pattern shearing interferometry in NDT, strain measurement and vibration analysis, *SPIE*, Bellingham, USA, ISBN 0-8194-4110-4.

Yang, L.X., Steinchen, W., Schuth, M. & Kupfer, G. 1995. Precision measurement and nondestructive testing by means of digital phase shifting speckle pattern and speckle pattern shearing interferometry. Measurement – *Journal of the International Measurement Confederation* 16: 149–160.

Emerging Technologies in Non Destructive Testing, Van Hemelrijck, Anastasopoulos & Melanitis (eds)
© 2004 Swets & Zeitlinger, Lisse, ISBN 90 5809 645 9

Determination of stochastic matrix cracking in brittle matrix composites

H. Cuypers, D.Van Hemelrijck, B. Dooms & J. Wastiels
Vrije Universiteit Brussel, Faculteit Toegepaste Wetenschappen,
Dienst Mechanica van Materialen en Constructies (MEMC), Belgium

ABSTRACT: Inorganic Phosphate cement (IPC) is a cementitious material developed at the "Vrije Universiteit Brussel" which can be reinforced with E-glass fibres. In order to improve its application in civil engineering constructions like wall and roof panels, the constitutive behaviour of this IPC composite should be fully understood and accurately modelled. In compression, the behaviour of IPC composite is nearly linear elastic up to failure. Unfortunately, in tension the behaviour of IPC composite is completely different. Since matrix cracks initiate and propagate at very low stress levels, the stress–strain relation is linear only in a very small region. Using a stochastic cracking theory, the averaged matrix crack distance can be calculated as a function of the applied tensile stress. This theory is experimentally verified using a stereomicroscope equipped with a video camera and image acquisition board. Both unidirectional and 2D-random oriented fibre reinforcement is used.

1 INTRODUCTION

Inorganic Phosphate cement (IPC) is a cementitious material developed at the "Vrije Universiteit Brussel, Department MEMC". IPC is a two-component system, consisting of a calcium silicate powder and a phosphate acid based solution of metal oxides. Like most cementitious materials, IPC has high compression strength but low tensile strength. Fortunately, after hardening IPC provides a non-alkaline environment, allowing E-glass fibres to be introduced for added stiffness, strength and toughness. Other advantages are the high fire resistance and the fact that in case of fire no toxic gasses are released.

For an economical and safe design without excessive safety factors the constitutive behavior of IPC reinforced with E-glass fibres must be accurately modelled. In compression, the behavior of this IPC composite is nearly linear elastic up to failure (see Cuypers & Wastiels 2002). Unfortunately, in tension the behavior of IPC is completely different. Since matrix cracks initiate and propagate at low stress levels, the stress–strain relation is linear in a very small region only. A first attempt to model the behavior of brittle matrix composites was done by Aveston et al. (1971) and Aveston & Kelly (1973). However, according to Bauweraerts (1998) and Gu et al. (1998), the main shortcoming of this theory, when applied to IPC composites, is found in the fact that in the ACK theory the matrix strength is assumed to be a deterministic variable.

Based on the work of Curtin et al. (1998) and Curtin (1999), presenting a statistical treatment of matrix crack evolution in ceramic micro-composites with unidirectional reinforcement, Cuypers & Wastiels (2002) presented a stochastic cracking theory for E-glass fibre reinforced IPC. Two Weibull parameters (σ_{Rc}, which is the composite reference cracking stress and m, which is the Weibull modulus) account for the stochastic cracking behavior of the matrix, which is highly dependent on the presence of fibres. Another parameter introduced into the model is the shear stress (τ_0) along the matrix-fibre interface. These three material properties are obtained by finding a "best fit" between the theoretical and experimental stress–strain curves.

Once the Weibull-parameters are known, the average crack spacing can be predicted theoretically as a function of the applied stress and compared with experimentally obtained data.

2 THEORETICAL BACKGROUND

2.1 *Introduction*

Figure 1 shows a typical stress–strain curve of a brittle matrix composite. Based on the ACK-theory (Aveston-Cooper-Kelly theory) of Aveston et al. (1971) and Aveston & Kelly (1973), we may divide this curve in Figure 1 into three zones: the pre-cracking zone

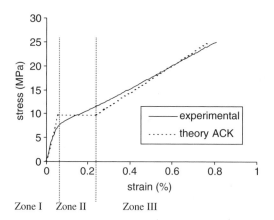

Figure 1. Stress–strain curve IPC composite (2D-random fibre reinforcement, fibre volume fraction 10%) in tension.

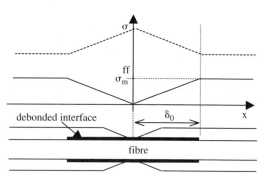

Figure 2. Evolution of the matrix (σ_m) and fibre (σ_f) normal stresses as function of the distance from the crack face.

(zone I), the multiple cracking zone (zone II) and the post-cracking zone (zone III).

Basic assumptions, used in the development of the ACK theory as well as in the development of the stochastic cracking theory, are:

- The fibres are only capable of carrying load along their longitudinal axis. They provide no bending stiffness.
- The matrix-fibre bond is weak. The adhesion shear bond strength τ_{au} is low. Propagation of matrix cracks leads to matrix-fibre debonding along a certain length (slip length, δ_0) in the vicinity of a matrix crack (see Fig. 2).
- Once the matrix and the fibre are debonded, a pure frictional shear stress τ_0 replaces the previously existing adhesion shear stress τ_a at the matrix-fibre interface.
- At the debonded matrix-fibre interface, this shear stress τ_0 is constant with slip.
- The frictional interface shear stress τ_0 is constant along the debonded interface.
- Poisson effects of the fibre and matrix are neglected.
- Global load sharing is used for the fibres.
- The matrix normal stresses are assumed to be uniform in a section, transverse to the loading direction.

2.2 Pre-cracking zone (zone I)

Zone I is often also called the "linear elastic zone". The composite obeys the law of mixtures. The composite stiffness E_{c1} is thus function of the fibre and matrix stiffness, E_f and E_m and the fibre and matrix volume fractions, V_f and V_m.

$$E_{c1} = E_f V_f^* + E_m V_m \qquad (1)$$

with

$$V_f^* = \eta_l \eta_\theta V_f \qquad (2)$$

η_l and η_θ are factors containing the influence of the fibre length and orientation respectively.

2.3 Multiple cracking zone (zone II)

2.3.1 Introduction

As the tensile strength of IPC is rather low, matrix cracks will appear in the composite at relatively low tensile stress levels. When a first matrix crack initiates and reaches a fibre, debonding of the matrix-fibre interface occurs, since it is assumed that the matrix-fibre interface is weak. At the vicinity of a crack face, the matrix is stress-free. Along the debonded interface, the assumption of the existence of a constant frictional matrix-fibre interface shear stress τ_0 is used. Away from the crack face, this frictional interface shear stress provides stress transfer from fibres to the matrix. Figure 2 illustrates the evolution of the normal stresses in matrix and fibres as a function of the distance from a crack face, provided the existence of other cracks is neglected. Slip length δ_0 is the distance from a crack face where the matrix stress reaches the maximum value $\sigma_m^{ff} \cdot \sigma_m^{ff}$ is the far field matrix stress (stress in the matrix at an infinite distance from the crack face):

$$\sigma_m^{ff} = \frac{E_m \sigma_c}{E_{c1}} \qquad (3)$$

It is shown in Figure 2 that the matrix stresses are lower at the vicinity of a crack face, within distance δ_0 on both sides of a crack. This means it becomes very unlikely that new matrix cracking will occur within distance δ_0 on both sides of this matrix crack. This zone is therefore often called exclusion zone. Debonding length

δ_0 can be determined from equilibrium of forces around a fibre (see Aveston et al. 1971), which gives:

$$\delta_0 = \frac{V_m r \sigma_m^{\mathit{ff}}}{V_f^* 2\tau_0} \tag{4a}$$

where r is the fibre radius.

Widom (1966) determined the average final crack distance $\langle cs \rangle_f$ equals:

$$\langle cs \rangle_f = 1.337 \delta_0 \tag{4b}$$

2.3.2 ACK-theory

In the ACK-theory it is assumed that all matrix cracking occurs at the same stress level.

2.3.3 Stochastic cracking theory

In the stochastic cracking theory the matrix failure strength is introduced through a probability distribution. Cracks can now occur gradually at various stress levels. Once the first matrix crack is initiated, we have fibres bridging this matrix crack. Consequently the external load can be further increased, leading to the initiation and propagation of new matrix cracks at higher stress levels outside the already existing exclusion zones. From three-point bending tests on pure IPC matrix specimens (see Cuypers & Wastiels 2002) it was concluded that the two-parameter Weibull probability distribution seems to be the most appropriate choice to model the IPC tensile strength. The shape of the Weibull cumulative distribution function, describing the failure of specimens in uniform tension, is (see Weibull 1952):

$$P = 1 - \exp\left[-\left(\frac{\sigma}{\sigma_R}\right)^m\right] \tag{5}$$

where: P = failure probability
   σ = uniform tensile stress in the material
   σ_R = reference failure stress
   m = Weibull modulus.

According to the stochastic cracking theory, the average crack spacing <cs> is a function of the applied composite stress. Equation (5) provides information about the percentage of inherent matrix flaws, which are able to propagate at a certain stress level. When multiple cracking has occurred completely, the average crack spacing <cs> equals $\langle cs \rangle_f$, being the final crack spacing. When only partial multiple cracking has occurred, <cs> is still larger than the final crack spacing. The average crack spacing <cs> at a certain composite stress is found by dividing the final crack spacing $\langle cs \rangle_f$ by the percentage of

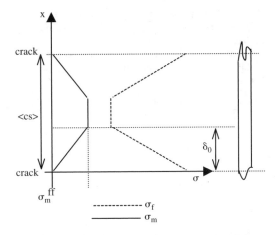

Figure 3a. Stresses in matrix (σ_m) and fibre (σ_f) along a cracked composite, <cs> > $2\delta_0$.

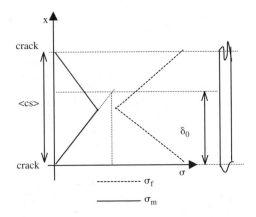

Figure 3b. Stresses in matrix (σ_m) and fibre (σ_f) along a cracked composite, <cs> < $2\delta_0$.

matrix cracks that already propagated, which is expressed by equation (6):

$$\langle cs \rangle = \langle cs \rangle_f \left[1 - \exp\left[-\left(\frac{\sigma_c}{\sigma_{Rc}}\right)^m\right]\right]^{-1} \tag{6}$$

where: σ_c = applied composite stress
   σ_{Rc} = composite reference cracking stress.

In case the stochastic cracking theory is used, the formulation of the stress–strain relationship should be done for two cases, as shown in Figures 3a and 3b.

Figure 3a shows the matrix (σ_m) and fibre stress (σ_f) in the composite in case the average crack spacing <cs> is larger than $2\delta_0$ (lower composite stress levels, occurrence of first matrix cracks). Figure 3b shows these stresses if $<cs>$ is smaller than $2\delta_0$ (higher composite stress levels, end of multiple cracking zone).

From Figure 3a, the composite stress–strain formulation can be determined in case $<cs> > 2\delta_0$ (see Curtin et al. 1998):

$$\varepsilon_c = \frac{\sigma_c}{E_{c1}}\left(1+\frac{\alpha\delta_0}{\langle cs \rangle}\right) \tag{7a}$$

with:

$$\alpha = \frac{E_m V_m}{E_f V_f^*} \tag{7b}$$

From Figure 3b, the composite stress–strain formulation can be determined in case $<cs> < 2\delta_0$:

$$\varepsilon_c = \sigma_c\left(\frac{1}{E_f V_f^*} - \frac{\alpha\langle cs \rangle}{4\delta_0 E_{c1}}\right) \tag{8}$$

Three parameters, which are used in equations (7) and (8) of the stochastic cracking theory, are difficult to obtain. These are (i) the composite reference cracking stress, σ_{Rc}; (ii) the Weibull modulus, m and (iii) the frictional interface shear stress, τ_0. Determination of these values will be discussed in paragraph 3.

2.4 Post-cracking zone (zone III)

Once full multiple cracking has occurred, the fibres only provide further stiffness. The stiffness in zone III can therefore be expressed as:

$$E_{c3} = E_f V_f^* \tag{9}$$

3 EXPERIMENTAL VERIFICATION

3.1 Optical stereo microscope

The use of an optical stereomicroscope is an alternative method for the determination of the Weibull parameters, which are directly related to the average distance between the cracks as a function of the tensile stress. The stereomicroscope (Leica MZ12.5, total magnification up to 150×) and analog video camera are mountable on the test frame and can be moved in both horizontal and vertical direction. A National Instruments image acquisition board (Imaq-PCI-1411) is used to digitize the images (8 bit, grayscale). With the aid of special software (Crack-D) written in the Labview version 6 programming environment and

using the IMAC Vision toolbox, the cracks are detected. In order to improve the detectability (contrast) of the cracks an ink solution is used after each loading step. Once all crack edges are detected, the picture is calibrated to obtain the distances between the detected cracks and the average crack distance can then be determined as a function of the applied stress.

3.2 Specimens

Two types of E-glass fibre reinforcements are used: chopped glass fibre mats ("2D-random") with a fibre density of $300\,g/m^2$ (Owens Corning MK12) and quasi-unidirectional woven roving ("UD") with a fibre density of $158\,g/m^2$. This quasi-unidirectional woven roving contains $141\,g/m^2$ of fibres as reinforcement in the main direction and $17\,g/m^2$ of the reinforcement is aligned along the transverse direction (Syncoglas Roviglas R17/141). The fibre diameter is $14\,\mu m$. The E-glass fibre reinforcement consists of fibre bundles rather than of single fibres. The number of fibres per bundle is ± 1500 for the UD-reinforcement and ± 540 for the 2D-random reinforcement.

The unidirectional fibres are continuous fibres. The length of the 2D-random oriented chopped fibres is $50\,mm$. The 2D-random oriented fibres used in the studied laminates are relatively long, such that $\eta_l = 1$ (see Gu et al. 1998). The value of η_θ as determined by Cox (1952) is used for the 2D-random oriented fibres. The value of η_θ is then 1/3. Values $\eta_l = 1$ and $\eta_\theta = 1/3$ will thus be used when determining V_f^* (equation 2) for 2D-random oriented fibres.

Two IPC laminates (UD1 & RD1) were fabricated using hand lay-up technique: UD1 contains unidirectional reinforcement, while RD1 contains 2D-random reinforcement. The IPC laminates were cured at ambient conditions for 24 hours and afterwards post-cured at 60°C for 24 hours.

The average thickness of the laminates is about $2\,mm$. Specimens with a length of $250\,mm$ and a width of $18\,mm$ were cut. The obtained fibre volume fractions of the plates are listed in Table 1.

The E-modulus of the fibres is $72\,GPa$ and the tensile strength is $1400\,MPa$. The properties of pure IPC are determined by Cuypers (2001). The E-modulus is $18\,GPa$ and the average failure strength σ_{mu} is $7.8\,MPa$. The Weibull parameters are also determined on these pure IPC specimens. When the specimens have a length

Table 1. Plate properties.

Plate	UD1	RD1
Reinforcement	Unidirectional	2D-random
Average V_f (%)	10.4	9.1
Stdev V_f (%)	0.6	0.3

56

of 250 mm, a width of 18 mm and a thickness of 2 mm, the Weibull modulus "m" is 9.3 and the reference cracking stress σ_R is 8.2 MPa. With a fibre volume fraction (V_f) of 10.4% the associated composite reference cracking stress σ_{Rc} of a unidirectional reinforced composite will equal 10.8 MPa (using equation 10a). Using equation 10a, the composite reference cracking stress of a 2D-random reinforced composite with V_f of 9.1% equals 8.4 MPa.

$$\sigma_{Rc} = \frac{\sigma_R E_{c1}}{E_m} \qquad (10a)$$

According to Purnell et al. (2000) the matrix-fibre interface shear stress τ_0 can be calculated using equation 10b, which was established using the ACK-theory.

$$\tau_0 = \frac{1.337 r \sigma_{mu} V_m}{2 V_f^* \langle cs \rangle_f} \qquad (10b)$$

The final average crack distance of both plates is determined. For laminate UD1 $\langle cs \rangle_f$ is 0.9 mm. For laminate UD1 the estimated τ_0, using equation 10b, then equals 0.35 MPa. For laminate RD1 $\langle cs \rangle_f$ is 1.0 mm. τ_0 is then 1.1 MPa.

3.3 Experimental results

The Weibull parameters obtained on pure IPC are listed in paragraph 3.2. However, one should keep in mind that these properties were obtained on pure IPC and might not represent IPC combined with fibre reinforcement. The Weibull properties are indeed very sensitive to the matrix flaw size distribution. These matrix flaws are mainly introduced during material processing. The introduction of fibres might influence this matrix flaw size distribution considerably. Also the value of the matrix-fibre interface shear stress is very sensitive on processing parameters (fibre bundle rearrangement, pressure applied during laminating, etc.) and is difficult to obtain.

The Weibull parameters and matrix-fibre interface shear stress are thus model parameters, used as input in the stochastic cracking theory, which are difficult to obtain. These three material properties can be obtained by finding a "best fit" between experimentally obtained stress–strain curves and theoretical predictions, with equations 7 and 8. This "parameter optimisation" methodology is described by Cuypers (2001). Once the optimized Weibull-parameters and matrix-fibre interface shear stress are known, the average crack spacing can be predicted as a function of the applied stress, using equation 6.

From each laminate, three specimens are loaded in tension up to fracture. The ultimate stress σ_{cu} and strain ε_{cu} are listed in Tables 2 and 3. Initial values of

the Weibull modulus (m) and composite reference cracking stress (σ_{Rc}) of the IPC matrix and the average matrix-fibre interface shear stress (τ_0) have been determined in paragraph 3.2. These values are listed in Tables 4 and 5. The "optimized" values are the values leading to a "best fit" between experimentally obtained stress–strain curves and theoretical predictions. They are also listed in Tables 4 and 5.

Using equation 6 we can obtain a theoretical curve for the averaged crack spacing $\langle cs \rangle$. The same value can be determined as a function of the applied tensile load, using the stereomicroscope and the software Crack-D.

Eight specimens from laminate UD1 have been tested in tension up to 60 MPa. Five specimens of plate RD1 have been tested up to failure. Along the length of the specimen pictures were taken and the average crack spacing was calculated.

Figure 4 shows the results of the average crack spacing for plate UD1, together with the theoretical curve given by equation 6, using the optimized model parameter values listed in Table 4. Correlation between experimental and theoretical curves is good.

Figure 5 shows the theoretical curve of the average crack spacing versus the experimental curves, when

Table 2. Ultimate composite stress and strain, UD1.

Specimen	σ_{cu} (MPa)	ε_{cu} (%)
UD1-1	127	1.77
UD1-2	117	1.96
UD1-3	103	2.00

Table 3. Ultimate composite stress and strain, RD1.

Specimen	σ_{cu} (MPa)	ε_{cu} (%)
RD1-1	18	0.926
RD1-2	21	0.891
RD1-3	19	0.976

Table 4. Laminate properties, UD1.

	τ_0 (Mpa)	m (-)	σ_{Rc} (MPa)
Initial	0.35	9.3	10.8
Optimized	0.52	1.4	14.8

Table 5. Laminate properties, RD1.

	τ_0 (MPa)	m (-)	σ_{Rc} (MPa)
Initial	1.1	9.3	8.4
Optimized	0.43	2.6	16.9

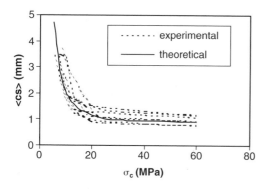

Figure 4. Average crack distance as a function of the applied load, UD1.

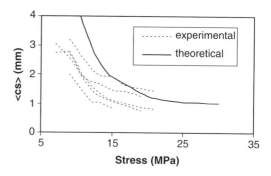

Figure 6. Average crack distance as a function of the applied load, RD1.

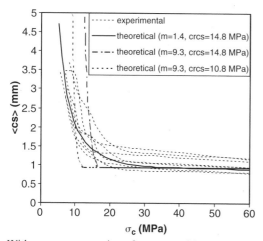

With: crcs = composite reference cracking stress

Figure 5. Influence of m and σ_R on theoretical curve average crack spacing versus composite stress, UD1.

the initial model parameter values (see Table 4) are inserted into equation 6 for the prediction of the average crack spacing. As can be noticed a higher value of m (9.3, see initial values Table 4) concentrates the multiple cracking process too much at one stress level. In case the composite reference cracking stress is 10.8 MPa (initial value, obtained on pure IPC specimens, see Table 4) instead of 14.8 MPa (optimized value, see Table 4) we only observe a shift to the left.

Figure 6 shows the average crack distance of the specimens of plate RD1, determined from the pictures, together with the theoretical curve given by equation 6, using the optimized values listed in Table 5. Correspondence between experimental and theoretical curves is less good for this plate. As can be seen in Figure 6, more cracks are observed experimentally than theoretically at lower stress levels. However, it

should be mentioned that the observed failure stress and strain (see Table 3) were relatively low, compared to previously obtained results (Cuypers et al. 2001). Moreover, the scatter between the experimental curves is rather high. The impregnation of the matrix between the fibre bundles of RD1 might have been poor. Extended experimentation on this type of plates is thus needed. However, the tendency of the experimental curves approaches the theoretical curve.

4 CONCLUSIONS

A stochastic cracking theory is used in this paper to describe the stress–strain behavior of E-glass fibre reinforced IPC composite specimens under monotonic tensile loading. In case the stochastic cracking theory is used, the composite reference cracking stress (σ_{Rc}), the Weibull modulus (m) and the matrix-fibre interface shear stress (τ_0) are model parameters, which are difficult to obtain. These parameters can be obtained from "best-fitting" of the theoretical stress–strain curve with an experimental curve, obtained from monotonic tensile loading, performed on a specimen of the studied laminate.

The Weibull parameters have been determined on pure IPC matrix and on IPC composite by finding the "best fit". It was noticed that the introduction of the fibres decreases the Weibull modulus (m) considerably. Probably the presence of the fibres broadens the matrix flaw length distribution.

One goal of this paper is to verify whether the value of σ_R, m and τ_0 obtained by "best fit" of the experimental and theoretical stress–strain curve under loading, can be inserted into a Weibull model for prediction of the average crack spacing as a function of the external load. The theoretical predicted value of the average crack spacing is compared with visual observations with microscope and digital camera on specimens. For the laminate with unidirectional reinforcement it was found that the model parameters (σ_R, m and τ_0) obtained by the "best fit" technique indeed lead to better prediction of the average crack distance as a function of applied stress, than the values obtained on the pure IPC without reinforcement. Further testing is needed on 2D-random oriented specimens.

REFERENCES

Aveston, J., Cooper, G.A. & Kelly, A. 1971. Single and multiple fracture, The Properties of Fibre Composites. In *Proc. Conf. National Physical Laboratories, IPC Science & Technology Press*: 15–24.

Aveston, J. & Kelly, A. 1973. Theory of multiple fracture of fibrous composites. In *J. Mat. Sci.*, Vol. 8: 411–461.

Bauweraerts, P. 1998. *Aspects of the Micromechanical Characterisation of Fibre Reinforced Brittle Matrix Composites*, Phd. Thesis VUB.

Cox, H.L. 1952. The elasticity and strength of paper and other fibrous materials. In *British Journal of Applied Physics*, Vol. 3: 72–79.

Curtin, W.A. 1999. Stochastic Damage Evolution and Failure in Fibre-Reinforced Composites. In *Advances in Applied Mechanics*, Vol. 36: 163–253.

Curtin, W.A., Ahn, B.K. & Takeda, N. 1998. Modeling Brittle and Tough stress–strain Behaviour in Unidirectional Ceramic Composites. In *Acta mater., No. 10*: 3409–3420.

Cuypers, H., Gu, J., Croes, K., Dumortier, S. & Wastiels, J. 2000. Evaluation of Fatigue and Durability Properties of E-glass fibre Reinforced phosphate cementitious Composite. In *Proc. Int. Symp. BMC6, A.M. Brandt, V.C. Li, I.H. Marshall eds., ZTUREK RSI and Woodhead Publ., Warsaw*: 127–136.

Cuypers, H. 2001. *Analysis and Design of Sandwich Panels with Brittle Matrix Composite Faces for Building Applications*, Phd thesis VUB.

Cuypers, H. & Wastiels, J. 2002. Application of a stochastic matrix cracking theory on E-glass fibre reinforced cementitious composites. In *Proc. Conf. ECCM10, June 3–7 2002, Brugge*: paper 305 (CD-ROM).

Gu, J., Wu, X., Cuypers, H. & Wastiels, J. 1998. Modelling of the tensile behaviour of an E-glass fibre reinforced phosphate cement. In *Proc Conf. Computer Methods in Composite Materials VI, proceedings CADCOMP 98*: 589–598.

Purnell, P., Buchanan, A.J., Short, N.R., Page, C.L. & Majumdar, A.J. 2000. Determination of bond strength in glass fibre reinforced cement using petrography and image analysis. In *J. of Materials Science*, 35: 4653–4659.

Weibull, W. 1952. A statistical distribution function of wide applicability. In *ASME J.*: 293–297.

Widom, B. 1966. Random sequential addition of hard spheres to a volume. In *J. Chem. Phys.*, 44: 3888–3894.

Emerging Technologies in Non Destructive Testing, Van Hemelrijck, Anastasopoulos & Melanitis (eds)
© 2004 Swets & Zeitlinger, Lisse, ISBN 90 5809 645 9

An extensometer based on Bragg-sensors for modal testing of large civil structures

W. De Waele & J. Degrieck
Ghent University, Faculty of Engineering, Dept. of Mechanical Construction and Production
Sint–Pietersnieuwstraat Gent, Belgium

S. Jacobs & G. De Roeck
KU Leuven, Dept. of Civil Engineering – Structural Mechanics Kasteelpark Arenberg
Heverlee, Belgium

S. Matthijs & L. Taerwe
Ghent University, Faculty of Engineering, Dept. of Structural Engineering Technologiepark-Zwijnaarde
Gent, Belgium

ABSTRACT: This paper discusses the design of an extensometer of which the sensing part is based on an optical fibre Bragg sensor. Major characteristics of the extensometer are high resolution, applicability for dynamic measurements, ease of use and sufficient robustness for in-field situations. Finite-element-calculations have been conducted during the design process to simulate the response to dynamic loading. Static and dynamic calibration experiments indicate a guaranteed resolution of approximately 0,05 to 0,1 $\mu\varepsilon$ for periodic loading up to a frequency of 150 Hz. Currently, the extensometer is being used in a research project concerning condition monitoring of large civil structures by means of modal testing. A short description and typical results of such experiments are given in this paper.

1 INTRODUCTION

Modal testing is considered as a valuable tool for the assessment of the global "health" condition of structures (Ewins 1984). Generally, modal testing consists of using a shaker to get a structure or structural element "in motion", upon which accelerations are measured using classical accelerometers. In this way one can determine the eigenfrequencies (and the modal curvatures) of the structure under consideration, which give valuable information concerning its health. The social relevance of such a technique for structures such as bridges, aeroplanes, ... is obvious. However, for a better interpretation of the results (localisation and determination of damage), modal strains are considered to be the most sensitive "dynamic system characteristic". These can change considerably in the damaged zones, but are however very difficult to measure on existing civil structures. Therefore suitable instrumentation has to be developed. The design proposed in this paper has been developed in the framework of a joined research project of Belgian universities (De Roeck et al. 2001).

2 PROTOTYPE DESIGN

2.1 Design requirements

In the research project mentioned above, large concrete beams with a total length of 17,6 m and height of 0,8 m have been choosen as representative structural elements (see Figure 1). Damage was introduced by means of successive steps of static loading, up to final fracture. After each of these loading steps, the beam was dynamically loaded by a falling weight. Accelerations were measured by accelerometers and strains by the extensometer and by internal Bragg-sensors glued onto the steel reinforcement of the concrete beam. In order to obtain a sufficiently good characterisation of the degree of damage, a number of prerequisites have been set forward for the design of the extensometer.

The modal strains should of course be measured for a sufficient number of eigenfrequencies. Finite-element-simulations (De Roeck & Jacobs 2002) indicated that measurements of the vibrations of the beam

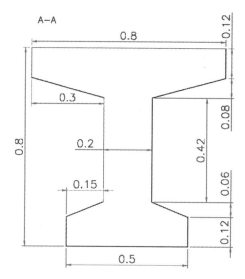

Figure 1. Cross-section of the tested concrete beams.

up to 50 Hz should be sufficient. Thus, the first eigenfrequency of the extensometer had to be sufficiently higher than the cited value. This was accomplished by a design with low weight and high stiffness. However, a high stiffness also means that great forces are needed in order to transfer the deformations of the concrete beam onto the extensometer. This could have severe implications for the attachment of the gauge onto the structure.

As mentioned above, the concrete beam was dynamically loaded by means of impacts. The energy of these impacts had to be restricted in order not to introduce additional damage. Finite-element-simulations showed that the maximum (dynamic) strain variations could be expected in the order of magnitude of $100\,\mu\varepsilon$ ($1\,\mu\varepsilon = 10^{-6}\,$m/m) and less. A design goal of $40\,\mu\varepsilon$ has been set forward. An accurate measurement of these small deformations could only be obtained if the extensometer has a high resolution, postulated in the order of 0,1 to $0,2\,\mu\varepsilon$.

Further, the extensometer needed to have a sufficiently large gauge-length. This is due to the fact that the extensometer must span at least one or two cracks of the damaged beam to get a "global" idea of the health state of the beam. It was supposed, from previous experience, that this goal could be obtained by a gauge-length of 500 mm.

At last, the extensometer must self-evidently be sufficiently robust so that it can be repeatedly mounted onto the concrete beams and can be used in a rather harsh environment. It could then also be used for in-situ condition monitoring of eg. existing bridges or other civil structures.

2.2 Proposed design

It should first be mentioned that an optical fibre Bragg-sensor (Meltz et al. 1989) has been choosen as the gauge type of the extensometer. A Bragg-grating is a periodic modulation of the refractive index inside the core of an optical fibre. Typical length of the grating for sensing applications is in the order of 10 mm. A Bragg-grating acts as a wavelength-selective mirror; it reflects a very small spectrum of the in-coupled broadband light. The central wavelength, or so-called Bragg-wavelength, shifts in a linear way with respect to applied axial stress. Main advantages of this optical fibre sensor are small size, linear dependence on strain, absolute measurements, immunity to electromagnetic interference and inherent mechanical strength. This sensor type is therefore considered to be a valuable alternative for the classical electrical resistance strain gauge, and is widely used for long-term monitoring of composite and concrete structures (Measures 2001). The authors have also been involved in other monitoring projects of which the results have been reported in (De Waele et al. 2001, Moerman et al. 2001).

The instrumentation currently used to demodulate the Bragg-sensors has a measurement range of approximately $8500\,\mu\varepsilon$ with an accuracy of $5\,\mu\varepsilon$ and a resolution of $1\,\mu\varepsilon$. The maximum measurement frequency is 1 kHz. Therefore the basic idea has been to design a device that transfers and enlarges the small deformations of the concrete structure onto the optical fibre Bragg-sensor. The sensing part of the Bragg-sensors used has a length of 10 mm. Therefore a minimal free length of the optical fibre containing the Bragg-grating of 20 mm has been presupposed in order to obtain a uniform straining of the sensor part. This means that a theoritical magnification factor of 25 could be obtained for an extensometer with a gauge-length of 500 mm.

A drawing of the proposed design of the extensometer is shown in Figure 2, with an indication of the main dimensions and the different components. A compromise between high structural stiffness, low weight and sufficient compliance of the device has been obtained by choosing an aluminium circular tube as the main structural element.

The extensometer consists of two concentric aluminium tubes (1 & 2) with outer diameters of 32 mm and 22 mm respectively and a wall thickness of 1,5 mm. Both tubes are glued to an aluminium end-piece (3) at one end. At the other end, only the outer tube is glued to a second aluminium end-piece (4), whilst the inner tube is somewhat shorter in length and is freely suspended inside the larger tube. The optical fibre (7) is glued in end-piece (4) and in a drill hole in a small aluminium piece (5) that is glued in the inner tube. For protection against possible buckling under compressive loading, the optical fibre with diameter of

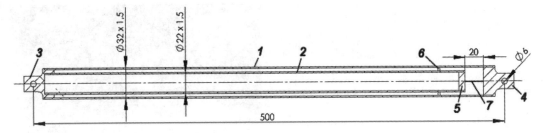

Figure 2. Prototype design of an extensometer with long guage-length for the measurement of dynamic deformations of large concrete structures. The entire deformation of the outer tube (1) is transferred to a short piece of optical fibre with Bragg-sensor (7).

125 μm is positioned in a glass capillary filled with epoxy resin. Finally, a small teflon ring (6) firmly attached to the inner tube prevents this inner tube from lateral movements without obstruction of longitudinal displacements. Lateral shaking due to the dynamic movements of the concrete structure could after all damage the fibre sensor.

The working principle of the extensometer can be clearly understood from Figure 2. Through the eyes in end-pieces (3) and (4) the outer tube of the extensometer is attached to steel styles in the concrete structure. Hereby it undergoes the same (local) deformation as the structure. This structural deformation will be transferred to the optical fibre that is at one end directly connected to the outer tube (1) and at its other end indirectly by means of the inner tube (2). Due to the large difference in stiffness between the Bragg-sensor and the inner aluminium tube, the structural deformation will almost entirely be transferred to the optical fibre. The ratio of strain measured by means of the optical fibre sensor (further called Bragg-strain) to the structural strain is further called the enlargement factor.

2.3 Finite-element-simulations

The mechanical behaviour of the extensometer has been extensively studied by means of finite-element-simulations. Calculations were performed using the software ABAQUS™. The material properties used in the simulations are summarized in Table 1.

Initially, the prototype has been modelled for the determination of its eigenfrequencies. Therefore the entire device has been modelled as a three-dimensional structure consisting of linear tetrahedral elements. The eigenfrequencies have been calculated with the boundary conditions as they appear during the experiments. For one eye a rotation around its axis is the only degree of freedom and for the other eye also a displacement in longitudinal direction is possible. The first eigenfrequency of the extensometer appears at a frequency of 288 Hz and corresponds with a bending mode, see Figure 3. At a frequency of 1867 Hz

Table 1. Material properties used in finite-element-simulations.

Aluminium	
E-modulus [MPa]	70.000
v [–]	0.33
ρ [kg/m^3]	2.700
Teflon	
E-modulus [MPa]	460
v [–]	0.46
ρ [kg/m^3]	2.200
Glass	
E-modulus [MPa]	69.000
v [–]	0.19
ρ [kg/m^3]	2.100
Epoxy	
E-modulus [MPa]	3.000
v [–]	0.3
ρ [kg/m^3]	1.400

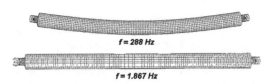

$f = 288$ Hz

$f = 1.867$ Hz

Figure 3. Graphical illustration of the first eigenmodes in bending and longitudinal deformation.

the first longitudinal eigenmode appears. Obviously, these eigenfrequencies are sufficiently higher than the frequencies that appear during the experiments.

The response to longitudinal harmonic excitation has been simulated using a two-dimensional axisymmetric model of the extensometer. A partial view on this model and on the applied mesh is shown in Figure 4.

Because the frequencies involved in the test program, beneath 50 Hz, are sufficiently smaller than the (longitudinal) eigenfrequencies of the extensometer, the mechanical behaviour can most probably be (quasi-)

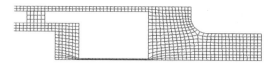

Figure 4. Partial view on the two-dimensional axisymmetric finite-element-model of the extensometer.

statically simulated. Therefore linear static analyses have been performed to calculate the deformation of the model under an imposed load uniformly distributed on one end-face. The enlargement factor of the extensometer was then defined as the ratio of the longitudinal strain in the middle element of the optical fibre to the longitudinal strain in the middle element of the outer tube. In this way an enlargement factor equal to 23 has been calculated.

The validity of the static analysis has been verified by a series of dynamic loading calculations. Harmonic excitations with frequencies from 10 to 150 Hz and with different load values have been simulated. These dynamic simulations lead in all cases to the same value of the enlargement factor as in the static simulation. A second reason (besides the high eigenfrequencies) for this good agreement is probably the fact that the mass of the extensometer is uniformly distributed along its length with the concentrated masses (of the aluminium end-pieces) in the close neighbourhood of the boundaries. The distribution of the masses (inertia) is indeed very important in the dynamic response of a structure.

3 EXPERIMENTS

3.1 Calibration experiments

The enlargement factor of the developed prototype design has been experimentally calibrated by a series of experiments described in this paragraph.

Based on the results of the finite-element-simulations simple static tensile tests were first choosen to determine the enlargement factor of the extensometer. These experiments have been performed on a universal testing machine INSTRON 4505.

From these experiments it was clear that the displacement of the clamps could not be used for calculation of the deformation of the extensometer. The natural variability in these displacement recordings was too high with respect to the very small imposed deformations. Therefore the universal testing machine was stopped at different forces and the Bragg-strains measured. The ratio of these Bragg-strains to the structural strain, calculated from the imposed forces, varied between 18,6 and 20,4. These values are fairly close to the theoretical value of 23, notwithstanding the uncertainty on the exact free length of the optical fibre and certain material properties.

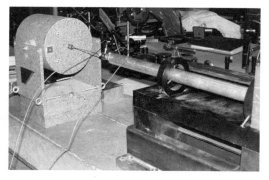

Figure 5. Overview of the experimental set-up used during the dynamic experiments. A shaker is used for harmonic excitation of the extensometer. Acceleration and load are measured by means of an impedance head glued on the left end-piece.

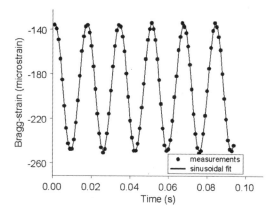

Figure 6. Measured Bragg-strain during a harmonic excitation of the extensometer at a frequency of 60 Hz with an applied tensile force of 30 N.

In the following, a series of dynamic experiments has been performed with the support of the Department of Civil Engineering of the Catholic University of Leuven. An overview of the experimental set-up is given in Figure 5. A harmonic excitation has been imposed by means of a shaker connected to one eye of the extensometer. At the other end, the second eye of the extensometer is fixed (all 6 degrees of freedom). In the middle of the extensometer a (security) device is placed to prevent lateral shaking due to an eccentrically applied load imposed by accident. The applied load and acceleration were measured using an impedance head (type PCB 288D01) that is glued on top of the free end of the extensometer.

A great number of experiments have been conducted in a frequency range up to 130 Hz for several values of the applied load. An example of a measured Bragg-signal due to a harmonic excitation of 30 N at a frequency of 60 Hz is given in Figure 6. The measured Bragg-signal shown during a time period of 0,1 s is very clearly represented by the measurement points. It should here be mentioned that the imposed force corresponds to a total deformation of the extensometer of only 1,5 μm (or 3με) in amplitude! Therefore the enlargement factor can be estimated from the figure as being almost equal to 20. Taking into account a resolution of 1 με

of the demodulation instrument, it can roughly be estimated that a resolution of almost 0,05 με can be obtained for the structural strain values. Also indicated on the figure is a calculated sinusoidal fit used for extraction of the amplitude of the Bragg-signal.

The structural strain has been derived from the measured force, thus following a quasi-static approximation. Based on the results of these dynamic experiments an enlargement factor of 19,3 was obtained. This value is a perfect confirmation of the values calculated from the static experiments.

3.2 Modal testing of a concrete beam: preliminary experiments

As already mentioned above, the extensometer has been designed for the measurement of deformations of large scale concrete beams (17,6 m long by 0,8 m height) during modal testing. The prestressed concrete girder has been produced and tested at the Magnel Laboratory for Concrete Research of the Ghent University. After initial dynamic testing of the yet unloaded girder, several quasi-static loading cycles have been executed, each time followed by dynamic testing. For the dynamic testing the beam is placed on air spring bellows in order to obtain free-free support conditions. To vibrate the girder a drop weight of 115 kg falling from a height of 1 m causes an (eccentric) impact on one end of the concrete beam.

During consecutive experiments, dynamic deformations are measured at 6 different positions on the side of the concrete girder. On Figure 7 the extensometer is

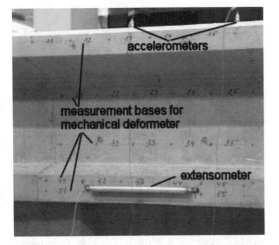

Figure 7. Instrumentation of the large scale concrete beam. Accelerometers and the extensometer are installed for measurements during the dynamic experiments. The measurement bases for mechanical deformeters are used to determine the permanent deformation due to the static loading cycles.

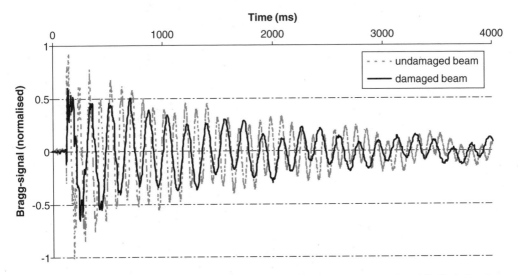

Figure 8. Modal Bragg-strain measured on a large prestressed concrete girder by means of specifically designed extensometer. Measurements on the undamaged beam are indicated by the dashed line and these for a (severely) damaged beam are indicated by the full line. The Bragg-strains clearly indicate the different mechanical behaviour of the concrete beam before and after damage was introduced.

installed on the lower flange of the concrete girder. Accelerometers are installed on top of the beam to measure the modal accelerations. On the side of the girder 5 bases have been installed for control measurements (with a mechanical deformeter) of permanent deformation due to the static loading cycles.

Preliminary results of measurements of Bragg-strains after the impact loading are shown on Figure 8 both for a healthy (or undamaged) beam and for a damaged beam due to loading up to plastic straining of the steel reinforcement.

The difference in the modal behaviour of the healthy beam and the (heavily) damaged beam is obvious. Both the global and local variations in the modal strain-signal are clear, in terms of frequency as well as amplitude.

4 CONCLUSIONS

An extensometer has been designed and developed, intended for the measurement of modal strains of large structural elements. A transducer transfers and enlarges the structural deformation to an axial deformation of an optical fibre Bragg-sensor. The mechanical behaviour of the extensometer has been extensively simulated by means of finite-element-calculations. High eigenfrequencies and a high enlargement fac-tor show the possible advantageous use of the proposed design. Static and dynamic calibration experiments using the instrument indicate the close agreement to the design intentions. The extensometer has been tested up to a frequency of 150 Hz and strains

with an amplitude of only $3\,\mu\varepsilon$ were measured with an approximate resolution of $0,05\,\mu\varepsilon$! At this moment, the extensometer is extensively being used to measure dynamic strains of a large concrete beam after impact by a falling weight. Preliminary results indicate the possibility of condition monitoring through the measurement of modal strains with the designed extensometer.

REFERENCES

De Roeck G., Sol H., Vantomme J., Taerwe L. & Degrieck J. 2001. Enhanced performance of dynamic monitoring of civil engineering structures by integration of optical fiber technology. *Research project* G.0266.01 funded by the Flemish Institute for Scientific Research FWO.

De Roeck G. & Jacobs S. 2002. Ansys-simulation of impact on prestressed concrete beam. *Internal report* FWO-project G.0266.01.

De Waele W., Degrieck J., Baets R., Moerman W. & Taerwe L. 2001. Load and deformation monitoring of composite pressure vessels by means of optical fibre sensors. *Insight* 43(8): pp 518–525.

Ewins D.J. 1984. *Modal testing: theory and practice*. Letchworth: Research Studies Press Ltd.

Measures R.M. 2001. *Structural monitoring with fiber optic technology*. San Diego: Academic Press.

Meltz G., Morey W.W. & Glenn W.H. 1989. Formation of Bragg gratings in optical fibers by a transverse holographic method. *Optics Letters* 14 (15): pp 823–825.

Moerman W., Taerwe L., Baets R., De Waele W. & Degrieck J. 2001. Application of optical fiber sensors for monitoring civil engineering structures. *Structural Concrete* 2(2): pp 63–71.

Emerging Technologies in Non Destructive Testing, Van Hemelrijck, Anastasopoulos & Melanitis (eds)
© 2004 Swets & Zeitlinger, Lisse, ISBN 90 5809 645 9

Damage in CFRP composite materials monitored with intensity modulated fiber optic sensors

M. Wevers & L. Rippert
Katholieke Universiteit Leuven, Department of Metallurgy and Materials Engineering, Research Group Materials Performance and Non-Destructive Testing

J.-M. Papy & S. Van Huffel
Katholieke Universiteit Leuven, Department of Electrical Engineering, Division SCD-SISTA

ABSTRACT: In this research study, intensity modulated fiber optic sensors, whose working principle is based on the microbending concept, are used to monitor the damage in C/epoxy laminates during tensile loading. The use of advanced signal processing techniques based on time-frequency analysis is explained in order to get information on the damage developing in the composite. The signal Short Time Fourier Transform (STFT) has been computed and several robust noise reduction algorithms have been applied. Principally, Wiener adaptive filtering, improved spectral subtraction filtering, minimum-phase FIR (Finite Impulse Response) filtering and Singular Value Decomposition (SVD)-based filtering have been used. An energy and frequency-based detection criterion is introduced to detect transient signals that can be correlated with the Modal Acoustic Emission (MAE) results and thus damage in the composite material. Hints are that time-frequency analysis and Hankel Total Least Square (HTLS) method can also be used for damage characterisation (delamination, matrix cracking and fiber breaking).

1 INTRODUCTION

The damage development in composite materials, which is much more complex than in metals, involves more than one damage type and can cause a degradation in the mechanical properties which jeopardyzes the functionality of a composite well before final fracture. Research studies are therefore focussed on the development of tools and techniques to monitor and assess the damage development in those materials while being in use. Thanks to the development of optical fibre communication technologies and the evolution in computer technology, new testing methods emerged. With optical fibres embedded in composite materials and advanced data processing techniques of the optical signals, a Non-Destructive Testing (NDT) system can be integrated into this complex material. In this approach fibre optic sensors may offer an alternative for the robust piezoelectric transducers used for Acoustic Emission (AE) monitoring.

Several kinds of optical fibre sensors have been developed, namely intensity-modulated sensors, phase-modulated sensors (interferometers), and Bragg grating sensors. The Michelson, Mach-Zenhder and Fabry-Perot interferometers are the most widely used configurations for phase- measuring of physical quantities like strain and temperature.

Intensity-modulated sensors detect variations in the intensity of the transmitted light caused by a perturbing environment. The main causes for intensity modulation are transmission, reflection and microbending. Intensity-modulated optical fibre sensors require only a low cost, simple and robust sensing system. The major limitation for these sensors is that any intensity fluctuation in the output not associated with the measurand produces erroneous results, so their repeatability and overall accuracy is not very high. The first intensity-modulated sensors developed, used the microbending concept to detect pressure, acceleration, displacement, temperature and strain (Berthold 1995, Lawson & Tekippe 1983, Yao & Asawa 1993). Several intensity-modulated sensors have been successfully used to measure damage but they usually rely on the optical fibre fracture (Glossop et al. 1990, Waite & Sage 1988).

In this study it will be shown that the optical signal, collected from an intensity-modulated sensor based on the microbending concept, contains information on the elastic energy and hence strain released whenever suddenly damage is being introduced in the host material.

Advanced filtering and signal processing techniques are applied to obtain these results, which are compared with those obtained from an AE monitoring system.

2 PRINCIPLE OF OPERATION

Bending an optical fiber locally reduces the critical reflection angle and thus a small amount of light leaks in the cladding. For a curvature radius in the order of centimeters this is called macrobending. Microbending is related to a curvature radius in the order of micro-meters. Microbends are axial fiber distortions having spatial wavelengths small enough to cause coupling between propagating and radiation modes (leaky modes).

Damage created in composite materials also produces mechanical waves propagating in the material. When a wave hits an optical fiber, the wave displacement bends it locally and so some coupling between propagating and radiation modes may appear. So, transient features in the measured optical signal could be related to stress waves released by matrix cracking, delamination or reinforcing fibers fracture phenomena. To reveal this, signal analysis tools such as filtering (denoising), time analysis and time-frequency analysis have to be used.

3 MATERIALS AND EXPERIMENTAL SET-UP

Laminates were produced from a VICOTEX carbon-epoxy prepreg. The prepreg was cut and stacked into a $[0_2°, 90_4°]_s$ lay-up. Two multimode optical fibers were embedded 60 mm apart in the 90° direction, in the middle plane of the specimen. A polymeric bore tube was put around the optical fibers at their exit point from the composite specimen. It shrank around the fiber during the cure and so protected this weak point. The tested specimens had a length of 150 mm, a width of 25 mm and a thickness of 1.2 mm. The gauge length of the sensor was 10 mm. (see Fig. 1)

Tensile tests were carried out on the 4505 INSTRON testing machine with a 100 kN loadcell to introduce damage in the composite. Aluminium end tabs were bonded to the specimens to prevent grip damage, using a two components ARALDITE 2011 epoxy glue. The load was applied continuously and the tensile machine was operating at a displacement rate of 0.5 mm/min.

A 10 mWatt HeNe laser has been used as light source. A beamsplitter has been used to connect both optical fibers. The optical fiber core diameter was 100 μm, the cladding diameter was 110 μm and the coating diameter was 125 μm. The optical fiber has been chosen to maximise the microbending and the injected light intensity. The polyimide coating max-imised the strain/stress transfer from the composite

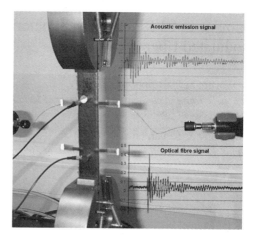

Figure 1. Photograph of instrumented specimen in tensile test machine.

material to the optical fiber and minimised the influence of the optical fiber embedment (Roberts et al. 1991, Jensen et al. 1992, Surgeon & Wevers 1999, Rippert et al. 2000). The small diameter difference between the core and the cladding increased the loss of light due to bending. Great care was also taken with the different optical connections to try to maximise the Signal to Noise Ratio (SNR).

The optical signal was collected in a photodiode, further amplified and then sent to an oscilloscope card with a 12 bits A/D Converter. A LABVIEW® program has been developed to control the acquisition card and the data collection. An AC-coupled amplifier was also designed. Basically, it is a first order High Pass (HP) filter with a 1.6 Hz cut-off frequency and an amplifier with a gain equal to 10. For each optical fiber, the output and the AC-coupled output optical signals are collected. The Sampling Rate (SR) was set to 10 kHz.

To detect and identify damage in the composite material specimens an AE system has been used (WAVE EXPLORER® from DIGITAL WAVE CORPORATION) with broadband sensors B1025 with a nearly flat frequency response in the 50–3000 kHz frequency range.

4 TESTING AND SIGNAL ANALYSIS

4.1 Testing and data acquisition

Tensile tests were performed on ten composite material specimens. A fourth order minimum-phase Low Pass (LP) Butterworth filter was applied (with a 5 Hz cut-off frequency) on the measured optical signal. A typical loading curve from the tensile machine (upper curve) and the corresponding filtered optical attenuation curve (middle curve) are shown in Figure 2.

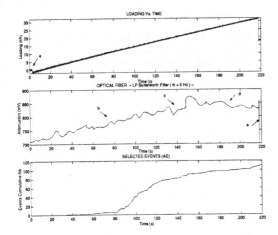

Figure 2. The loading (upper), the measured optical signal (middle) and the cumulative number of MAE events (lower).

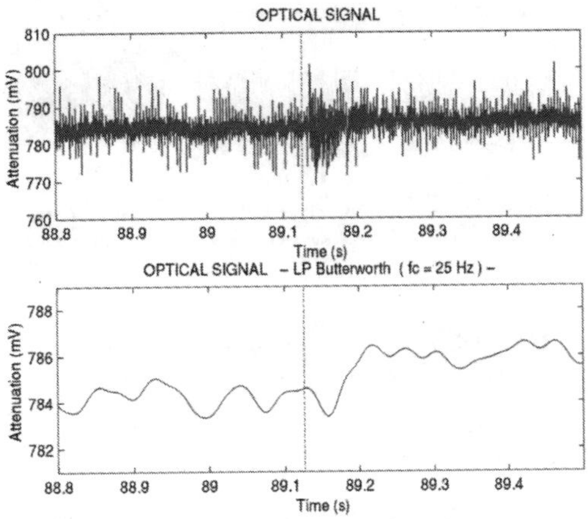

Figure 3. The measured optical signal (upper) and the filtered optical signal (lower).

Events detected by the MAE system are also illustrated in the figure (lower curve). The test can be separated in several parts (a–e).

At the beginning of the test (part b) the stress is uniformly distributed in the sample. Next, the first types of damage appear, namely transverse matrix cracking and delamination, in and around the 90° plies. With the increasing damage, the stress is more and more sustained by the 0° plies.

So, close to the optical fiber in the 90° plies, the stress is not increasing any more and some stress relief may even appear. When the Characteristic Damage State (CDS), i.e. a saturation of the transverse matrix cracks, is obtained nearly all the stress is carried by the 0° plies. It explains why after 132 seconds the optical signal does not follow the loading curve. This is also corroborated by the MAE results. The cumulative number of AE events has been plotted (with the very low energy events being filtered out). The shape of the curve is quite typical and it shows that some substantial damage has occurred (and very probably the CDS is obtained) before 132 seconds.

A close look at the optical signal shows some particularities as illustrated by Figure 3.

The upper curve shows the attenuation signal whereas the lower curve displays the optical signal filtered by the same LP Butterworth filter (but with a 25 Hz cut-off frequency). The occurrence of this transient feature is correlated with the occurrence of a MAE event (the vertical dotted line) that can be related to damage. So this feature can be related to damage as well. This optical event (i.e. transient detected in the optical signal) is clearly the sum of a step and a high frequency transient feature. The main limitation is the sensitivity; the events detected by this very simple method are only the most energetic ones (Papy et al. 2001). So more advanced noise reduction and signal processing methods are required to proceed further.

4.2 The numerical postprocessing of the optical signals

Several steps were taken to increase the SNR and to look for small transient features in the optical signal. A hardware filter has been designed and several numerical filters were implemented in software (using MATLAB®). There is a high static (DC) component in the measured optical signal, therefore a home-made first order analog HP filter has been designed. The cut-off frequency is 1.6 Hz. As the signal is also amplified (by a factor of 10), this kind of system is typically referred as an AC-coupled amplifier circuit. To further filter the static component, several standard FIR and IIR filters were tested. Finally, a third order minimum-phase Chebyshev HP filter has been chosen (with a 50 Hz cut-off frequency) for its slightly better frequency behaviour.

One of the main noise sources was found to be the laser power supply. It appears in the frequency domain as a high component at 50 Hz and its harmonics at 100 Hz, 150 Hz, etc. in the measured signal. A two taps adaptive filter, based on Wiener theory and the Least Mean Squares (LMS) method, was used to remove this noise (Proakis & Manolakis 1996). This kind of filter is very efficient to remove specific frequencies but the algorithm is quite consuming in terms of computation time. Therefore, since this noise above 250 Hz was small enough, the filtering was stopped at this frequency. Next, an improved spectral subtraction method has been applied. The spectral subtraction is a filtering technique that has been developed by S.F. Boll for speech processing (Boll 1979).

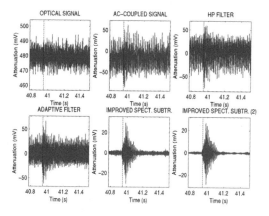

Figure 4. The different filtering techniques that are applied on the optical signal to detect a pencil lead break.

It is assumed that significant noise reduction is possible by removing the effect of noise from the magnitude spectrum only. The noise is estimated in a window of the signal just before an interesting transient. To overcome this limitation, an improved version of the spectral subtraction method has been used (Sovka et al. 1996). The noise estimate is computed at each time frame using an adaptive Wiener filter. The main requirements are that the useful signal and the noise are uncorrelated and that changes in the optical signal due to noise are slower than the ones due to damage. This technique happens to be quite fast and robust.

An example of the application of the filtering techniques is given by Figure 4. The AC-coupled amplifier first filters the optical signal. Both signals are then digitised and are displayed in the upper row (left and middle) of the figure. Then the AC-coupled signal is filtered by the Chebyshev HP filter (upper right curve), the resulting signal is filtered by the adaptive filter (lower left curve). Then the improved spectral subtraction filter is applied twice (lower middle and right curves). During the time period displayed by the figure, a pencil lead break test had been performed and its occurrence is pinpointed by the vertical dotted line. The SNR improvement is significant, mostly due to the improved spectral subtraction filter. Applying it twice increases slightly the SNR at the expense of a slight increase in computation time.

After the application of these filters, the main source of fluctuations in the optical signal is some noise with a time-varying frequency content. (Papy et al. 2001) The source of this non-linear phenomenon is most probably the laser. These fluctuations appear in the time domain as transient features on the signal. Studies are going on to filter this noise but other robust signal processing methods can also be used to distinguish between damage related events and false detections.

4.3 Damage detection

A detection method based on energy tracking has been developed (Papy et al. 2001). Basically, a smooth estimate of the energy is compared to the instantaneous energy estimate. Based on their ratio, a decision factor is calculated during a training period (i.e. a sequence where there is no event, typically a few seconds at the very beginning of the test). This method is quite robust if the noise does not evolve too much during a test. The improved spectral subtraction computes and also uses these energy estimates. So, the additional computation cost of this method is quite low. This method does not require setting a threshold that would depend of the overall light intensity, the decision factor is automatically computed from the statistical properties of the noise in the signal.

Additionally, the Hankel Total Least Squares (HTLS) method has been used (Van Huffel 1993). This very robust method is used to filter data that are arranged in an Hankel matrix. The algorithm uses the Singular Value Decomposition (SVD) (Golub & Van Loan 1996, Doclo et al. 1998) and the Total Least Squares (TLS) (Van Huffel et al. 1994) methods to estimate the parameters of a damped exponential model. So every detected event is modelled as a sum (up to a chosen order K) of damped exponentials and the related parameters (frequencies, damping coefficients, amplitudes and phases) are obtained with a very good accuracy. The method can be used on complex or real signals. Then a classification of the detected transients has been attempted to separate false detections and correct events. The results of this classification have been correlated with the MAE results.

The detection method has been evaluated first on pencil lead break tests performed at several locations on a composite specimen. For the pencil lead break tests a good sorting out of the good and false detections could be obtained. (Fig. 5) After the HTLS computation, it appears clearly that some transients have common features such as three particular frequency bands (around 840 Hz, 1055 Hz and 1290 Hz). Correlation with MAE indicated that these events correspond to the pencil lead breaks while the other events are false detections.

The method has also been applied on detected events from the tensile tests but the results are not conclusive yet. The frequency content of damage events is quite widespread and so more advanced methods are required to separate events and false detection. Statistical methods or neural networks may be of use to solve this problem.

4.4 Time-frequency analysis (STFT) and damage identification

The STFT can be applied on the filtered signal or after the application of the HTLS on the reconstructed

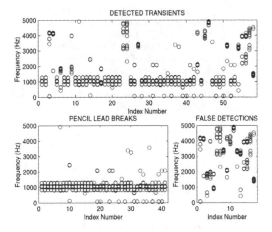

Figure 5. The frequency content (HTLS) of the detected transients.

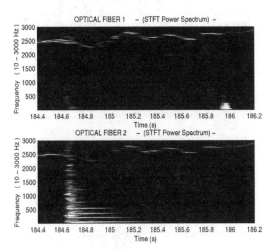

Figure 6. The STFT analysis.

signal. Since the pencil lead breaks were performed on the surface of the sample, the predominant mode in these tests should be the flexural mode whereas during tensile tests both modes should be present with relative magnitudes depending on the type of damage that produced the wave. It appears that most of the frequency content that can be identified with the STFT is from the flexural mode. The extensional mode appears at higher frequencies where the level of noise makes it more difficult to identify. The so-called optical events can be clearly localised and characterised in the time domain and in the frequency domain, but no final conclusive results can be produced so far to correlate the wave packages with the type of damage. An example of a damage related event is illustrated in the time-frequency domain by Figure 6.

5 CONCLUSION

It has been shown that an intensity modulated optical sensor based on the microbending concept can be used for continuous damage monitoring. Its sensitivity is less than those of interferometers but it is simple and robust. It requires some advanced signal analysis tools like adaptive filtering, spectral subtraction filtering, exponential data modelling (HTLS) and time-frequency analysis (STFT). Correlation methods between several embedded fibers (or between an embedded fiber and a reference fiber) and frequencies tracking methods are currently under study to enhance the system and more especially when the time-varying frequency content of the noise is in the same frequency range as the damage related events.

The principle has been proven to work. The sensor can detect damage initiation and characterise its frequency content. The similarities between optical and MAE signals and the use of time-frequency analysis (STFT or wavelets for better resolution) should permit damage identification.

ACKNOWLEDGEMENTS

The authors would like to thank Ing. J. Vanhulst for his precious help with the data acquisition system. This work was supported by the F.W.O. (Project no. G.0200.00) and by the Belgian Programme on Interuniversity Poles of Attraction (IUAP-V-10-29), initiated by the Belgian State, Prime Minister's Office for Science, and by a Concerted Research Action (GOA) project of the Flemish Community, entitled "Mathematical Engineering for Information and Communication Systems technology".

REFERENCES

Berthold, J.W., 1995. Historical review of microbend fiber-optic sensors, *J. Lightwave tech.* 13(7), 1193–1199.

Bhatia, V., Schmid, C.A., Murphy, K.A., Claus, R.O., Tran, T.A., Greene, J.A. & Miller, M.S., 1995. Optical fiber sensing technique for edge-induced and internal delamination detection in composites, *Smart Mater. Struct.* 4, 164–169.

Boll, S.F., 1979. Suppression of acoustic noise in speech using spectral subtraction, *IEEE Transactions on acoustics, speech, and signal processing* ASSP-27(2).

Claus, R.O., Gunther, M.F., Wang, A. & Murphy, K.A., 1992. Extrinsic Fabry-Perot sensor for strain and crack opening displacement measurements from −200°C to 900°C, *Smart Mater. Struct.* 1, 237–242.

Doclo, S., Dologlou, I. & Moonen, M., 1998. A novel iterative enhancement algorithm for noise reduction in speech, Proceeding of the *5th International Conference on spoken Language Processing (ICSLP98)*, Sydney, Australia, 1435–1438.

Glossop, N.D.W., Dubois, S., Tsaw, W., Leblanc, M., Lymer, J., Measures, R.M. & Tennyson, R.C., 1990. Optical fibre damage detection for an aircraft composite leading edge, *Composites* 21(1).

Golub, G.H. & Van Loan, C.F., 1996. Matrix Computations, 3rd edn. The Johns Hopkins University Press, Baltimore, Maryland.

Jensen, D.W., Pascual, J. & August, J.A., 1992. Performance of graphite/bismaleimide laminates with embedded optical fibres. Part I: uniaxial tension, Part II: uniaxial compression, *Smart Material Structure* 1, 24–30 & 31–35.

Lawson, C.M. & Tekippe, V.J., 1983. Environmentally insensitive diaphragm reflectance pressure transducer, Proceedings of SPIE, Vol. 412, 96–103.

Measures, R.M., 1992. Advances toward fiber optic based smart structures, *Optical engineering* 31(1), 34–47.

Papy, J.-M., Van Huffel, S., Rippert, L. & Wevers, M., 2001. Spectral subtraction method applied to damage detection in composite materials with embedded optical fibers, Proceedings of the *SAFE/ProRISC/SeSens Benelux Workshop on Circuits, Systems and Signal Processing*, paper on CD-ROM, Veldhoven, The Netherlands.

Proakis, J.G. & Manolakis, D.G., 1996. Digital signal processing: principles, algorithms, and applications, 3rd edn, Prentice Hall, Upper Saddle River, NJ.

Rippert, L, Wevers, M. & Van Huffel, S., 2000. Optical and acoustic damage detection in laminated CFRP composite materials, *Composite Science & Tech.* 60, 2713–2724.

Roberts, S.S.J., Davidson, R. & Paa, R., 1991. Mechanical Properties of Composite Materials Containing Embedded Fibre-Optic sensors, *Fibre Optics Smart Structures and Skin IV*; Proceedings of the Meeting, Boston, MA, Bellingham, WA, SPIE Vol. 1588, 326–341.

Sovka, P., Pollak, P. & Kybic, J., 1996. Extended Spectral Subtraction, Proceedings of European Signal Processing Conference, *EUSIPCO-96*, Trieste, Italia.

Surgeon, M. & Wevers, M., 1999. Static and dynamic testing of a quasi-isotropic composite with embedded optical fibres, *Composites Part A* 30(4), 317–324.

Tsuda, H., Ikeguchi, T., Takahashi, J. & Kemmochi, K., 1998. Damage monitoring of carbon-fiber-reinforced plastics using Michelson interferometric fiber-optic sensors', *Journal of materials science letters* 17, 503–506.

Tran, T.A., Miller III, W.V., Murphy, K.A., Vengsarkar, A.M. & Claus, R.O., 1991. Stabilized extrinsic fiber optic Fabry-Perot sensor for surface acoustic wave detection, Proceedings of SPIE *Fiber optic and laser sensors IX* Vol. 1584, 178–186.

Van Huffel, S., 1993. Enhanced Resolution Based on Minimum Variance Estimation and Exponential Data Modeling, *Signal Processing* 33(3), 333–355.

Van Huffel, S., Decanniere, C., Chen, H. & Van Hecke, P., 1994. Algorithm for time-domain NMR data fitting based on total least squares, *Journal of Magnetic Resonance, Series A* 110, 228–237.

Waite, S.R. & Sage, G.N., 1988. The failure of optical fibres embedded in composite materials, *Composites* 19(4).

Yao, S.K. & Asawa, C.K., 1983. Microbending fiber optic sensing, Proceedings of SPIE, Vol. 412, 9–13.

Ultrasound

Emerging Technologies in Non Destructive Testing, Van Hemelrijck, Anastasopoulos & Melanitis (eds)
© 2004 Swets & Zeitlinger, Lisse, ISBN 90 5809 645 9

Numerical simulations of ultrasonic polar scans

Nico F. Declercq & Joris Degrieck
Soete Laboratory, Department of Mechanical Construction and Production, Ghent University, Gent, Belgium

Oswald Leroy
Interdisciplinary Research Center, KULeuven Campus Kortrijk, Kortrijk, Belgium

ABSTRACT: Numerous experiments (Van Dreumel et al 1981, Degrieck et al 1994, Degrieck 1996, Declercq et al 2001, Degrieck et al 2003) have convinced the authors that ultrasonic polar scans are a promising tool for non-destructive characterization of fiber reinforced composites. However, because of the requirement to invert experimental data to extract quantitative information (Degrieck et al 2003), numerical simulations are mandatory. Such simulations have been developed before for single layered fiber reinforced composites. Nevertheless, since the vast majority of composites are multi-layered, the development of extended numerical models is needed. Such model is presented, together with a presentation of numerical simulations of ultrasonic polar scans for multi-layered composites. It is also shown that the polar scan of a fabric reinforced composite is quite different from a polar scan of $(0°/90°)$-stacked unidirectional layers.

1 INTRODUCTION

The principle of lowest possible mass for best suited stiffness and strength is the driving force behind the tailoring of high quality composites and the manufacturing of composite structures. This objective can only be achieved if it comes together with nondestructive characterization during manufacturing and service life. It has been shown before (Van Dreumel et al 1981, Degrieck et al 1994, Degrieck 1996, Declercq et al 2001) that the ultrasonic polar scan is a highly recommended and easy to use technique for the characterization of the fiber direction, the orthotropic stiffness, the extraction of information on porosity, resin fraction, etcetera. The present paper focuses on the theoretical modeling of ultrasonic polar scans. The last part of the paper presents some numerical examples of ultrasonic polar scans.

2 THE EFFECT OF ORTHOTROPY ON ELASTICITY

In what follows, we apply the double suffix notation convention of Einstein. The dynamics of an anisotropic material is described by (Naefeh 1995)

$$\frac{\partial \sigma_{ij}}{\partial r_j} = \rho \frac{\partial^2 u_i}{\partial t^2} \qquad (1)$$

while the generalized Hooke's law, taking into account symmetry properties which are due to the analytical feature of the strain energy and the symmetric nature of the stress and strain tensors, is given by

$$\begin{bmatrix} \sigma_{11} \\ \sigma_{22} \\ \sigma_{33} \\ \sigma_{23} \\ \sigma_{13} \\ \sigma_{12} \end{bmatrix} = \begin{bmatrix} C_{11} & C_{12} & C_{13} & C_{14} & C_{15} & C_{16} \\ C_{12} & C_{22} & C_{23} & C_{24} & C_{25} & C_{26} \\ C_{13} & C_{23} & C_{33} & C_{34} & C_{35} & C_{36} \\ C_{14} & C_{24} & C_{34} & C_{44} & C_{45} & C_{46} \\ C_{15} & C_{25} & C_{35} & C_{45} & C_{55} & C_{56} \\ C_{16} & C_{26} & C_{36} & C_{46} & C_{56} & C_{66} \end{bmatrix} \begin{bmatrix} e_{11} \\ e_{22} \\ e_{33} \\ 2e_{23} \\ 2e_{13} \\ 2e_{12} \end{bmatrix} \qquad (2)$$

Further symmetry considerations due to orthotropy result in

$$C_{14} = C_{24} = C_{34} = C_{15} = C_{25} = C_{35}$$
$$= C_{16} = C_{26} = C_{36} = C_{45} = C_{46} = C_{56} = 0 \qquad (3)$$

3 THE PROPAGATION OF BULK PLANE WAVES

The reverse Voigt procedure transforms the stiffness tensor C_{mn} of rank 2 to the stiffness tensor c_{ijkl} of rank 4 as

$$1 \rightarrow 11 \qquad 2 \rightarrow 22 \qquad 3 \rightarrow 33$$
$$4 \rightarrow 23 = 32 \quad 5 \rightarrow 13 = 31 \quad 6 \rightarrow 12 = 21 \qquad (4)$$

Equation (1) then becomes

$$\rho \frac{\partial^2 u_i}{\partial t^2} = c_{ijkl} \frac{\partial^2 u_l}{\partial x_j \partial x_k} \qquad (5)$$

A plane wave solution of (5) is of the form

$$u_i = U_i \exp i\left(n_j r_j - \omega t\right) \qquad (6)$$

where $\mathbf{n}$ is the wave vector. If this is entered in (5), straightforward calculations then result in

$$\left(\frac{1}{\rho} c_{ijkl} n_k n_j - \omega^2 \delta_{il}\right) U_l = 0 \qquad (7)$$

The latter equation is called Christoffel's equation. It relates the slowness n/ω and the polarization $\mathbf{U}$ to the propagation direction and is solved by demanding nontrivial solutions, followed by the determination of the corresponding eigenvectors.

4 THE SCATTERING OF PLANE WAVES

4.1 Snell's law

If sound inside the bulk of the composite laminate results from impinging plane waves (denoted by superscript "inc"), Snell's law must be taken into account which for interfaces perpendicular to n_3 states that

$$n_1 = n_1^{inc} \text{ and } n_2 = n_2^{inc} \qquad (8)$$

Then, requiring nontrivial solutions, (7) leads to a sixth degree polynomial equation of the form

$$\beta^6 + B_5 \beta^5 + A_4 \beta^4 + B_3 \beta^3 + A_2 \beta^2 + B_1 \beta + A_0 = 0 \qquad (9)$$

in which β represents n_3.

Furthermore, it can be shown (Naefeh 1995) that the presence of symmetry higher than or equal to monoclinic symmetry (as is the case for ortotropy) results in

$$B_j = 0 \qquad (10)$$

whence 3 solutions

$$\beta_2 = -\beta_1 \quad \beta_4 = -\beta_3 \quad \beta_6 = -\beta_5 \qquad (11)$$

exist.

4.2 Continuity of normal stress and displacement

For a plate immersed in water, two different continuity conditions are involved. The water/solid interface is determined by continuity of normal stress and normal displacement, hence along the interface,

$$u_3^{water} = u_3^{solid} \qquad (12)$$

and

$$\sigma_{i3}^{water} = \sigma_{i3}^{solid}, \quad i = 1,2,3 \qquad (13)$$

The solid/solid interface(s) are determined by the continuity of the displacement vector and continuity of normal stress, hence along the interface(s)

$$u_i^{solid} = u_i^{solid}, \quad i = 1,2,3 \qquad (14)$$

and

$$\sigma_{i3}^{solid} = \sigma_{i3}^{solid}, \quad i = 1,2,3 \qquad (15)$$

By taking into account the appropriate continuity conditions and by applying the discussion of previous subsection 4.1, one is able to simulate a polar scan.

5 THE PRINCIPLE OF A POLAR SCAN

A classical C-scans is formed by registering the reflected or transmitted signal on many spots on the laminate surface, by using normal incidence. Most often, C-scans are used to detect material defects and eventually to find out the 3D locations of such defects. Lately, some attempts have been undertaken to unveil the fiber direction by means of C-scans. However, the reported methods can only be used if sufficient microscopic material defects exist on the fiber/matrix interface or if the fibers are not equally distributed within the matrix.

A polar scan differs from a classical C-scan in that the transducer is not constantly hold perpendicular to the interface.

On the contrary, a polar scan exploits oblique incidence and measures the reflected/transmitted specular sound resulting from sound that is subsequently incident from all possible directions from above the plate. As is seen in Figure 1, the incidence direction is defined by (θ, φ). The reflected/transmitted amplitude is then registered in a polar diagram where each spot corresponds to a certain $(\theta + \pi, \varphi)$ and represents the amplitude for that direction. Physically, the process of sound impinging the plate and traveling inside the plate, being scattered once and again by the different

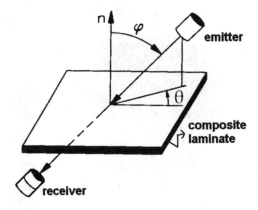

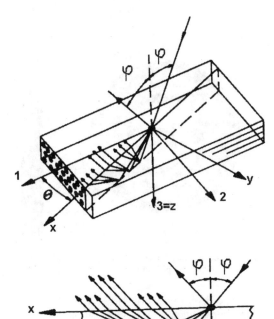

Figure 1. The position of the transducers in a polar scan.

Figure 2. The complicated interaction of an incident plane wave with a single layered composite plate. Each scattering generates 3 propagation modes in the plate and 1 in the liquid.

interfaces, is a very complicated phenomenon, cf. Figure 2. However, for harmonic incident waves (we may always consider a pulse as a superposition of harmonic waves), a "standing wave pattern" is formed inside the plate, which is modeled by demanding only 6 modes (the 6 coming from Christoffel's equation (7)) propagating in each layer. This standing wave

pattern may result in some kind of an eigen-vibration of the plate. If this occurs, this pattern is called a quasi Lamb wave. It is characterized by a reflection/transmission coefficient tending to zero. This results in "dark regions" in the registered polar scan.

The position and the characteristics of these regions are determined by the physical parameters of the plate, such as the thickness of the layers, the density, the stiffness coefficients and the damping. Ultrasonic polar scans form therefore an excellent tool for monitoring these physical properties. The interpretation of a polar scan is a difficult task. However, in the case of thick plates, the only patterns that do appear are due to bulk critical angles. For reasons of explanatory simplicity we solely focus on this case. Snell's law for critical waves is as follows

$$\sin(\varphi_{crit})|_\theta = v_l / v_{crit}(\rho, C_{ij}, \theta) \qquad (16)$$

where v_l is the plane wave velocity in the liquid and v_{crit} is the velocity of the critical bulk wave.

If a certain contour in one direction is wider than in other directions, this means that the velocity in the "wider" direction is lower than in the other direction. For example the velocities of quasi longitudinal waves (corresponding to the inner contour of a polar scan) traveling along the in-plane axes of orthotropy are given by

$$\sqrt{C_{11}/\rho} \text{ and } \sqrt{C_{22}/\rho} \text{ respectively.}$$

Hence, regarding equation (16), the directions of highest stiffness produce the smallest critical angles for quasi longitudinal waves. Even though the contours of polar scans for thin plates are much more difficult to interpret, the basic idea remains unchanged.

6 SOME NUMERICAL EXAMPLES

6.1 Single layered unidirectional fiber reinforced composites

Hereafter, each polar scan is simulated for a 1 mm thick carbon/epoxy fiber reinforced plate and an ultrasonic sound frequency of 5 MHz.

Figure 3 presents the numerical simulation of an ultrasonic polar scan (in reflection) for a single layered unidirectional fiber reinforced composite. The fibers are oriented along the 0° polar direction. In that direction, it is indeed verified that the inner contour (which corresponds to the quasi longitudinal plane wave critical angle) is smallest. Figure 4 is equivalent to Figure 3, except that here the ultrasonic polar scan in transmission is plotted. Due to damping, the overall amplitude is smaller, but characteristic contours are still visible.

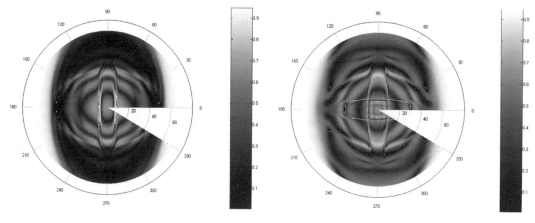

Figure 3.   Ultrasonic polar scan (in reflection) of a single layered unidirectional fiber reinforced composite.

Figure 5.   Ultrasonic polar scan (in reflection) of a double layered (0°/90°) cross-ply laminate.

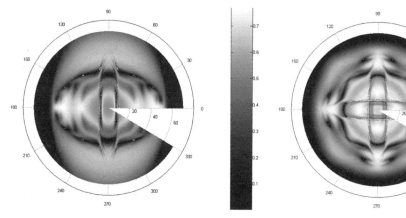

Figure 4.   Ultrasonic polar scan (in transmission) of a single layered unidirectional fiber reinforced composite.

Figure 6.   Ultrasonic polar scan (in transmission) of a double layered (0°/90°) cross-ply laminate.

6.2   Double layered cross-ply laminate (0°/90°)

Figures 5 and 6 represent the numerical simulations of ultrasonic polar scans on a double layered fiber reinforced composite with the top layer consisting of fibers in the 0° direction and the bottom layer built up by fibers in the 90° direction.

The presence of both symmetries (one on top of the other), is clearly visible in the characteristic contours if compared to Figures 3 and 4. Moreover, it is seen that the reflected pattern in the 0° polar direction is not equivalent to the pattern in the 90° polar direction.

6.3   Single layered fabric reinforced composites

Figures 7 and 8 are the simulations of ultrasonic polar scans on a fabric reinforced composite. Even though the laminate also consists of carbon fibers in epoxy resin, the actual stiffness is different from the cross-ply

composite described above. Hence only qualitative comparison is allowed. It is seen that the pattern in the 0° polar direction matches perfectly to the pattern in the 90° polar direction. It is clear that the patterns qualitatively differ strongly from those of Figures 5 and 6. Hence, a single layered fabric fiber reinforced composite is quite different from a (0°/90°) double layered fiber reinforced composite.

6.4   Multi-layered cross-ply laminate (0°/90°)

In order to check the relevancy of the above statement that a fabric fiber reinforced composite is different from a (0°/90°) stacked composite, we have decided the number of layers up to 10. The numerical results are seen in Figures 9 and 10. Even though the number of layers is much larger, still there is a difference between the polar scans of Figures 9 and 10 and the ones of

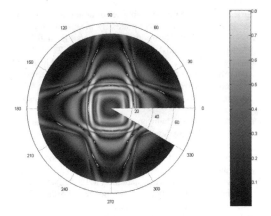

Figure 7. Ultrasonic polar scan (in reflection) of a single layered fabric reinforced composite.

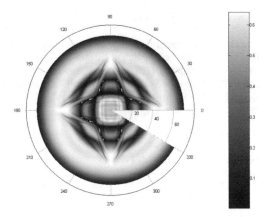

Figure 10. Ultrasonic polar scan (in transmission) of a cross-ply reinforced composite consisting of 10 (0°/90°) stacked layers of unidirectional fiber reinforced material.

Figures 7 and 8. Even here, with a much increased number of stacked layers, the reflected pattern in the 0° polar direction still differs significantly from the pattern in the 90° polar direction. This finding is important since it is common for many researchers to model a fabric fiber reinforced composite as (0°/90°) stacked layers of unidirectional material. It is obvious from the presented preliminary results that this should be avoided.

7 CONCLUDING REMARKS

It is shown how numerical simulations of polar scans are performed, starting from simple principles of mechanics and wave motion. Even though simulations on single layered fiber reinforced composites already existed, the theoretical model has been extended to multi-layered composites. As an excellent example, it has been shown that fabric fiber reinforced composites cannot be modeled sufficiently accurate by means of a model in which unidirectional fiber reinforced layers are stacked in large numbers in the 0° polar direction and the 90° polar direction. In the near future, the applied model will be further extended to crystals and also to pre-stressed composites. These numerical simulations, together with the upgraded and highly modernized experimental set up may become a means of characterizing the stiffness of anisotropic plates. It is the authors' purpose to develop an automated tool that applies a numerical/experimental inversion technique and to rebuild the experimental apparatus to meet in-field requirements. Due to the strong connection to stiffness, the developed technique is intended to monitor fatigue damage on composites, porosity, and resin fractions. Furthermore, it will be used to verify micro-mechanical models.

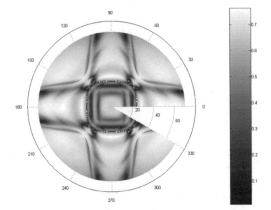

Figure 8. Ultrasonic polar scan (in transmission) of a single layered fabric reinforced composite.

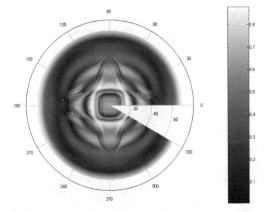

Figure 9. Ultrasonic polar scan (in reflection) of a cross-ply composite consisting of 10 (0°/90°) stacked layers of unidirectional fiber reinforced material.

ACKNOWLEDGEMENTS

The authors are thankful to "The Flemish Institute for the Encouragement of the Scientific and Technological Research in Industry (I.W.T)" for financial support and to the "Fund for Scientific Research – Flanders-Belgium (F.W.O)" for offering a travel grant.

REFERENCES

Declercq N. F., Degrieck J. & Leroy O., 2001. Characterization of Layered Orthotropic Materials Using Ultrasonic Polar Scans. Proceedings of the 2nd FTW PhD symposium, Ghent University.

Declercq N. F., Degrieck J. & Leroy O., 2003. Numerical simulations of ultrasonic polar scans on fiber reinforced composites. Proceedings of the 29 Deutsche Jahrestagung für Akustik DAGA 2003, 18–20. March 2003, Aachen, Germany.

Degrieck J., 1996. Some possibilities of nondestructive characterization of composite plates by means of ultrasonic polar scans. Non Destructive Testing. Van Hemelrijck & Anastasopoulos (eds), Balkema, Rotterdam, 225–236.

Degrieck J., Declercq N. F. & Leroy O., 2003. Ultrasonic Polar Scans as a possible means of nondestructive testing and characterization of composite plates. Insight – The Journal of the British Institute of Non-Destructive Testing, 45(3), 196–201.

Degrieck J. & Van Leeuwen D., 1994. Simulatie van een Ultrasone Polaire Scan van een Orthotrope Plaat. Proceedings of "The 3rd Belgian National Congres on Theoretical and Applied Mechanics", Liege University, 39–42 (in dutch).

Naefeh A. H., 1995. Wave propagation in layered anisotropic media with applications to composites, North Holland series in Applied Mathematics and Mechanics.

Van Dreumel W. H. M. & Speijer J. L., 1981. Nondestructive Composite Laminate Characterization by Means of Ultrasonic Polar Scan, Materials Evaluation, 39(10), 922–925.

Emerging Technologies in Non Destructive Testing, Van Hemelrijck, Anastasopoulos & Melanitis (eds)
© 2004 Swets & Zeitlinger, Lisse, ISBN 90 5809 645 9

A new nonlinear-modulation acoustic technique for crack detection

V.Yu. Zaitsev & V.E. Nazarov
Institute of Applied Physics RAS, Nizhny Novgorod, Russia

V.E. Gusev & B. Castagnede
Université du Maine, Le Mans Cedex 09, France

ABSTRACT: A new nonlinear-moduation method for crack detection is discussed. It is based on the so-called cross-modulation effect consisting of the transfer of modulation from an intensive, initially slowly amplitude-modulated stronger (pump) excitation to the probe signal. Advantage of this technique is a very flexible independent choice of the frequencies for both carriers and the modulation, since their ratio may be rather arbitrary. This in its turn makes possible to effectively use the sample resonances in order to achieve the necessary level of the pump excitation and to ameliorate conditions for detection of the modulation sidelobes for the probe wave. Unlike harmonic-generation methods the initial nonlinear distortions of the pump and probe excitations (e.g. due to nonlinearities in the electronics) are not critical for this technique. The performed test experiments indicated high sensitivity of the new technique.

1 INTRODUCTION

Exploitation of nonlinear vibro-acoustic effects for diagnostics and, in particular, for early crack detection is an emerging technique, which is rapidly progressing during last several years after pioneering encouraging results (e.g. Gitz, Guschin & Konyukhov 1973, Sessler & Weiss 1975, Buck, Morris & Richardson 1978, Antonets, Donskoy & Sutin 1986, Achenbach, Parikh & Sortiopoulus, 1989). Among other nonlinear effects such as, for example, observation of higher harmonics generation in the damaged sample, the use of modulation vibro-acoustical interactions yields several advantages (see e.g. Korotkov & Sutin 1994, Zaitsev & Sas 2000). In the conventional variants (Korotkov & Sutin 1994, Zaitsev & Sas 2000, Van Den Abeele, Johnson & Sutin 2000) of this technique the increase of modulation depth of a weaker probe wave induced by another intensive, low-frequency (pump) wave or vibration is used as a sign of damage appearance.

The modulation approach was used in the first (to our knowledge) patented nonlinear method of crack detection (Sessler & Weiss 1975), whose principle was clearly described in the following formulation: "The material specimen or structure having a crack to be detected is subjected to tensile or compressive forces due to excitation caused by low-frequency sound waves or mechanical vibrations from a generator, thus changing the effective size of the crack in the specimen. An ultrasonic search unit is used to follow modulations of reflected energy at the crack interace due to variation of the effective size of the crack". The method principle is schematically shown in Figure 1.

Thus this method is very close to the conventional linear pulse-echo technique supplemented with an additional pump source in order to produce the crack modulation. The latter provides the possibility to discriminate the echo-signals produced by highly compliant cracks from other scattering sources (boundaries, near spherical or cylindrical technological holes, etc.) whose reflection strength is not influenced by the pump excitation. However, the method principle implies that the defects are strong enough obstacles

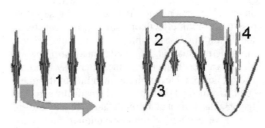

Figure 1. Principle of the pulse-echo modulation technique for crack detection. 1 – incident probing pulses, 2 – reflected pulses, 3 – intensive low-frequency excitation (pump), 4 – crack "breathing" under the action of the pump.

for the propagating waves in order to produce noticeable reflection (scattering). The additional pump serves only for better discrimination of the crack echo-signal from that produced by other scatterers.

More recently it was also demonstrated that even a rather weak damage, whose scattering strength is very small since its size is much smaller than the length of the sounding probe wave, may also induce strong modulation of a probe excitation under the action of an additional low-frequency pump (Korotkov & Sutin 1994, Zaitsev & Sas 2000a, Van Den Abeele, Johnson & Sutin 2000). In particular, both the probe and pump waves could be continuous and excited at eigenfrequencies of the sample. The method principle is schematically shown in Figure 2. In the latter case, the defects are so small in the scale of the wavelength that the reflection or scattering from the defects in the aforementioned pulse-echo sense cannot be singled out. However, for detection of even such a small crack-like damage in a sample this modulation technique has proven to be very sensitive. The defect localization for (quasi-) continuous excitations, in principle, can be possible using additional analysis of differences in the modulation response for different spatial distributions of the pump and probe waves. As a disadvantage of this method it may be noted that the necessity to produce an intensive low-frequency pump action often causes complication of feasibility of the technique. For example, essentially different types of the actuators may be required (for example, a piezo-source for the higher-frequency probe wave and a powerful vibration bench for the low-frequency pump). Another disadvantage is connected to the fact that normally it is convenient to excite both the pump and the probe waves tuned to own resonances of the sample. However, the combination frequencies of the modulation components may get at a rather arbitrary position with respect to the sample resonance peaks. This fact complicates the observation of the modulation sidelobes and estimation of the modulation depth, whose apparent magnitude is strongly affected by the sample resonance structure.

Recently, a new variant (Zaitsev, Gusev & Castagnede 2002a,b) of the nonlinear-modulation technique was successfully tested for detecting a weak damage. The new method is based on the so-called cross-modulation effect consisting in the transfer of modulation from an intensive, initially slowly amplitude-modulated pump excitation to the probe signal. The carrier frequency of the latter may be either lower or comparable, or higher than that of the pump signal. The method principle is schematically shown in Figure 3. This variant of the modulation technique keeps all positive features of the conventional modulation schemes. An additional advantage of this technique is the very flexible, independent choice of both carrier frequencies as well as the modulation frequency. This in its turn makes possible to effectively use the sample

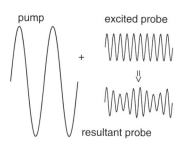

Figure 2. Schematically shown conventional nonlinear modulation of a probe wave caused by a low frequency pump excitation.

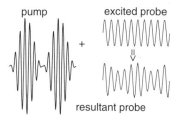

Figure 3. Schematically shown cross-modulation of a probe wave caused by an amplitude-modulated pump excitation.

resonances in order to achieve the necessary level of the pump excitation and to ameliorate conditions for detection of the modulation sidelobes. In particular, the modulation frequency may be readily chosen low enough in order to place the sidelobes within the same resonance peak to which the probe wave is tuned. This provides favorable conditions for detection of the modulation sidelobes around the probe wave carrier and eliminates strong distortion of the apparent modulation depth. Besides, since the pump and probe carrier frequencies should not be strongly separated in this scheme, similar (e.g. piezoceramic) types of the actuators may be effectively used for the both waves. Unlike harmonic-generation methods the initial nonlinear distortions of the pump and probe excitations (e.g. distortion of the sounding waves spectra due to nonlinearities in the electronics) are not critical for this technique. Test experiments with a sample containing small cracks indicated very high sensitivity of the new technique.

2 WHY STRUCTURE-SENSITIVITY OF NONLINEAR EFFECTS IS SO HIGH?

Many publications related to the observation of nonlinear effects in damaged solids report that "structure-sensitivity" of nonlinear properties (that is magnitude of their variations induced by a damage in the sample) is often much higher than the complementary variations in the linear elastic parameters. However, the origin of such a high contrast between variations in linear versus nonlinear properties in many cases seems to be understood not clearly enough. This situation has an essential physical reason: nonlinear features are indeed less rude compared to the linear ones and, besides, apparently similar nonlinear effects may be caused by essentially different physical nonlinear mechanisms. Nonlinearity is often considered in terms of nonlinear elasticity only, whereas actually in real cases the similar effects may be caused either by nonlinear elasticity or, for example, amplitude-dependent dissipation with essentially different physics, which may be of a non-hysteretic or a hysteretic type. The hysteretic nonlinearity comprises both nonlinear elasticity and the nonlinear dissipation. Ample understanding of the actual underlying mechanisms is necessary for optimization of practical implementation the nonlinear methods. Here we briefly elucidate the reasons of high "structural sensitivity" sensitivity of the nonlinear properties of damaged samples containing high-compliant crack-like defects.

2.1 Crack-induced elastic nonlinearity

We start with the case of purely elastic nonlinearity. It is well known that real cracks are normally not simple planar cuts inside the intact material, but contain inner contacts. The contribution of the contacts to the elastic stress is often described by the Hertzian nonlinearity (Landau & Lifschitz 1986), which in simplest case leads to the following stress–strain relation at the contacts:

$$\sigma = B\varepsilon^{3/2}H(\varepsilon),\qquad(1)$$

where factor B depends on elastic moduli of the individual asperities and their sizes, and the Heaviside function $H(\varepsilon)$ indicates that only compressed (here we assume sign $\varepsilon > 0$ for the compression) contacts contribute to the stress in the material. There is always some mean static contact strain ε_0 determining linear elastic modulus $d\sigma(\varepsilon_0)/d\varepsilon$ for the wave perturbations, whose strain $\tilde{\varepsilon}$ is significantly smaller, $\tilde{\varepsilon} << \varepsilon_0$ and stress $\tilde{\sigma} << \sigma_0 = \sigma(\varepsilon_0)$. In real materials, however, the contact loading is essentially different, so that along with contacts bearing some average loading there are essentially weaker loaded contact, which also give their contributions to the resultant $\sigma(\varepsilon)$. Singling out explicitly the static and wave parts in stress and strain the contributions of an average loaded and an initially unloaded contact to the stress in the material may be rewritten as:

$$\sigma_0 + \tilde{\sigma} = B(\varepsilon_0 + \tilde{\varepsilon})^{3/2}H(\varepsilon_0 + \tilde{\varepsilon}) + \\ B(\mu\varepsilon_0 + \tilde{\varepsilon})^{3/2}H(\mu\varepsilon_0 + \tilde{\varepsilon})\qquad(2)$$

here coefficient $|\mu| << 1$ characterizes the extent of unloading. We may allow for $\mu < 0$ in order to describe the contacts with initial gaps, so that they may be activated at wave strains $|\tilde{\varepsilon}| > |\mu|\varepsilon_0$. For initially weakly compressed contacts ($0 < \mu << 1$, $|\tilde{\varepsilon}| << \mu\varepsilon_0$) differentiation of Equation (2) with respect to $\tilde{\varepsilon}$ readily yields the following expression for the contribution of these contacts to the linear elastic modulus:

$$d\tilde{\sigma}/d\tilde{\varepsilon}\big|_{\varepsilon_0} = (3/2)B(\varepsilon_0)^{1/2} + (3/2)B(\mu\varepsilon_0)^{1/2}\qquad(3)$$

It is evident from the second term in the left-hand side of Equation (3) that the relative contribution of the weak contacts to the linear elasticity is essentially smaller (by a factor of $\mu^{1/2} << 1$) compared to the contribution of more stiff, stronger compressed contacts. In contrast, for the higher-order nonlinear moduli, which are characterized by the higher-order derivatives, the contribution of the weak unloaded contacts strongly dominates over the nonlinear coefficients related to the stronger pre-compressed contacts. This statement is illustrated by the following expressions for the 2nd and 3rd derivatives of $\tilde{\sigma}$ with respect to $\tilde{\varepsilon}$ corresponding to the quadratic and cubic (in the vibrational strain) nonlinear terms in the power expansion of the stress–strain relation:

$$d^2\tilde{\sigma}/d\tilde{\varepsilon}^2\big|_{\varepsilon_0} = (3/4)B(\varepsilon_0)^{-1/2} + (3/4)B(\mu\varepsilon_0)^{-1/2}\qquad(4)$$

$$d^3\tilde{\sigma}/d\tilde{\varepsilon}^3\big|_{\varepsilon_0} = -(3/8)B(\varepsilon_0)^{-3/2} - (3/8)B(\mu\varepsilon_0)^{-3/2}\qquad(5)$$

These expressions demonstrate that for the nonlinear moduli of the order $n \geqslant 2$ the relative contribution of the weak contact is greater than that of the average-loaded contact by a large factor of $\sim\mu^{3/2-n} >> 1$. Thus presence of highly compliant crack-like defects containing weakly pre-compressed contacts (whose contribution to the linear elasticity is very small) may strongly increase the sample elastic nonlinearity. The above simple arguments clearly indicate that, in contrast to the almost negligible contribution to the linear elasticity, the weak contacts may play dominant role in increase of the magnitude of nonlinear effects determined by the higher-order nonlinear elastic moduli.

In particular, the factor described by Equation (4) corresponds to the quadratic in the vibrational strain contribution to the total stress and, correspondingly, to the linear-in-strain correction to the elastic modulus. This contribution is responsible for the instantaneous variations of the elastic modulus under an intensive vibrational excitation and may cause modulation of a weak probe wave by an intensive low-frequency pump (of the type schematically shown in Figure 2).

The factor described by Equation (5) corresponds to the quadratic-in-strain nonlinear correction to the elastic modulus. Via this term, a periodic pump induces period-averaged variations in the sample elasticity. Thus this effect may, in principle, be responsible for the appearance of the effect of the cross-modulation of a probe wave by an amplitude-modulated pump excitation (of the type shown schematically in Figure 3). The transformation of the signal phase modulation corresponding to the variation of the elasticity into the amplitude modulation may be obtained by tuning the probe-wave frequency near a resonance peak of the sample.

2.2 Crack-induced amplitude-dependent dissipation

The above arguments concerning the role of the crack inner contacts in the increase of the sample nonlinearity are rather instructive and support the simplest representation of the nonlinearity as the elastic nonlinearity. However, the same cracks with contacts may produce strong modulation-type interaction between the probe and pump waves via an essentially different mechanism. Namely, pronounced amplitude modulation of the probe signal may be produced by the pump wave via its direct influence on the probe signal dissipation. Conventionally, this dissipation is mostly attributed to friction or adhesion hysteresis at crack interfaces (Zaitsev & Sas 2000b). However, it is physically clear and recently corroborated by direct atomic-scale experiments (Mate et al. 1987) that for manifestation of adhesion and friction, mutual displacement at interfaces should exceed the atomic size a. In this context, for a crack with diameter L, the average compressional or shear strain ε can produce maximal lateral or normal interfacial displacement (Zaitsev, Gusev & Castagnede 2002b) $D \sim \varepsilon L$. The requirement $D > a$ determines the threshold strain $\varepsilon_{th} > a/L$, below which the interfacial displacement is of sub-atomic scale. For a typical atomic size $a \sim 3 \times 10^{-10}$ m and a macroscopic crack with $L \sim 10^{-3}$ m, this yields $\varepsilon_{th} \sim 0.3 \times 10^{-6}$, which should be exceeded in order to activate frictional and adhesional hysteretic losses. However, even at much smaller strains, the defects can efficiently dissipate elastic energy due to locally enhanced thermoelastic coupling at cracks and especially at the inner crack contacts. These losses are thresholdless and are essential even for rather weak probe waves with strains $\varepsilon < 10^{-8}$. The corresponding mechanism was recently considered in (Zaitsev, Gusev & Castagnede 2002b), in which we argued additionally that, in crack-containing solid samples, quite moderate pump strains $\varepsilon \sim 10^{-6}-10^{-5}$ may significantly affect the probe wave dissipation at the inner contacts in cracks. In this context, it is well known that cracks with ratio of the opening d to characteristic crack diameter L,

may be completely closed by average strain $\varepsilon \sim d/L$, typical d/L for cracks being $10^{-3}-10^{-4}$. However, at small loosely separated regions, local separation (or inter-penetration) d_{loc} of crack interfaces is much smaller than average separation d. Such contacts are extremely stress-sensitive, since due to the described geometry they are strongly perturbed by the average strain, which can be orders of magnitude smaller (roughly $d/d_{loc} >> 1$ times) than the typical magnitude $\varepsilon \sim 10^{-3}-10^{-4}$ required to close the whole crack. For the magnitude of the dissipation it is important that inner contacts in real cracks are normally not point-like, but rather strip-like due to the quasi-2D character of crack initiation. Such a shape of contacts agree with direct electron- and atomic-force microscopy images of cracks normally exhibiting wavy corrugated structure of the interfaces. Due to this geometry (and the resultant strong local concentration of stress and increased temperature gradients), the inner contacts may very efficiently dissipate elastic wave energy via the thermoelastic coupling. In order to estimate the magnitude of this dissipation we applied an approximate approach similar to that (Landau & Lifschitz 1986) used for estimates of thermoelastic losses in polycrystals and additionally took into account stress-concentration at the inner contacts. Thus we derived the following approximate expressions for thermoelastic losses in the low-frequency limit, in the high-frequency limit and at the relaxation maximum, when the thermal wave length coincides with the width of the contact:

$$W_{LF}^{dis} = 2\pi\omega T(\alpha^2 K^2/\kappa)l^2 \tilde{L}L^2 \varepsilon^2, \qquad (6)$$

$$W_{HF}^{dis} = (2\pi/\omega)\kappa T(\alpha K/C\rho)^2 \tilde{L}(L/l)^2 \varepsilon^2, \qquad (7)$$

$$W_{cont}^{max} = 2\pi T(\alpha^2 K^2/\rho C)\tilde{L}L^2 \varepsilon^2,$$
$$\omega = \omega_l \approx \kappa/(\rho C l^2), \qquad (8)$$

where ω is the wave cyclic frequency, T is the temperature, α is the temperature expansion coefficient of the solid, K is the bulk elastic modulus, ρ is the density, C is the specific heat, ε is the average strain, κ is the thermal conductivity, ω is the relaxation frequency for contact width l, L is the characteristic crack diameter and $\tilde{L}$ is the contact length. Note that for strip-like contacts with $\tilde{L} \sim L$ the maximal losses at the narrow contacts are roughly the same as at the whole crack. However, the relaxation frequencies for millimeter-scale cracks are fractions of Hz for most of metals, glasses or rocks, whereas for narrow contacts, this frequency can reach kHz and even MHz band. Quantitatively the estimates based on Equations (6)–(8) indicate that the considered thermoelastic losses at the inner contacts can be rather strong and may produce pronounced modulation of the probe wave under the action of quite a moderate-amplitude pump excitations. The relevant experimental examples will be considered below.

3 EXPERIMENTAL DEMONSTRATION

In order to demonstrate high sensitivity of the cross-modulation effect to the presence of defects in a solid sample we implemented an instructive experiment in the form of interaction of two longitudinal modes in a glass rod 30 cm in length and 8 mm in diameter.

The rod contained three corrugated thermally-produced cracks 2–3 mm in size (Figure 4). The probe and pump wave were tuned to different resonance peaks of the rod (for example, to the 2nd and 1st longitudinal resonances about 3.8 and 11 kHz respectively). The modulation frequency was chosen equal to a few Hertz, so that it was convenient to observe the modulation sidelobes together with the carrier-frequency lines for the pump and the probe waves within the respective resonance peaks. In a reference rod without cracks, the modulation sidelobes (existing due to residual parasite nonlinearities) were 25–40 dB lower than shown in Figure 5. In principle, variation in the amplitude of the probe wave may be attributed to the pump-induced mismatch between the frequencies of the probe wave and the sample resonance frequency that should be affected by the stronger pump wave via the crack-induced nonlinear elasticity (see the discussion in section 2.1). Alternatively, the probe wave modulation could be caused by direct influence of the pump wave on the dissipation of the probe signal, and certainly the two effects could be mixed. In order to clarify the actual mechanism we obtained resonance curves for the probe waves for different amplitudes of the pump. The curves presented in Figure 6 indicate that primarily the dissipation, not the elasticity, is affected by the stronger wave in the discussed case. Magnitudes and frequencies, at which the observed amplitude-dependent variations in dissipation were observed, agree well with estimates based on Equations (6)–(8). Quantitatively, the estimates for typical parameters of glass show that even a single contact-containing crack of a few millimeters in size suffices to explain the observed 10–12% variation in the initial magnitude of the quality-factor of about 300–350 for the probe wave.

Another experimental demonstration of the high-sensitivity of the cross-modulation effect was obtained for a thick glass plate $230 \times 190 \times 14.5$ mm in sizes.

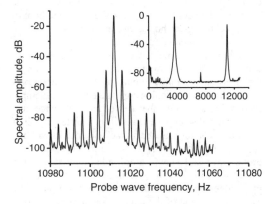

Figure 5. Experimental spectrum of the cross-modulation in the rod. In the inset the lower-resolution spectra of the pump and probe waves are displayed.

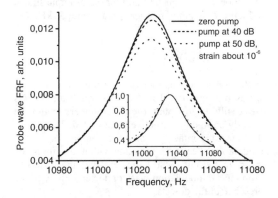

Figure 6. Probe wave frequency-response functions obtained at different pump amplitudes. The inset displays widening of the normalized resonance curves.

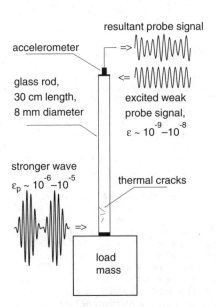

Figure 4. Schematically shown experimental configuration.

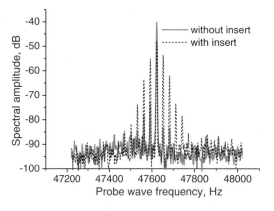

Figure 7.   Probe wave spectra without the insert in the cut (no crack) and with the insert (with the crack-like defect).

In order to demonstrate high sensitivity of the nonlinear effects to the presence of crack-like defects in the sample and, in contrast, to prove that complex boundaries, technological holes, etc. are not important for the nonlinear interaction, we prepared the sample in a special manner. Namely, in order to model technological holes and cracks we made two cuts about 1 mm in thickness and 15 mm in depth in the directions perpendicular to two neighboring plate edges. The sample prepared in such a manner served as a reference one. When a small steel plate was tightly inserted in one of the cuts, the latter imitated a crack-like defect with the apparent interface area of 2–4 mm^2. The sinusoidal probe wave and slowly amplitude-modulated pump waves were excited by two piezo-disks glued to the plate edges opposite to the cuts. The insertion of the steel piece in one of the cuts caused appearance of a rather pronounced modulation of the probe signal as shown in Figure 7. The pump-wave frequency for this example could be chosen either lower or higher than the probe-wave frequency, and for the example shown in Figure 7 the pump frequency was about 54 kHz.

4   CONCLUSION

In the considered instructive experimental examples, we used a transparent material in which the defects are easily visible and their parameters could be directly and non-destructively estimated. The use of the reference intact rod and the comparison of the modulation spectra for the plate with and without the insert in the cut clearly indicate that the observed pronounced modulation was connected only with the presence of the defects.

As indicate both the experimental results and the theoretical arguments the new variant of the modulation technique is very sensitive to the presence of the crack-like defects and offers significant technical advantages (possibility to effectively use sample resonances both for the primary excitations and the modulation sidelobes; flexibility and independence in the choice of the working carrier and modulation frequencies; possibility to use piezo actuators at arbitrary low modulation frequencies). The obtained results indicated that, in contrast to conventionally considered role of the elastic part of the sample nonlinearity, the proposed mechanism related to the amplitude-dependent dissipation may play the dominant role for the modulation effects in weakly damaged samples. This understanding is important for the optimization of the parameters of the sounding pump and probe waves. Certainly this modulation mechanism is not the unique one, and nonlinear-elastic or hysteretic mechanisms may be also important in other conditions.

Note finally that in this communication we focused primarily on the physical features of the cross-modulation effect and discussion of its relation to the conventional modulation technique concerning the problem of crack detection. For obtaining additional information on the defect localization the dependence of the cross-modulation effect on the spatial structure of the probe and pump waves may be used. The understanding of the effect physical mechanism is also important for the localization procedure, since the working frequencies of the sounding signals essentially influence not only on the spatial distribution of the sample normal modes, but may essentially affect the modulation depth via the frequency dependence of the dissipation of the probe signal as discussed in section 2.2.

The work was partially supported by RFBR (grants No 02-02-16237 and 02-02-08021-INNO).

REFERENCES

Achenbach, J.D., Parikh, O.K. & Sortiopoulus, D.A. 1989. Nonlinear effects in reflection from adhesive bonds, *Review of Progress in QNDE* (eds. D.O. Thompson and D.E. Chimenti) 8B: 1401–1407.

Antonets, V.A., Donskoy, D.M. & Sutin, A.M. 1986. Nonlinear vibrodiagnostics of flaws in multilayered structures. *Mechanics Compos. Mater.* 5: 934–937.

Buck, O., Morris, W.L. & Richardson, J.M. 1978. Acoustic harmonic generation at unbounded interfaces and fatigue cracks, *Appl. Phys. Lett.* 33(5): 371–373.

Gitz, I.D., Guschin, V.V. & Konyukhov, B.A. 1973. Measurements of nonlinear distortions of acoustic waves in polycrystalline aluminium in fatigue tests, *Akust. Zh.* (*Sov. Phys. Acoust.*) 19(3): 335–338.

Korotkov, A.S. & Sutin, A.M. 1994. Modulation of ultrasound by vibrations in metal constructions with cracks, *Acoust. Lett.* 18: 59–62.

Landau, L.D. & Lifschitz, E.M. 1986. *Theory of Elasticity*, 3rd (revised) English edition: Pergamon, New York.

Mate, C.M. et al. 1987. Atomic-scale friction of a tungsten tip on a grafite surface, *Phys. Rev. Lett.* 59(17): 1942–1945.

Rudenko, O.V. 1993. Nonlinear methods in acoustic diagnostics, *Russian J. Nondestructive Testing* 29(8): 583–589.

Sessler, J.G. & Weiss, V. 1975. Patent US3867836 Crack detection apparatus and method.

Van Den Abeele, K., Johnson, P.A. & Sutin, A.M. 2000. Nonlinear elastic wave spectroscopy (NEWS) techniques to discern materialdamage. Part I: nonlinear wave modulation spectroscopy. *Res. Nondest. Eval.* 12(1): 17–30.

Zaitsev, V., Gusev, V. & Castagnede, B. 2002a. Observation of the "Luxemburg–Gorky effect" for elastic waves, *Ultrasonics* 40: 627–631.

Zaitsev, V., Gusev, V. & Castagnede, B. 2002b. Luxemburg–Gorky effect retooled for elastic waves: a mechanism and experimental evidence, *Phys. Rev. Lett.* 89(10) 105502.

Zaitsev, V.Yu. & Sas, P. 2000a. Nonlinear response of a weakly damaged metal sample: a dissipative mechanism of vibro-acoustic interaction, *J. Vibration and Control* 6: 803–822.

Zaitsev, V.Yu. & Sas, P. 2000b. Dissipation in microinhomogeneous solids: inherent amplitude-dependent attenuation of a non-hysteretical and non-frictional type, *Acustica-Acta Acustica* 86: 429–445.

Emerging Technologies in Non Destructive Testing, Van Hemelrijck, Anastasopoulos & Melanitis (eds)
© 2004 Swets & Zeitlinger, Lisse, ISBN 90 5809 645 9

Ultrasonic spectroscopy of adhesively bonded multi-layered structures

T. Stepinski

Signals and Systems, Uppsala University, Uppsala, Sweden

ABSTRACT: Ultrasonic spectroscopy utilizes information in the frequency domain obtained due to the constructive and destructive interference of elastic waves for nondestructive evaluation of adhesively bonded structures. In resonance inspection (RI) an ultrasonic toneburst with a sweeping frequency is applied to an ultrasonic transducer and a resonance spectrum of the inspected structure is acquired. A theoretical model is developed in the paper to predict the modal shapes and resonance frequencies of the thickness mode resonances occurring in multi-layered structures. The model includes the piezoelectric transducer used for sensing the resonances. Modeling results are verified by the measurements performed using a network analyzer. The principles of narrowband and wideband RI are explained, and an example of the solution to the spectrum classification issue in wideband test is presented.

1 INTRODUCTION

The use of resonance has long been recognized as the key to an ultimate method for nondestructive testing because all of the mechanical properties of a part are contained in the resonance spectrum. Global *resonance inspection* (RI) is a whole body measurement and its result is a measure of the structural integrity of the part itself. Local resonance tests are focused on a limited volume within a test object with a well defined geometry (for instance, a layered structure).

Most NDT techniques focus on defects. The classic questions are "What type of defects can be detected?" and "What is the minimum detectable defect size?" Global resonant inspection focuses on the part. It defines the resonant pattern for acceptable parts and rejects all parts that do not fit that pattern. Parts are rejected if they are structurally defective. Local resonant inspection is capable of detecting local structure variations based on small variations in the acquired spectrum.

The earliest resonance techniques involved striking the part and listening to the resonance. Subsequent improvements substituted electronic filters or FFT (Fast Fourier Transforms) for the human ear. Using swept frequency measurement techniques has improved the accuracy of the RI and extended its bandwidth with short ultrasonic waves indicating the presence of small defects.

However, a more fundamental limitation on resonance measurements is that normal production variations in dimensions and material properties mask the effect of defects on the resonances for all but the largest defects. As a result, resonance testing has been limited to applications where production variations are non-existent (cf. Quasar (2000), Migliori & Sarrao (1997)).

Solution to the problem of normal production variations is a key to the successful application of RI.

Ultrasonic RI is widely used for the inspection of aerospace structures and its application field is likely to increase rapidly with the growing application of layered structures in modern aircraft (e.g. GLARE that is being considered as the principal structure material of the A380 from Airbus). Unfortunately, very limited information concerning the operation principles of the commercially available RI instruments is available in the literature.

The aim of this paper is to fill this gap and to enlighten the potential and the limitations of the RI techniques. We start from explaining the principles of RI of multi-layered structures and presenting a model enabling simulation of the test setup. Then, we address the main issue limiting RI applications, the ways of reliable interpretation of the acquired spectra.

2 RESONANT INSPECTION OF MULTI-LAYERED STRUCTURES

An ultrasonic inspection technique commonly used for aircraft multi-layered structures is based on ultrasonic spectroscopy. Commercially available instruments (bond testers) used for this test operate on the principle of mechanical resonance in a multi-layered structure. A piezoelectric probe, excited by a swept frequency signal is coupled to the surface of the inspected structure. A frequency spectrum in the range of some tens of kHz to several MHz is acquired by the instrument.

A resonance in the layered structure occurs when echoes between two boundaries travel back and forth due to differences in acoustic impedances at the boundaries. For multi-layered structures, a number of resonances can be observed depending on their geometry and condition. A characteristic resonance pattern, an ultrasonic signature, obtained for each particular defect-free structure and given transducer can be used as a reference.

During the inspection of an unknown object, its surface is scanned with the probe and ultrasonic spectra are acquired for a number of frequencies. Disbond detection is performed by the operator based on some simple features of the acquired spectra, such as, center frequency and amplitude of the highest peak in a preselected frequency range. This means that the operator has to perform a spectrum classification task based on primitive features extracted by the instrument.

Below, we present theory and experiments illustrating this principle.

3 MODELING MULTI-LAYERED STRUCTURE

To simulate different structures, a software tool was developed in Matlab for calculating the input acoustical impedance of sandwich structures consisting of a number of semi-infinite parallel layers. Material constants, such as density and sound velocity are used as parameters. The software uses transmission line concepts for calculating impedances seen on the top of the successive layers (see Brekhovskikh (1980) for details). It is assumed that a longitudinal plane wave is incident on the top boundary of the structure. Material attenuation is not included in the model.

The software has been used for simulating laminated airspace structures consisting of metal layers bonded by fibre reinforced resin layers. Variations in spectra due to disbonds and voids between the aluminum plate and the adhesive could be modeled and evaluated. As an example we present results obtained for a simple laminated structure consisting of two aluminum plates bonded with an adhesive, see Table 1.

Acoustical impedance seen from the top was calculated as a function of frequency for the following cases:

- perfect structure
- perfect structure with variable adhesive thickness
- disbond between upper layer and adhesive
- disbond between lower layer and adhesive

Modeling results are presented in the figures below.

From Figures 1 and 2, it can be seen that the disbond results in a regular spectrum typical for a single plate. Resonance peaks of this spectrum are shifted with respect to the peaks of the spectrum corresponding to a perfect structure. However, no regularity could be observed in the shifts as a function of frequency, as the shifts corresponding to different resonance peaks may have even different signs (some of the peaks are shifted towards the lower frequencies and the other to the higher frequencies, cf. Figure 2).

The disbonds would be relatively easy to detect if variations in the spectra measured for a defect free structure (*nominal spectra*) were small. Unfortunately variations of the adhesive parameters may result in quite large shifts of the resonance peaks. To illustrate this in Figure 3 we present modeling results obtained for 3 different values of adhesive thickness.

It is apparent that variations in adhesive thickness may mask peak shifts due to disbonds. The influence of adhesive thickness on peak positions in all frequency bands is summarized in Figure 4.

The examples presented above illustrate the main problem encountered in RI applications: *natural variations of spectra mask the variations due to flaws.*

Table 1. Parameters of the laminated structure analyzed in the paper.

Layer	Thickness [mm]	Parameters	
		c [m/s]	ρ [kg/dm^2]
Top	1.27	6140	2.77
Adhesive	0.2	2400	1.16
Bottom	1.00	6140	2.77

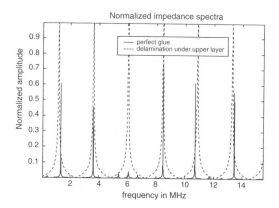

Figure 1. Normalized impedance spectra for perfect structure and a structure with disbond under upper Al layer.

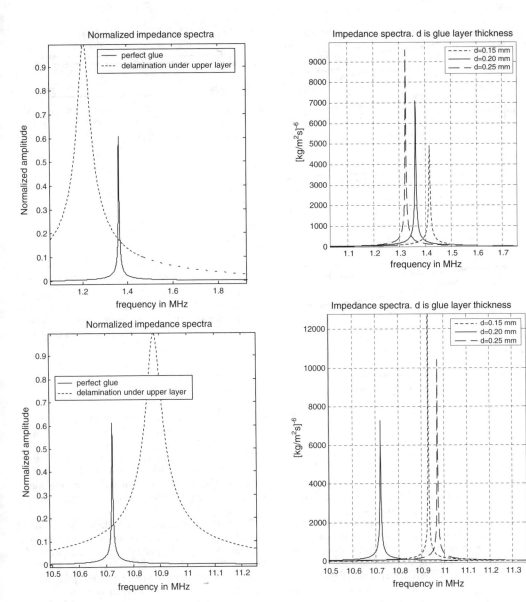

Figure 2. Details of the normalized impedance spectra shown in Figure 1 for the resonance peaks respectively at approx. 1.4 MHz (top) and 10.7 MHz (bottom).

Figure 3. Position of the resonance peaks for the sandwich structure defined in Table 1 for the resonance peaks respectively at approx. 1.4 MHz (top) and 10.7 MHz (bottom).

At least two solutions to this problem are possible:

- measurement in a carefully selected narrow frequency band using a specially designed resonance transducer, and
- wide band measurement followed by a advanced pattern recognition scheme.

The first solution is applied in bond testing instruments from *Fokker* and *Staveley NDT Technologies*

while the second alternative is used in RI systems from *Quasar International Inc*. Below we explain in more detail how bond testing instruments operate.

4 NARROWBAND RI OF BONDS

A piezoelectric low frequency (below 1 MHz) transducer is a vital part of narrowband bond testing instruments. Since the transducer is undamped, it has a distinct

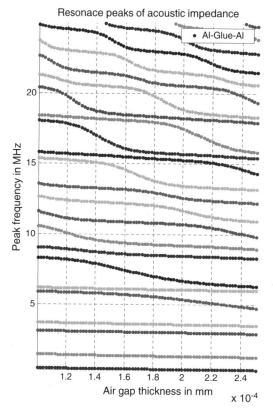

Figure 4. Position of the resonance peaks for the sandwich structure defined in Table 1 as a function of adhesive thickness.

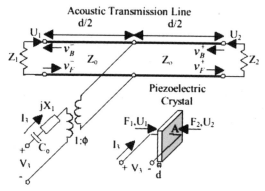

Figure 5. KLM model of the transducer used in the simulations.

frequency peak corresponding to its thickness mode (cf. Guyott et al. (1986)). When the transducer is loaded (e.g. coupled to a sandwich structure) its spectrum changes depending on the acoustic impedance of the load. These changes can be detected measuring transducer's electrical impedance. Therefore the transducer is excited with a swept frequency signal and its electrical impedance is measured in some frequency band or for a user-selected frequency.

To illustrate this we present an experiment verifying our modeling result. In the experiment we use a piezoelectric element (cylinder with diameter approx. 13 mm and thickness 25 mm). The electrical admittance of this element was measured using *4395A Network/Spectrum/Impedance Analyzer from Agilent Technologies*.

Parameters of the KLM model corresponding to the transducer were estimated from the *4395A* measurements. The KLM model (Krimholtz et al. (1970)), which is based on transmission line concepts was chosen since it enables modeling transducers under various load conditions.

The acoustical side of the KLM model is coupled to the electrical side by a transformer with ratio $(1:\phi)$ (cf. Figure 5). Thus, the input impedance of this model, which depends explicitly on the acoustic impedances (Z_1 and Z_2) seen from the each side of the piezoelectric crystal can be expressed as:

$$Z_{in} = \frac{1}{j\,\omega C_0} + jX_1 + \frac{Z_a}{\Phi^2}$$

where Z_a = the acoustic impedance seen from the secondary side of the transformer; X_1 = the input electrical reactance; C_0 = the input capacitance.

The transducer's electrical admittance measured in air in the frequency range around its thickness resonance is shown in Figure 6 together with the admittance of the KLM model. The admittance shows a resonance peak at 61.4 kHz.

The next measurement was made for the transducer placed on 1.5 mm aluminum plate (a thin layer of water was used for coupling). The theoretical admittance for this condition was calculated using the KLM model. The results presented in Figure 7 indicate that the resonance frequency has been shifted down to 59.6 kHz.

A reasonable agreement between the KLM model and the measurement has been achieved. Other measurements and modeling results show that the transducer's resonance shift depends on the plate thickness despite that the plate resonance frequency is substantially higher than that of the transducer (which is approx. 1 MHz for the 1.5 mm Al plate). This is due to variations in the plate acoustic impedance in the region of the transducer's resonance.

The spectrum obtained for a perfect sandwich structure has a complex form and a considerably lower amplitude. This means that the position of the resonance peak observed in the transducer's admittance can be used for detecting disbonds under the upper

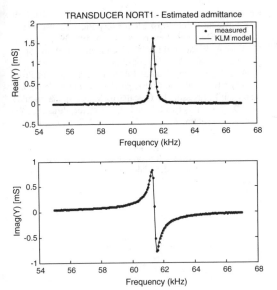

Figure 6. Electrical admittance of the transducer used in the experiment. Measured values (dots) and admittance of the KLM model (solid line).

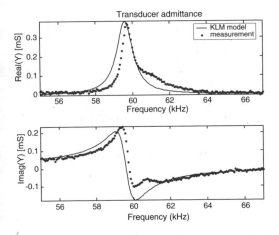

Figure 7. Electrical admittance of the transducer coupled to 1.5 mm Al plate. Measured values (dots) and admittance of the KLM model (solid line).

plate. Disbonds located at other interfaces can also be detected based on the resonance shift and shape of the transducer characteristic (Stepinski et al. (1998)).

Transducer admittance is a complex number that can be represented in a complex plane. It is obvious that transducer admittance measured for some frequency close to resonance underlies considerable variations depending on condition of the inspected material. This feature is used in the *Bond Master* from *Staveley NDT Technologies*.

5 BROADBAND RI

Broadband RI is applied most often as a global integrity test for relatively small objects. The test is performed using separate transducers for exciting and receiving oscillations in the test object.

In broadband RI an acquisition of several resonance peaks in a broad frequency range or in a number of narrow frequency bands is performed using a well damped transducer. The result is then processed by a special software tool with built-in *self learning* capacity, for example, using neural network concepts (cf. Stepinski et al. (2000)).

QUASAR International, Inc., USA has developed a technique referred to as Quantitative Resonant Inspection (QRI). QUASAR has developed an advanced software tool that assists the user during developing set-up for a new test object as well as during production (cf. Quasar (2000)).

The key to QRI defect detection is the Sorting Module, a data file in the RI computer, which includes the set of patterns that describes the characteristics of an acceptable part. Each pattern is the mathematical relationship of the frequency measurements, not the shape of the frequency peaks or anything visual.

The QRI system is used to measure selected resonances of a *learning set* consisting of known parts (both good and defective). The results are input to a proprietary pattern recognition program, which generates the Sorting Module and translates it into instructions for the QRI factory system. The learning set should contain examples that represent the broad range of acceptable production variation. This learning set should also contain structurally defective parts from the same production lots.

However, even when certain defect types are not included in the sample set, the QRI system generally rejects these structural defects because they break out of the pattern consistent with structurally sound parts. It is important to recognize the ability of QRI to recognize a pattern deviation in a part that has, for example, an inclusion of a foreign object or a void. Even when the specific defect was never represented in the learning set, QRI will reject that part in the factory sorting if there is a significant structural deficiency (i.e. its pattern is outside the set of sound parts).

6 CONCLUSIONS

Principles of narrowband and wideband RI inspection was introduced and explained in some detail. A theoretical model of the multi-layered structure including the piezoelectric transducer was presented. Simulation results obtained using this model were used to explain the narrowband inspection of adhesively bonded structures. The simulation results were verified by measurements performed using a network analyzer. The issue of natural variation in dimensions and material properties for the sensitivity of RI to defects was addressed. Example solutions employing self-learning techniques were briefly presented.

REFERENCES

Brekhovskikh, L.M. 1980. Waves in Layered Media, Academic Press.

Guyott, C.C.H., Cawley, P. & Adams, R.D. 1986. Vibration characteristics of the Mk II Fokker Bond Tester probe, Ultrasonics, vol. 24, pp. 318–324.

Krimholtz, R., Leedom, D.A. & Matthaei, G.L. 1970. New Equivalent Circuits for Elementary Piezoelectric Transducers, Electronic Letters, vol. 6, pp. 398–399.

Migliori, A. & Sarrao, J.L. 1997. Resonant Ultrasound Spectroscopy, John Wiley & Sons, Inc.

Quasar, 2000. Using Quasar Resonant Inspection in Production Environment, *Copyright Quasar International, Inc.*

Stepinski, T., Ericsson, L., Gustafsson, M. & Vagnhammar, B. 1998. Neural network based classifier for ultrasonic resonance spectra, *Proc of the 7th ECNDT*, Copenhagen, pp. 2363–2370.

Stepinski, T., Ericsson, L., Vagnhammar, B. & Gustafsson, M. 2000. Classifying Ultrasonic Resonance Spectra Using Neural Network, Materials Evaluation, vol. 58, no. 1, pp.74–79.

Emerging Technologies in Non Destructive Testing, Van Hemelrijck, Anastasopoulos & Melanitis (eds)
© 2004 Swets & Zeitlinger, Lisse, ISBN 90 5809 645 9

Air-coupled ultrasonic testing of textile products

E. Blomme, D. Bulcaen, F. Declercq & P. Lust
KATHO Dept. VHTI, Kortrijk, Belgium

ABSTRACT: An air-coupled ultrasonic system is applied to examine several kinds of textile products with regard to various types of errors. The feasibility is demonstrated to detect and visualize fluctuations in the weight and thickness of coatings, the penetration of a coating into a textile substrate, some common fabric defects, density variations of a non-woven and the quality of an impregnation on a fibre web product. In all cases the textile is inspected in through-transmission mode at frequencies between 0.25 and 2 MHz, the ultrasonic transducers being separated by an air gap of 3 cm.

1 INTRODUCTION

Ultrasonic methods never have been popular for use in textile inspection systems. The main reason for this is that conventional ultrasonic NDT-methods operating at frequencies higher than 100 kHz are either immersion techniques or require physical contact with the medium under investigation. This is changing. Nowadays it is possible to produce high frequency ultrasound up to 3 MHz in air over distances varying from some mm to several cm (depending on the frequency) without needing a coupling gel or a liquid as a propagation medium. This can be achieved by either classical piezo-transducers to which one or more matching layers are applied on the piezo-electric element in order to overcome the large acoustic impedance gap with air (e.g. Bhardwaj 2002) or by capacitive devices which essentially are vibrating membranes (e.g. Schindel et al. 1995). Although the useful distance between transmitter and receiver is limited, e.g. 3 to 4 cm at 1 MHz, this small spatial area is sufficient to enable automated inspection methods which can be competitive or complementary with existing quality control systems such like weight sensors based on β-rays. In particular the air-coupled technique provides a solution for the detection of errors which do not induce weight variations.

This paper describes some experiments involving weight and thickness variations of a coating, the penetration of a coating into a textile substrate, density variations of a non-woven and the quality of an impregnation on a fibre web product. In addition the feasibility is demonstrated to detect some common weaving faults.

2 EXPERIMENTAL SET-UP

All results in this report have been obtained in through-transmission mode (Figure 1). The distance between the transmitter and the textile is 2 cm while the distance between the textile and receiver was ca. 1 cm. Fluctuations in these distances hardly effect the measurement as long as the distance between the transmitter and the receiver is kept constant. The apparatus is conceived in such a way that the transducers are kept in fixed position while the sample is attached to a PC-controlled XY-stage. The step is typically 1 or 2 mm in both the X- and Y-direction. Piezo-transducers with matching layers as well as capacitive devices can be implemented in the set-up. As no resonance phenomena take place in most textile products, the applied frequency in general can be chosen rather arbitrarily but preferably between 250 kHz and 2 MHz. In all examples the transmitting transducer was fed by a continuous sinusoidal signal of 18 V_{pp}.

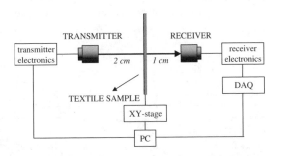

Figure 1. Scheme of the experimental arrangement.

At receiver side the signal first was amplified (gain of 34 dB) and then measured by a homemade super-heterodyne receiver (see also Blomme et al. 2002a). Each point composing a scan was obtained from the acquisition of the non-averaged signal level (received signal strength indicator RSSI). The signal levels were imaged by a logarithmic detector on an arbitrarily scale ranging from 1 to 4096 ($= 2^{12}$) which is related to a dB-scale. The total dynamic range was about 120 dB. The final data were post-processed in MATLAB for visualisation.

3 EXPERIMENTAL RESULTS

3.1 Weight variations of a coating

An important issue in the field of textile inspection is the monitoring of a coating process. Table 1 shows the (dry) coating weights of two series of cotton samples on which up to 4 layers of protective coating (BEMICOAT STR) were applied. The transmitting piezo-transducer ($\varnothing$ 12 mm) was activated at 0.7 MHz.

Figure 2 shows the average transmission values of a matrix of 25 $\times$ 25 point measurements (C-scan) of an area of 5 $\times$ 5 cm^2. The 0-level on the scale refers to the transmission in air. In absence of a sample an attenuation is recorded of 66 dB with respect to the input signal, which is an accumulated loss due to the acoustic impedance mismatch between air and transducer surface and the absorption in air. Inserting the samples, additional losses increase from 11 dB in case of the non-coated sample of series A up to 93 dB in the case of the heaviest sample of series B. It is clear that the most significant ultrasonic transmission drops occur at coating weights between 0 and 50 g/m^2, a range frequently met in textile coating applications. A further increase of coating weight results into a stabilization of the transmitted signal, due to saturation of the substrate by the coating. In these experiments continuous ultrasound has been used but comparable results can be obtained in pulse mode (Blomme et al. 2002c).

The feasibility of monitoring fluctuations of a coating weight by air-coupled ultrasound is also demonstrated by the next example. Figure 3 shows the case of a 180 g/m^2 polyester/acryl substrate (ratio

Table 1. Weights of a coating applied on two series of five cotton samples (in g/m^2).

Coating layers	0	1	2	3	4
Series A (117 g/m^2 substrate)	0	25	39	53	65
Series B (212 g/m^2 substrate)	0	42	58	76	95

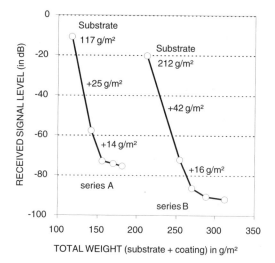

Figure 2. Air-coupled through-transmission measurements of the samples of series A and B (see Table 1). The weight of the first two coating layers is indicated.

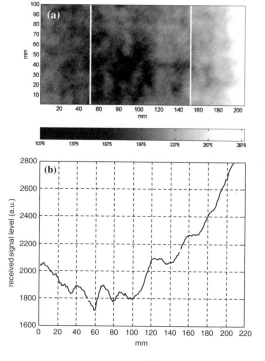

Figure 3. Air-coupled through-transmission scan at 700 kHz of a polyester/acryl substrate (180 g/m^2) + flame-retardant coating (30 g/m^2 averaged) in arbitrary units. The figure below shows an averaged line scan from left to right. The difference between maximum and minimum received signal level corresponds to ca. 30 dB and a coating weight difference of 9 g/m^2.

65/35) on which a flame retardant coating with an average weight of 30 g/m^2 is applied on one side.

However the coating weight is irregularly distributed over the substrate and is higher at the right than at the left side. The picture is a composition of three overlapping scanned areas from the material. Figure 3b represents the average of the Y-transmission values of Figure 3a in the X-direction. Although the maximum difference in coating weight is not more than 9 g/m^2, the fluctuation in the received ultrasonic signal level is considerable and corresponds to a variation of 30 dB.

3.2 *Thickness of a PVC-coating*

In contrast to the previous examples we consider here rather thick (>0.5 mm) and heavy (>300 g/m^2) coatings, such like various types of PVC-coatings which are commonly used to produce artificial leather. An interesting situation occurs if the thickness of the coating is varying while its weight remains constant. In that case a thickness variation is directly related to a density variation of the coating. As ultrasound is very sensitive to density changes, it is to be expected that these variations can be monitored ultrasonically. Figure 4 shows three fragments of a (non processed) line scan taken of a 680 g/m^2 PVC-sample of 150 cm width with capacitive transducers (Ø 10 mm) at a frequency of 250 kHz. Due to irregularities in a heating process, density changes may occur resulting into thickness variations. From Figure 4, it can be seen that the level of transmitted ultrasound is significantly higher in the middle part where the smallest thickness is measured (1.55 mm against 2.10 mm left and right). This can be understood from the higher porosity of the coating towards the borders.

It should be remarked that this kind of thickness variations cannot be monitored by conventional weight sensors, as the weight remains constant along the whole width of the sample. In view of the rather high acoustical impedance of the PVC-layer ($Z \approx 3 \times 10^5$ kg/m^{2}s, $Z_{air} = 420$ kg/m^2/s) which results into a very low transmission coefficient – more than 99% of the incident ultrasound is reflected from the PVC-coating – the experiment reveals promising perspectives with respect to the on-line monitoring of thick coatings.

3.3 *Penetration of a coating*

In many situations not only the weight of a coating is important but also the penetration of the coating into the textile. Too deep or insufficient penetration may result into quality errors or may cause difficulties in a next step of a finishing process. Figure 5 shows the result of a scan at 700 kHz with Ø 12 mm piezo-transducers

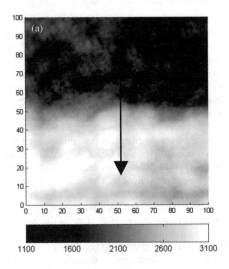

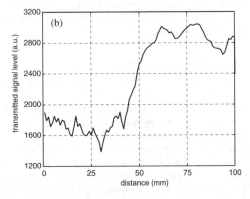

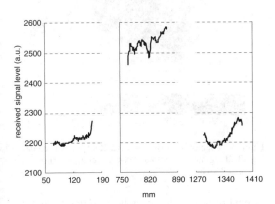

Figure 4. Air-coupled ultrasonic monitoring of thickness variations of a PVC-coating at 0.25 MHz. Left and right parts: 2.10 mm; middle part: 1.55 mm.

Figure 5. (a) Air-coupled ultrasonic image of a coated cotton sample with varying degree of penetration (deeper penetration on the upper half); frequency = 0.7 MHz. (b) Line scan in the direction of the arrow.

over a $10 \times 10 \, cm^2$ surface of a $130 \, g/m^2$ cotton sample. The cotton substrate was prepared with a protective coating of $34 \, g/m^2$ but on the upper half more pressure was applied than on the lower half. The grey scale in the picture clearly reveals the region of increased penetration. Due to the higher pressure exercised on the upper half, the air holes between the yarns of the fabric structure are more filled up with coating than in the lower half, which results into higher density and decreased ultrasonic transmission (difference of ca. $30 \, dB$). A line scan in the direction of the arrow is sufficient to detect the penetration change. Again it should be noted that the error cannot be detected by conventional gravimetric methods as the same amount of coating is applied to both the upper and lower part of the sample. For another example, see Blomme et al. 2000b.

3.4 Density changes and errors in a fibre web

The air-coupled approach also opens new perspectives for testing on line the quality of certain types of non-wovens like fibre web products. A homogeneous density is an important quality factor of a web product, whether it is made of metal or synthetic fibres. It should contain no compressed areas, rarefactions or inclusions. Figure 6a shows a photographic picture of a non-transparent metal web. The $10 \times 8 \, cm^2$ sample has a thickness of 3 mm (not compressed) and is composed of $8 \, \mu m$ thick fibres. It contains an area with a clustered accumulation of fibres. Figure 6c shows an air-coupled ultrasonic scan obtained with $\emptyset \, 3 \, mm$ piezo-transducers at 2 MHz. The bad area can be seen very clearly, which can be explained by the higher density with respect to the remaining part of the sample. The ultrasonic image is as clear as the X-ray picture shown in Figure 6b.

In this case the error could be detected visually. In general however inhomogeneities also may occur in the interior of a web, especially in case of thick fibre webs, but this makes no difference for the air-coupled ultrasonic technique.

3.5 Impregnation on a non-woven product

If a non-woven like a fibre web is impregnated with some liquid substance, the homogeneity of the impregnation is an important quality factor. As variations in the impregnation result into local density variations, it is to be expected that these can be detected ultrasonically. The possibility to perform the detection without any contact with the web opens new perspectives for the on-line monitoring of an impregnation process. To illustrate this we consider a sample of a $647 \, g/m^2$ needle felt of thickness 4 mm, impregnated with ca. $300 \, g/m^2$ synthetic resin (see inset in Figure 7). The impregnation on the upper half of the sample has acceptable quality, but the lower half is badly impregnated: the

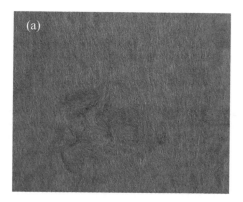

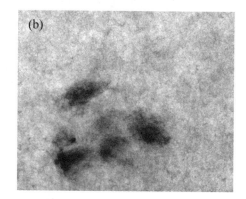

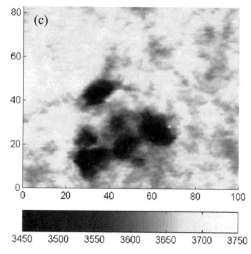

Figure 6. Sample of a $10 \times 8 \, cm^2$ metal fibre web containing a bad region: (a) photo; (b) X-ray picture; (c) ultrasonic image at 2 MHz.

amount of resin is lower (only ca. $200 \, g/m^2$) and the resin has not fully penetrated into the web. Figure 7 shows the result of a $10 \times 8 \, cm^2$ transmission scan at 400 kHz with capacitive transducers of $\emptyset \, 1 \, cm$.

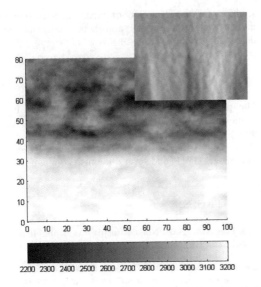

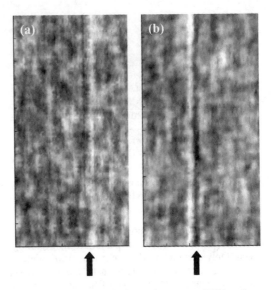

Figure 7. Impregnated needle felt sample of $10 \times 8\,cm^2$ and thickness 4 mm with acceptable (upper half) and insufficiently impregnated (lower half) region. Ultrasonic through-transmission scan at 0.4 MHz in arbitrary units.

The bad area has a significantly higher ultrasonic transmission (due to the smaller amount of resin) and can clearly be distinguished from the good region.

3.6 Some common fabric defects

Air-coupled ultrasound also opens new perspectives with regard to the detection of some common weaving faults. Most fabric defects induce local density variations and hence affect the transmission of ultrasound. In principle defects in both the warp and pick direction can be detected although it should be remarked that the detection of errors of small size (in both X- and Y-direction) require high speed scanning in combination with fast data acquisition, which is not yet easy to establish. In view of the small size of yarns, we would recommend a rather high frequency and a receiver of small diameter. The following experiments have been carried out at a frequency of 2 MHz with a transmitter of $\varnothing\,25\,mm$ and a receiver of $\varnothing\,3\,mm$. Exceptionally the step of the XY-stage was doubled in the direction of the picks (2 steps/mm in the X-direction).

Figure 8 shows scans of two $180\,g/m^2$ polyester samples with 36 warps/cm and 18 picks/cm. The first picture shows the influence of a missing warp (reed mark) on the ultrasonic through-transmission image, the second one the consequence of a shifted warp, i.e. a missing yarn immediately followed by a double one. Non-filtered line scans averaged over 13 lines, spaced

Figure 8. Air-coupled ultrasonic scan at 2 MHz of two polyester samples with weaving errors: (a) missing warp; (b) shifted warp.

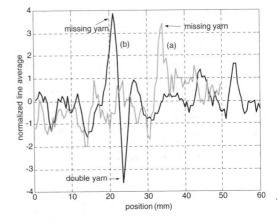

Figure 9. Air-coupled ultrasonic scan at 2 MHz of two polyester samples with weaving errors: (a) missing warp; (b) shifted warp (see Figure 8). The record represents an average of 13 lines.

8 mm apart, clearly reveal both types of defects and the discriminating feature between them: whereas in the first case only a peak transmission occurs, also a fall back appears due to the double yarn (see Figure 9).

The next example demonstrates the possibility to detect another kind of error. Figure 10 shows the result of a spinning fault on a weft thread. A thickening of the weft over a 3 cm distance is clearly imaged by the transmitted ultrasound. However, although the experiment shows that ultrasonic detection of this

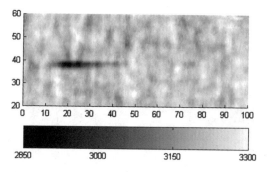

Figure 10. Air-coupled ultrasonic scan at 2 MHz of a woven sample containing a weft yarn with a spinning fault.

kind of error in the pick direction in principle is possible, in practice it mainly depends on the scanning speed. Only at high speed of scanning, 100% detection is possible but simultaneously the resolution decreases. The use of an array of transducers could provide a solution (Chien et al. 1999).

4 CONCLUSION

The availability of air-coupled ultrasonic transducers operating at frequencies between 0.25 and 2 MHz opens interesting perspectives for the on-line inspection of many kinds of textile products and with regard to various types of errors. In this paper the feasibility has been demonstrated to detect and visualize some common fabric defects, fluctuations in the weight and thickness of coatings, the penetration of a coating into a textile substrate, density variations of a non-woven and the quality of an impregnation on a fibre web product. In particular the air-coupled technique provides a solution for the detection of errors which do not induce weight variations and hence not can be detected by weight sensors.

ACKNOWLEDGEMENT

The present work has been performed within the framework of the HOBU-Fund projects (Nr. 000197/20082) supported by the Institute for the Promotion of Innovation by Science and Technology in Flanders (IWT) and several Flemish companies.

REFERENCES

Bhardwaj, M.C. 2002. Non-destructive Evaluation. In Mel Schwartz (ed), *Encyclopedia of Smart Materials*: 690–714. New York: John Wiley & Sons.
Blomme, E., Bulcaen, D. & Declercq, F. 2002a. Air-coupled ultrasonic NDE: experiments in the frequency range 750 kHz–2 MHz. *NDT&E Int*. (35): 417–426.
Blomme, E., Bulcaen, D. & Declercq, F. 2002b. Recent observations with air-coupled NDE in the frequency range of 650 kHz to 1.2 MHz. *Ultrasonics* (40): 153–157.
Blomme, E., Bulcaen, D., Declercq, F. & Lust, P. 2002c. Air-coupled ultrasonic evaluation of coated textiles. In *Proc. IEEE Int. Ultras. Symp. 2002 (Munchen)*: Session 2A-2.
Chien, H.-T., Sheen, S.-H., Lawrence, W.P., Razazian, K. & Raptis, A.C. 1999. On-loom real-time non-contact detection of fabric defects by ultrasonic imaging. In B. O. Thompson & D. E. Chimenti (eds), *Review of Progress in Quantitative Nondestructive Evaluation (18B)*: 2217–2224. New York: Plenum Press.
Schindel, D.W., Hutchins, D.A., Zou, L. & Sayer, M. 1995. The design and characterization of air-coupled capacitance transducers. *IEEE Trans. Ultrason. Ferroelect. Freq. Cont.* (42): 42–50.

Emerging Technologies in Non Destructive Testing, Van Hemelrijck, Anastasopoulos & Melanitis (eds)
© 2004 Swets & Zeitlinger, Lisse, ISBN 90 5809 645 9

On some applications of the non-linear acoustic phenomena for the evaluation of granular materials

V. Tournat, Ph. Béquin & B. Castagnède
Laboratory of Acoustics at University of Maine, Université du Maine, Le Mans Cedex 09, France

V.E. Gusev
Laboratoire de Physique de l'État Condensé, Université du Maine, Le Mans Cedex 09, France

V.Yu. Zaitsev & V.E. Nazarov
Institute of Applied Physics, RAS, Nizhny Novgorod, Russia

ABSTRACT: Granular materials are widely used in different branches of industry from the building construction to pharmaceutics and production of the nanostructured materials. Classical acoustical methods for their diagnostics are well known to be very useful. In the present communication we would like to attract the attention to the possible applications of the emerging nonlinear acoustical methods for the non-destructive evaluation of granular materials. Three potential applications are discussed. Firstly, due to high nonlinearity of the granular materials the parametric emitting antenna can operate with sufficient efficiency in these materials. Thus, similarly to the advantages achieved in underwater applications of the parametric sonar, compact sources of highly directive low-frequency sound waves can be created for the acoustic spectroscopy, tomography and depth-profiling of the granular piles and columns. Secondly, the (low frequency) acoustic signal, emitted by the parametric antenna, carries information on the processes of the absorption and scattering of high frequency waves, and also on their dispersion related to the discrete nature of the granular assemblages. It contains information on the transition in the high-frequency wave transport from ballistic to diffusive regime, and from propagative to evanescent. Thirdly, due to high sensitivity of the nonlinear phenomena to the state of the inter-grain contacts, the processes involving polarized (shear) waves are sensitive to the anisotropy in the force chains network induced by uniaxial static loading.

1 INTRODUCTION

Acoustic waves are useful when they penetrate deeper than other waves in the media to be tested. This seems to be the case for dense granular materials, for the scales from several grains to large distances. Many physical phenomena are known to play a role in the propagation of an elastic perturbation in a granular material. Among them are absorption, scattering (due to randomness of the grain positions and contact positions between the grains), velocity dispersion (a manifestation of the micro-structure), and also nonlinearity (mainly due to the soft contacts). The equation of state describing the behavior of the contact between two spherical beads is known as the Hertz relation and is fundamentally nonlinear (Johnson 1954).

A lot of studies, both theoretical and experimental, have been performed in idealized media (typically 1-D chains of beads), but also in the more realistic 3-D granular materials (unconsolidated assemblages of beads) (Liu and Nagel 1992; Jia et al. 1999; Zaitsev et al. 1999a; Moussatov et al. 2001). For regularly assembled beads of the same dimension, the equivalent quadratic nonlinear parameter, which characterizes the importance of the quadratic nonlinear effects on the acoustic propagation, was evaluated (Belyaeva et al. 1993; Zaitsev 1995). The obtained values were between 10^2 and 10^4, to be compared to the classical nonlinear parameters of homogeneous media: 3 for water, 1.2 for air and 1.4 for glass. As a consequence, it is relatively easier in such media to generate nonlinear effects, even with moderate amplitudes of excitation, in comparison with the homogeneous media.

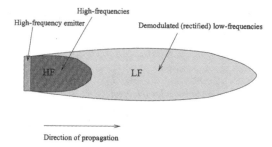

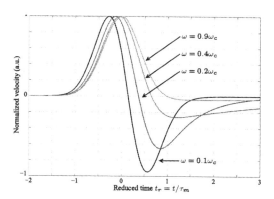

Figure 1. Qualitative scheme of the parametric antenna.

Figure 2. Transformation of the demodulated signal due to the transition from the regime dominated by HF wave ballistic propagation to the regime dominated by their diffusion. Profiles of demodulated signals are plotted as a function of the reduced time $t_r = t/\tau_m$ (where τ_m is the characteristic time of the Gaussian modulation envelope), for four frequencies of carrier wave. ω_c is the so-called cut-off frequency of the medium.

2 PARAMETRIC EMITTING ANTENNA IN SCATTERING MEDIA

The parametric antenna emits low-frequency (LF) acoustic signals due to the demodulation of amplitude-modulated high-frequency (HF) waves. In Fig. 2, regions of existence of HF primary waves and LF demodulated waves are schematically represented.

The main interests in applying the parametric antenna method are coming from several inherent features already observed in underwater acoustics such as low frequency emission with a high directivity (Novikov et al. 1987). The HF waves (wavelength comparable to several bead diameters) are sensitive to the micro-structure through the velocity dispersion or the scattering, but are usually attenuated before arriving to the receiver. In contrast, LF waves (with very large wavelengths compared to the beads diameter) are less dispersive or attenuated and can be detected far from the emitter. It has been shown that the LF

waves, radiated due to the demodulation of initially modulated HF waves, contain information on the HF waves absorption (Zaitsev et al. 1999a; Zaitsev et al. 1999b).

The high directivity of the radiated LF waves allows to perform experiments in samples that have smaller lateral dimensions than the LF wavelength and to avoid at the same time reflections of these waves on the sample lateral boundaries.

A theoretical model on the parametric antenna operation in granular materials has been recently developed, including velocity dispersion, HF absorption and scattering (Tournat et al. 2002). We have shown that velocity dispersion influence on the demodulated LF profile can lead to an important change of its shape. It is also demonstrated that the transition from propagative to diffusive HF primary waves transport manifests itself by an integration of the demodulated temporal profile, under some conditions. Figure 2 shows an example of the influence on the demodulated LF signal of the transition from ballistics to diffusion in the HF primary waves transport.

This model, developed for the one-dimensional propagation problem, does not take into account the real three-dimensional effects such as diffraction of the acoustic beams. This latter may mask the theoretically predicted effects of dispersion or diffusion.

Experimental results obtained recently (Tournat et al. 2003) are in qualitative accordance with the developed theoretical model. To obtain a quantitative agreement, it is necessary to solve the 3-D problem numerically. By fitting the temporal demodulated signals with the model, we are trying to solve the inverse problem, i.e. to determine the free parameters of the model (Tournat et al. 2002).

3 PARAMETRIC EMISSION BY EVANESCENT MODES

We have also developed a model that describes the nonlinear self-demodulation process in a one-dimensional chain of identical elastic beads. Linear (but also nonlinear) elastic propagation in such media has been actively investigated. It has been verified that in such chains, where beads interact due to the Hertzian contacts, the propagation is non-linear (Coste and Gilles 1999; Nesterenko 1983; Lazaridi and Nesterenko 1985). Also, a well-known dispersive feature of periodic lattices, is the existence of a cutoff frequency ω_c for the elastic waves, above which the acoustic modes become evanescent (Maradudin 1971). This cut-off frequency ω_c is related to the bead radius, the bead's material properties and the longitudinal static applied stress on the chain. As a result, the cut-off frequency is a source of information on the properties of the granular chain.

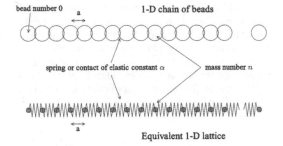

bead number 0

1-D chain of beads

spring or contact of elastic constant α

mass number n

Equivalent 1-D lattice

Figure 3. Problem under consideration.

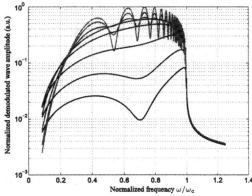

Figure 4. Demodulated signal amplitude as a function of the frequency of primary wave for different absorptions. From the dark line to the bright, the constant C in the absorption coefficient $\alpha(\omega) = C\omega_n = C\omega/\omega_c$ is equal to 230, 91, 36, 14, 6, 0.9, 0.36 respectively. The normalized demodulated frequency $\Omega_n = \Omega/\omega_c$ is equal to $3{,}3\cdot10^{-3}$, the distance of observation corresponds to the bead number $n = 2500$, and the bead diameter is $a = 2\cdot10^{-4}$m.

When considering elastic waves with a periodicity much longer than the elastic wave travel time along a bead diameter (the latter is less than 10^{-6}s for 2 mm diameter glass beads), the chain can be modeled by masses linked with non-linear springs (Fig. 3). To begin with, the potential energy of the chain E_p is expanded up to the cubic term in bead displacement $U(n)$ (where n is the bead number):

$$E_p = E_{p0} + \frac{a}{2}\sum_n [U(n) - U(n+1)]^2$$

$$+ \frac{\beta}{6}\sum_n [U(n) - U(n+1)]^3 + \ldots \tag{1}$$

where the cubic non-linear term induces the quadratic nonlinearity in the stress–strain relationship, and β denotes the quadratic nonlinear parameter.

Then, the following discrete non-linear equation of motion is derived for each mass m:

$$m\frac{\partial^2 U(n)}{\partial t^2} - \alpha[U(n+1) - 2U(n) + U(n-1)] \cong$$

$$\cong -\frac{\beta}{2}\frac{m}{\alpha}\frac{\partial^2 U(n)}{\partial t^2}[U(n+1) - U(n-1)]$$

Using the successive approximation method to solve this equation in the context of the self-demodulation process (i.e. strong modulated HF waves and weak demodulated LF wave), we obtain for the boundary conditions,

$$U_{HF}(n = 0) = A_0 \cos(\omega_1 t) + A_0 \cos(\omega_2 t)$$

$$U_{LF}(n = 0) = 0$$

the following solution,

$$U_{LF}(n) = \Im m\left\{\frac{\beta}{2}\frac{m}{\alpha}\left[\omega_2^2 \sin(k(\omega_1)a) - \omega_1^2 \sin(k*(\omega_2)a)\right]\right.$$

$$\left. \cdot A_{hf}^{(1)}(0)A_{hf}^{(2)}(0)\frac{\left[e^{-ik(\Omega)an} - e^{-i\Delta kan}\right]}{4\alpha \sin^2(\Delta ka/2) - m\Omega^2}e^{i\Omega t}\right\} \tag{2}$$

In this solution $k*$ denotes the complex conjugate of the wave number k, $\Delta k = k*(\omega_2) - k(\omega_1)$ and $\Omega = \omega_2 - \omega_1$. In order to derive this solution, we accounted for the dispersion relation of both propagative and evanescent modes:

$$k(\omega) = \begin{cases} \frac{2}{\pi}k_c \arcsin(\omega/\omega_c) & ,0 \leq \omega \leq \omega_c \\ k_c - i\frac{2}{\pi}k_c \operatorname{arccos}h(\omega/\omega_c) & ,\omega_c \leq \omega \leq +\infty \end{cases} \tag{3}$$

where $k_c = \pi/a$ is the maximum (real) wave number.

Figure 4 represents the demodulated signal amplitude U_{LF} (at the distance corresponding to the bead number $n = 2500$) as a function of the frequency of primary wave for different absorptions, i.e. different additional imaginary parts $\alpha(\omega)$ for the wave numbers.

For the highest absorptions (darkest lines), the dynamics of the amplitude as a function of pump frequency exhibits a minimum around the normalized frequency $\omega_n = \omega/\omega_c = 2^{-(1/2)}$. This is an effect of dispersion of nonlinearity; the non-linear quadratic term of the equation of motion (or equivalently the cubic term in the potential energy) has a minimum at this frequency. For the lowest absorptions (brightest lines of

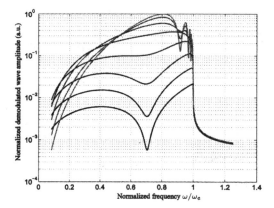

Figure 5. Demodulated signal amplitude as a function of the frequency of primary wave for different absorptions. From the dark line to the bright, the constant C in the absorption coefficient $\alpha(\omega) = C\Omega_n$ is respectively equal to 230, 91, 36, 14, 6, 0.9, 0.36. The normalized demodulated frequency $\Omega_n = \Omega/\omega_c$ is equal to $8,3 \cdot 10^{-4}$.

Fig. 4), the amplitudes have more and more pronounced minima at some fixed frequencies. These oscillations are due to the velocity dispersion effects between HF waves (sources) and LF wave, being sometimes in phase (maxima) and sometimes out of phase (minima) in the interaction region. This phenomenon is also a possible cause of saturation for the demodulated amplitude.

For the same conditions except of a different demodulated frequency $\Omega_n = 8,3 \cdot 10^{-4}$, the results are plotted on Fig. 5. The feature occurring for the frequency $\omega_n = \omega_c/2^{1/2}$ is still present and is independent of Ω_n. However, as could be anticipated, the phenomenon of beating depends on Ω (minima and maxima observed for low absorption – bright lines – are shifted compared to Fig. 4).

In each case, the transition from propagative ($\omega_n < \omega_c$) to evanescent primary waves ($\omega_n > \omega_c$) manifests itself by a strong decrease in the self-demodulation process efficiency. Obviously, this feature is of high interest for experimental purposes because it allows detecting the cut-off frequency of the medium.

4 PROBING FORCE CHAINS AND INTERGRAINS CONTACTS BEHAVIOUR WITH NON-LINEAR SHEAR WAVES

One of the conclusions, concerning an uniaxial externally applied static stress, is that it is transfered through the forced chains of beads that are preferentially oriented along the direction of this applied stress. A lot of recent studies have been performed on the properties of the force chains network in 2-D or 3-D granular

assemblages (Jaeger et al. 1996). We found that when anisotropy of the chain force network is realized, it influences the acoustical properties of the material due to the induced anisotropy in the contact pre-strains.

Velocity is sensitive to this preferentially oriented chain forces through its dependence on the static applied stress. This stress is applied vertically with a piston placed at the top of a 0.41 m diameter rigid cylindrical container, filled with 2 mm diameter glass beads. As a consequence, chain forces should be oriented along the vertical direction. We performed experiments to determine longitudinal wave velocities dependences on the applied static stress for two different directions of wave propagation: vertical and horizontal. When fitting the results with the following power law $v_\ell \propto P_0^\alpha$ (where v_ℓ is the longitudinal velocity and P_0 the applied stress), we obtained $\alpha \cong 1/4$ for the vertical direction of propagation and $\alpha \cong 1/9$ for the horizontal one. The first result was already observed experimentally several times and explained theoretically (Goddard 1990). This behavior is attributed to the Hertzian contact law between beads and to the creation of new contacts during the increase of the static stress (for low values of stress). The $\alpha \cong 1/6$ power law has been observed for higher static stresses, when there is no more creation of new contacts between the beads in the medium. Our observed $\alpha \cong 1/9$ behavior can be interpreted as a manifestation of the anisotropy of the applied static stress transmission in the medium. The increasing externally applied stress is mainly supported by vertically oriented contacts; in contrast, horizontally oriented contacts are not affected so much by this uniaxial static stress. This acoustical experiment confirms the observed results on anisotropy of force chains in granular materials.

It is well known that, for a fixed dynamic strain in the acoustical wave, the nonlinearity of Hertzian contacts is higher for weakly pre-strained contacts in comparison with strongly pre-strained contacts (Belyaeva et al. 1993). Moreover, when the dynamic elastic strain is higher than the contact pre-strain, clapping of the contact occurs (Zaitsev 1995). This type of nonlinear behavior is different from the one related to the quadratic non-linearity. As a consequence, non-linear acoustic phenomena can be applied for the evaluation of internal static stress of granular assemblages.

Several experiments have been performed on the shear wave self-demodulation effect. A strong HF carrier wave of 80 kHz, modulated in amplitude at 6 kHz, is radiated by a shear wave transducer. Due to nonlinearity, this wave is self-demodulated with mode conversion into a 6 kHz longitudinal wave (this feature has been checked precisely using different polarization experiments and checking the role of the membranes orientations of the transducers).

In order to compare, for the same configuration, this result with previous ones (Zaitsev et al. 1999a; Zaitsev

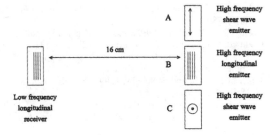

Figure 6. Configuration of the experiment. The two shear wave emitters are polarized orthogonally.

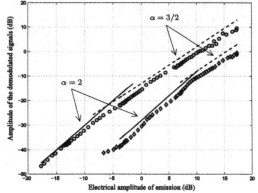

Figure 7. Circles correspond to the demodulated longitudinal signals from horizontally polarized shear HF waves, and diamonds, to demodulated longitudinal signals from vertically (along the force chains) polarized shear HF waves. The applied static pressure is $6 \cdot 10^{-4}$ Pa. Continuous lines corresponds to a quadratic law and dashed lines to a 3/2 power law.

et al. 1999b), we also emitted HF waves with a longitudinal transducer.

A longitudinal transducer is thus used for the detection, after 16 cm of propagation in the unconsolidated assemblage of 2 mm diameter glass beads. The usual configuration for the transducers disposition geometry is shown in Fig. 6. Similar to the previous velocity experiment, in order to enlight the anisotropy of the force chains network in the medium, we performed this experiment with horizontal propagation. The vertical externally applied stress is thus orthogonal to the plane of the Fig. 6. As a consequence, one shear emitter (denoted C) is polarized along the force chains and the other (denoted A) orthogonally to these force chains.

Amplitudes of the received LF demodulated longitudinal signals are plotted as a function of the amplitudes of the emitted HF shear waves on Fig. 7. Different features are noticeable. First, the signal demodulated from the horizontally polarized shear waves is up to 15 dB stronger than the signal demodulated from the vertically polarized shear waves. Second, the two signals are growing quadratically for low amplitudes of excitation and then there is a kink to a 3/2 power law for higher amplitudes of excitation. This transition from a quadratic to a 3/2 power law behavior, occurs around -5 dB for the horizontal polarization and around $+8$ dB for the vertical polarization.

The same experiment has also been performed for a vertical direction of propagation. Amplitudes difference between the two signals demodulated from the orthogonally polarized shear waves was less than 2 dB, ensuring that real efficiencies of the two shear transducers are very close.

The static strain in the medium, corresponding to the applied static stress, can be estimated using the following stress–strain relation (Belyaeva et al. 1993):

$$\sigma = \frac{\tilde{n}(1-\alpha)E\varepsilon^{3/2}}{3\pi(1-v^2)} \qquad (4)$$

with σ the applied stress, ε the strain, $\tilde{n}$ the average number of contacts per bead (3–5 for random packing),

E the Young modulus and v the Poisson ratio of the bead's material, and α the porosity of the medium.

Using the known values for these parameters, we estimate the static strain around 10^{-4} for our experimental configuration. Obviously, there exists a little dispersion of the static strain values, from one contact to another, around the estimated mean value. At the same time, calibration of the shear transducers using a vibrometer, provides an estimation of the dynamic applied strain in the medium at full input electrical amplitude around the same strain value, which is 10^{-4}.

When the dynamic strain is higher (or at least of the same order) than the static one, the contact has the opportunity to "open" and to close again. This is usually called the regime of clapping. Detailed considerations about this phenomenon and its consequences can be found in (Zaitsev 1995; Belyaeva et al. 1994). In fact, it manifests itself by the observed 3/2 power law behavior. Thus in Fig. 7, the quadratic behavior is associated with the classical pre-stressed Hertzian contacts, and the 3/2 behavior to the clapping of the contacts. Consequently, at the transition between these two regimes, the amplitude of the dynamic strain is roughly equal to the amplitude of the static strain of the contacts participating in the transmission of the dynamic perturbation.

Difference in the strain amplitude for the transition from quadratic to 3/2 power law behavior observed on Fig. 7 is thus associated with the difference of the contacts pre-strain. Shear waves polarized (vertically) along the force chains (vertical) feel an assembly of contacts with a higher pre-strain than the shear waves polarized (horizontally) orthogonally to the force chains. This difference in the static pre-strain can be estimated using the difference in the measured dynamic strains at the transitions 2 to 3/2. This provides the difference about 13 dB, which characterizes the level of the anisotropy of the force chains network in the granular material.

5 CONCLUSIONS

Several nonlinear methods have been proposed and applied to characterize granular materials through their acoustic wave absorption, scattering, velocity dispersion (section 2), through the cut-off frequency and various dispersive effects (section 3), or through the anisotropy of force chains network and contacts behavior (section 4). The method of the parametric antenna should find some applications in the granular materials characterization but also in other complex media, where scattering and absorption are strong for usual frequencies of ultrasonic waves.

REFERENCES

Belyaeva, I., V. Zaitsev, and L. Ostrovsky, 1993. *Acoust. Phys.* 39, 11.

Belyaeva, I., V. Zaitsev, and E. Timanin, 1994. *Acoust. Phys.* 40(6), 789.

Coste, C. and B. Gilles, 1999. *Eur. Phys. J. B* 7, 155–168.

Goddard, J., 1990. *Proc. R. Soc. Lond. A* 430, 105–131.

Jaeger, H., S. Nagel, and R. Berhinger, 1996. The physics of granular materials. *Physics Today.*

Jia, X., C. Caroli, and B. Velicky, 1999. *Phys. Rev. Lett.* 82, 1863.

Johnson, K., 1954. *Proc. Roy. Soc. A* 230, 531.

Lazaridi, A. and V. Nesterenko, 1985. *J. Appl. Mech. Tech. Phys.* 26, 405.

Liu, C. and S. Nagel, 1992. *Phys. Rev. Lett.* 68, 2301.

Maradudin, A. 1971. Theory of lattice dynamics in the harmonic approximation. New York: Academic Press.

Moussatov, A., B. Castagnède, and V. Gusev, 2001. *Phys. Lett. A* 283, 216.

Nesterenko, V., 1983. *J. Appl. Mech. Tech. Phys.* 24, 567.

Novikov, B., O. Rudenko, and V. Timochenko, 1987. *Nonlinear Underwater Acoustics.* New York: ASA.

Tournat, V., V. Aleshin, V. Gusev, and B. Castagnède, 2002. *16th International Symposium on Nonlinear Acoustics.*

Tournat, V., B. Castagnède, V. Gusev, and P. Béquin, 2003. *C.R. Mécanique*, 331.

Tournat, V., V. Gusev, and B. Castagnède, 2002. *Phys. Rev. E* 66, 041303.

Zaitsev, V., 1995. *Acoust. Phys.* 41(3), 385.

Zaitsev, V., A. Kolpakov, and V. Nazarov, 1999a. *Acoust. Phys.* 45, 202.

Zaitsev, V., A. Kolpakov, and V. Nazarov, 1999b. *Acoust. Phys.* 45, 305.

Emerging Technologies in Non Destructive Testing, Van Hemelrijck, Anastasopoulos & Melanitis (eds)
© 2004 Swets & Zeitlinger, Lisse, ISBN 90 5809 645 9

Slanted transmission mode of air-coupled ultrasound: new opportunities in NDT and material characterisation

I. Yu. Solodov, K. Pfleiderer, H. Gerhard, S. Predak & G. Busse
Institute for Polymer Testing and Polymer Science (IKP) – Nondestructive Testing (ZFP), University of Stuttgart, Stuttgart, Germany

ABSTRACT: The slanted transmission mode (STM) of air-coupled ultrasound is based on the "resonance" transmission of an acoustic wave through the sample due to Lamb mode excitation and re-radiation. It was shown that in-plane elastic anisotropy of a plate-like sample could be obtained by measurements of the optimal angles of incidence or the amplitude and phase of the STM-output signal as functions of the azimuth angle. Such a technique is demonstrated to be sensitive in discerning the directions of reinforcement in composites and orientation of crystalline substrates. Substantial contrast enhancement is demonstrated for the focused STM-C-scan imaging of the cracked defects and delaminations in polymers and composites. The STM phase measurements are used for monitoring variations in local stiffness due to tensile deformations. The non-contact STM-measurements of flexural wave velocity during tensile tests enabled to detect structural changes and damage in nonlinear and plastic zones for composite materials.

1 INTRODUCTION

Air-coupled ultrasound inspection is a challenging technique for non-contact NDT that was shown to be a successful alternative to conventional contact and water-immersion methods (e.g. Rogovsky 1991). Usually, it is based on the scattering of longitudinal waves generated in the sample by an ultrasonic beam incident normally onto the sample surface (normal transmission mode, NTM). In this case, acoustic energy transmitted through the sample for the NTM is extremely small due to a severe impedance mismatch on the boundaries of the sample. To increase the output, one can use an oblique incidence of the acoustic wave at an angle of "total" transmission (Brekhovskikh 1960):

$$\sin \theta_0 = v_{air} / v_L, \qquad (1)$$

which is caused by Lamb mode generation in the sample (Luukkalla et al. 1971). This brings about two attractive opportunities for material characterisation and NDT: firstly, one can determine the phase velocity v_L of Lamb modes and thus monitor remotely variations in a local stiffness of the material. Secondly, the plate modes can be scattered by the defects (especially, cracked defects) more efficiently than longitudinal waves and thus provide a higher sensitivity of

acoustic NDT and better contrast of acoustic imaging of flaws.

However, the technique is critically dependent on two factors: (a) precise measurements of the angles of incidence and (b) a requirement of an incident plane wave that inevitably stipulates for the use of wide-aperture transducers. The latter deteriorates the spatial resolution and prevents the development of the scanning STM-versions for air-coupled NDT systems.

In this paper, we develop a focused ultrasonic beam STM to enhance resolution of the technique and use it for a high-contrast acoustic imaging of various flaws. The FSTM version is also applied to the measurements of variations in a local material stiffness associated with either in-plain elastic anisotropy or damage induced by a nonlinear tensile load.

2 EXPERIMENTAL METHODOLOGY

In our experiment (Fig. 1), commercially available piezoelectric 1-3 composite material transducers (Hillger et al. 1998) were used to radiate and receive a focused ultrasound beam (fundamental frequency 450 kHz, diameter of the transducers 18 mm, full angular transducer aperture $\cong 20°$, focus distance $\cong 40$ mm, and focus neck $\cong 2$ mm). A 15-cycle burst

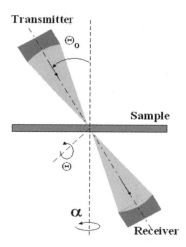

Figure 1.   Experimental set-up for air-coupled FSTM.

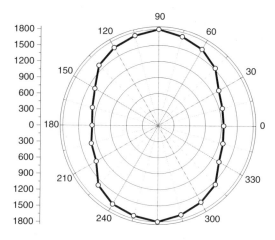

Figure 2.   Phase velocity of a_0 - mode (m/s) as a function of azimuth angle of 30 wt.% GFRPP sample.

of ≈1400 V amplitude was applied to the transducer to generate the acoustic wave; the two-step amplification of the receiver provided ≅ 140 dB total dynamic range of the PC-operated system.

To approach the plane-wave case for precise measurements of the angle of incidence, highly damping ring diaphragms were inserted into the transducers reducing the initial beam diameter to ≅ 7 mm and the beam angular aperture down to ≅ 5° (weak focusing). After 8 bit-A/D-conversion at 12.5 MHz sampling frequency the transmitted pulse can be observed and measured at the PC-display. A scan of the angle of incidence (θ) was implemented by a sample rotation using a line-scan of the PC-operated scanning table. Such angular B-scans usually demonstrated more than ≅ 20 dB rise in the transmitted wave amplitude under the "resonance" phase matching conditions (1) for the zero-order anti-symmetrical a_0-mode. This mode was chosen for further experiments where the accuracy of θ_0 measurements was about 0.5°.

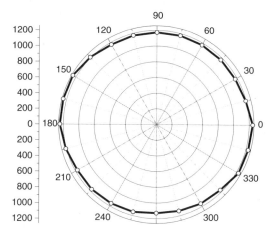

Figure 3.   Phase velocity of a_0-mode (m/s) as a function of azimuth angle in a pure PP sample.

3   STM-NDT OF IN-PLAIN ANISOTROPY

For a weakly focused beam, the wave vector of the excited plate modes lies predominantly in the plane of incidence (Fig. 1). By rotating the sample around the normal to its surface one can change the azimuth angle (α) of the in-plane wave propagation. In anisotropic materials, this rotation results in a velocity variation of the modes involved in the STM that may be monitored by the corresponding change in $\theta_0(\alpha)$ and subsequent $v(\alpha)$ calculations from Equation (1).

Figure 2 shows the results of such a methodology applied to a 2 mm thick sample of glass-fibre-reinforced (GFRP) polypropylen containing 30% wt.

short fibres. A strong elastic anisotropy due to fibre reinforcement is observed and the angle with the maximum velocity ($\alpha = 90°$) indicates the direction of glass fibre orientation.

Other examples of high sensitivity of the STM to the in-plain elastic anisotropy are shown in Figures 3–4. According to Figure 3, even without fibre vreinforcement a reasonable in-plain anisotropy ($\Delta v/v \cong 5\%$) exists in a polypropylen sample. It is, apparently, caused by molecular orientation of the polymer induced during the manufacturing process. Hence the acoustic anisotropy might be used to monitor the injection moulding process of polymers.

Figure 4 is concerned with the case of even lower anisotropy observed for the (001)-plane of a Si-wafer.

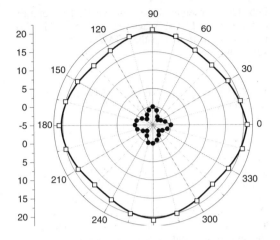

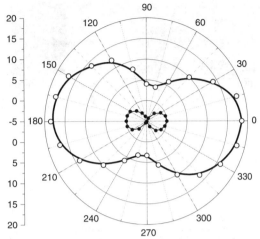

Figure 4. Amplitude (□) (rel. units) and phase (●) (% of 2π) of FSTM output signal as a function of azimuth angle in (001)-plane of Si-wafer.

Figure 5. Amplitude (○) (rel. units) and phase (●) (10% of 2π) of FSTM output signal as a function of azimuth angle in 30 wt.% GFRPP sample.

The in-plane variation of the a_0-mode phase velocity was found to be so small that within the error of experiment ($\approx 0.5°$) measurements of $\theta_0(\alpha)$ were not possible. Instead, Figure 4 shows the angular dependence of amplitude and phase of the STM output signal obtained for the incident beam of a wide angular aperture ($\Delta\theta \approx 20°$). For the Si-wafers, θ_0 values are in the range of $10°$–$15°$ so that the FSTM output comprises not only the a_0-mode but also the acoustic wave transmitted normally to the sample surface. As the angle α changes, the interference pattern formed by these two components varies due to the change in the phase difference $\Delta\varphi$ between them caused by variation of the a_0-mode velocity. This results in an amplitude variation and phase shift of the FSTM output signal. Therefore, by using well-focused air-coupled acoustic waves the information about elastic anisotropy can also be obtained from the amplitude and phase of the FSTM-output signal as functions of the azimuth angle α.

Data presented in Figure 4 illustrate the presence of the 4-fold axis of symmetry, which determines the $90°$-in-plane symmetry of elasticity. The maximum negative phase shift (smaller delay) of the output signal in Figure 4 corresponds to the maximum of v_{a0}. Therefore, one can conclude that the directions $\alpha = \pi/4 \pm m\pi/2$ ($m = 1, 2$) indicate the positions of natural "reinforcement" of the Si-crystal lattice.

Figure 5 is a further proof for the viability of the FSTM-interference version. In the case of strong anisotropy of the GFR polypropylen sample, both amplitude and phase variations exhibit a 2-fold symmetry and are very evident: $\Delta u/u \approx 400\%$; $\Delta\varphi \cong \pi$. By direct measurements of $\theta_0(\alpha)$ it was shown that the positions of the minima in Figure 5 ($\alpha \cong 80° \pm 180°$)

correspond to the maxima of v_{a0} and indicate the directions of reinforcement in the composite. Thus, the interference version of the FSTM eliminates measurements of the angles θ_0 that contributed significantly to the experimental error. Instead, the measurements of the azimuth amplitude and phase characteristics of the FSTM enhance substantially the sensitivity and accuracy of the technique in discerning the in-plane elastic anisotropy.

4 DEFECT IMAGING WITH FSTM

Besides an obvious contribution of the plate modes involved in the FSTM into ultrasound scattering by defects, they can also play an important role in the interference mechanism of image formation and thus affect the imaging of internal defects in the sample. Acoustic beams of the full angle aperture ($\Delta\theta \approx 20°$; focus nec $\cong 2$ mm) were used to obtain a high spatial resolution in the experimental results on imaging given below for both NTM and FSTM. In the FSTM-imaging, the optimal angle θ_0 was first measured for the intact part of the sample and then set for the subsequent C-scan images.

For the NTM through a very small crack, the transmitted acoustic field is strongly affected by diffraction (similar to light diffraction at a slot). The amplitude of the main diffraction maximum decreases rapidly with crack thickness thus making such defects practically invisible in the NTM (Fig. 6a). On the contrary, the plate modes propagate along the sample plane and the scattering efficiency is expected to be non-zero even

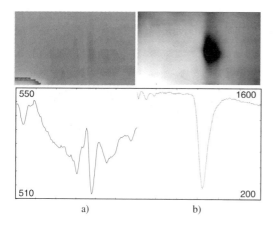

a)   b)

Figure 6. (top): NTM (a) and FSTM (b) C-scans of 60 ×
30 mm area of PMMA plate with a small closed crack;
(bottom): distribution of transmitted wave amplitudes along
the center line of the images (relative units inside frames).

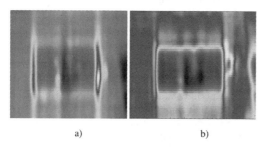

a)   b)

Figure 7. NTM (a) and FSTM (b) C-scans of an actuator
imbedded into epoxy-based composite.

for very tight cracks. Comparison of Figures 6a, b
demonstrates the much higher sensitivity of the
FSTM for small (closed) crack detection: the contrast
of the crack image is $\Delta u/u \cong 80\%$ with FSTM (Fig.
6b) while it is $\Delta u/u \cong 2\%$ with NTM (Fig. 6a).

In the case of delaminations, disbonds or inclusions
over the sample thickness, the local velocity of the
plate modes within the defect area may change substantially either due to variations of local thickness
(delaminations) or material elasticity (strongly inhomogeneous inclusion). Besides the increase of acoustic
scattering in those areas, an additional FSTM contrast
enhancement is to be expected due to the following
factors: (a) local θ_0-variation that should cause the
"resonance" drop of the transmitted wave amplitude;
(b) alteration of the interference pattern between the
plate modes and the acoustic wave transmitted normally to the sample surface, as discussed in the pre-
ceding section.

Such benefits of the STM acoustic imaging are
shown in Figures 7a, b for the piezoelectric actuator

imbedded into epoxy-based composite. The contrast
of the actuator FSTM-image ($\Delta u/u \cong 45\%$) (Fig. 7b)
is far superior to that of the NTM, which is $\Delta u/u \cong$
15% (Fig. 7a).

5   STM FOR NONLINEAR NDT OF DAMAGE INDUCED IN COMPOSITES

It has been shown quite recently that elastic nonlinearity of imperfect solids increases greatly due to
super-molecular structures, like dislocations, cracks,
delaminations, etc. (e.g. Solodov & Maev 2000). As a
result, nonlinear acoustic methodology can be used
for discerning damage development and prediction of
ultimate strength and life span of materials and industrial products. This section presents a phenomenology
and an experimental study of the STM application to
monitor elastic nonlinearity in a wide range of tensile
stress (up to a fracture limit) applied to thin plates of
fibre-reinforced composites.

5.1   Phenomenology of nonlinear STM-NDT

A deviation from a linear relation between stress (σ)
and strain (ε) is given by the generalised Hooke's law:

$$\sigma(\varepsilon) = C^{II}(1 - \beta_2\varepsilon - \beta_3\varepsilon^2 + ...)\varepsilon , \qquad (2)$$

where C^{II} is the linear stiffness; β_n is the nonlinearity
parameter of the n-th order.

The following mechanisms of nonlinearity can
contribute to the value of the nonlinear parameter:

1. Intrinsic lattice nonlinearity (nano-scale)
2. Micro(meso)-scale structural changes (micro-
   cracking, delaminations, etc.)
3. Macro-scale defect generation.

These types of nonlinearity can result from both
static and dynamic (acoustic) deformations. In the
former case, they bring about the well-known nonlinear stress–strain curve. For acoustic waves, the nonlinearity in Equation (2) results in the wave velocity
variation with static strain (Zheng et al. 1999):

$$c(\varepsilon) = c_0(1 - \beta_2\varepsilon - \frac{3}{2}\beta_3\varepsilon^2 - ...), \qquad (3)$$

where c_0 is the wave velocity in an unstrained body.

The lattice ("classical") nonlinearity (type 1) is
usually rather small and constitutes a background
nonlinearity; types 2 & 3 result in much higher values
of nonlinearity parameters. Thus, the increase in nonlinearity is an indication of damage (cracking, delamination, etc.) induced in a material.

To evaluate material acoustic nonlinearity we
retain only the second-order term in Equation (2) and

from Equation (3) obtain the following expression for the dynamic nonlinearity parameter:

$$\beta_2^d = -\frac{\partial c}{\partial \varepsilon}/c_0.$$ (4)

According to Equation (4), to determine the dynamic nonlinearity parameter one has to measure the acoustic wave velocity as a function of ε.

5.2 STM-measurements of nonlinearity parameters under tensile load

In experiments, to determine the nonlinearity parameters we measured the stress–strain relation and flexural wave velocity up to the fracture threshold for glass fibre-reinforced composites by using a standard automated tensile machine with maximum static stress 200 MPa. Computer software "Origin" was used to obtain velocity derivatives after a smoothing of the experimental data.

A pulse mode FSTM of air-coupled ultrasound (see Fig. 1) was adapted for generation and detection of flexural waves (frequency 450 kHz) in the samples. In the unstressed sample, the value of c_0 was found by determining the angle of optimal transmission from (1). The variation of the flexural wave velocity under stress was then determined by measuring the change in acoustic pulse delay $\Delta\tau(\varepsilon)$ from the following relation derived:

$$c(\varepsilon) = c_0(1 - \frac{c_0\Delta\tau(\varepsilon)}{c_0\Delta\tau(\varepsilon) + D}),$$ (5)

where $D \approx 4$ mm is the estimated distance of the flexural wave propagation within the beam.

Non-contact FSTM-measurements of local acoustic wave velocity were used to calculate the nonlinearity parameter (4) as a function of static strain and study its behaviour through a loading cycle. Figures 8 and 9 show the results of experiments for 30%-short glass fibre-reinforced polypropylene (GFRP) samples with fibres, respectively, parallel and normal to the stress applied direction.

For both materials, the acoustic wave velocity decreases over the entire deformation range that makes β_2 positive and shows local weakening at meso-scale as the strain increases. However, the nonlinearity parameter displays quite different behaviour for the GFRP with fibres normal and parallel to the stress. In the latter case (Fig. 8), 2 maxima of nonlinearity are observed: the first peak corresponds to an initial stage of deformation while the second one takes place in a plasticity zone. Apparently, for the stress applied to this composite material a major load is distributed along the glass-fibres. Since the fibre material is far more rigid, the initial deformation must, first,

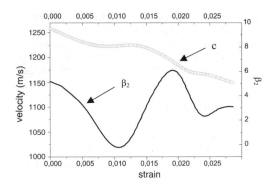

Figure 8.   Velocity and nonlinearity parameter as functions of strain for GFRP with fibres parallel to the stress applied.

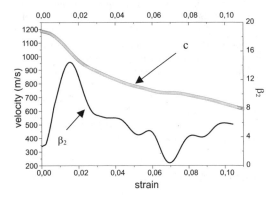

Figure 9.   Velocity and nonlinearity parameter as functions of strain for GFRPP with fibres normal to the stress applied.

concentrate in the matrix around the ends of the fibres and then expand until it hits the wall of a neighbouring fibre. Such predominantly matrix micro-cracking may result in the maximum of β_2 at low ε in Fig. 8. Further increase in ε is mostly due to opening of the matrix cracks which does not have a substantial impact on the stiffness (low nonlinearity). In the plasticity zone, the crack opening gap becomes too big to be accommodated by the extension of the fibers that results in the fiber cracking and pull-out (Ashbee 1994). This leads to a drop of the composite stiffness and appears as the second peak of β_2 in Figure 8.

The situation is different for the GFRP with fibres normal to the stress (Fig. 9). In this case, the load is apparently applied to the areas of low stiffness PP-matrix. As a result, a single peak of nonlinearity parameter is observed which is, apparently, associated with the matrix cracking in the vicinity of the yield strain point. Such a single-stage weakening of the composite material demonstrates a much lower contribution of fibres to material strength in this case. By comparing the absolute values of nonlinearity parameters in Figures 8 and 9 one can make a reasonable conclusion: the higher the material nonlinearity (Fig. 9), the lower its strength (fibres normal to stress).

6 CONCLUSIONS

The "resonance" angle behaviour of the weakly focused STM allows to determine quantitatively the in-plane elastic anisotropy by measuring the angles of maximum transmission as functions of azimuth directions. The much more sensitive interference version of the FSTM is a simple and robust technique for non-contact detection of the fibre orientation in composite materials and of minor anisotropy in crystalline samples.

Due to an additional scattering of the plate modes and a local alteration of the interference pattern of the output signal, the FSTM exhibits a far superior contrast to conventional NTM technique in imaging of cracks, delaminations, disbonds and other inhomogeneities.

The nonlinear STM-acoustic approach enables to discern structural changes and monitor damage induced in materials subjected to high mechanical loads. A correlation between material nonlinearity and strength is a basis for remote NDT of durability and failure prediction using the STM of air-coupled ultrasound.

ACKNOWLEDGEMENTS

The authors are grateful to the German Science Foundation (DFG) for financial support (collaborative research centre SFB 381), to Dipl.-Ing. R. Aoki (DLR; Stuttgart), and to Dipl.-Ing. J. Duerr (EADS Dornier GmbH; Friedrichshafen) for kindly providing the samples.

REFERENCES

Brekhovskikh, L.M. 1960. *Waves in Layered Media.* New York: Academic Press.

Hillger, W., Gebhardt, W., Dietz, M. and May, B. 1998. Ultraschallpruefungen beruehrungslos mit Ankopplung ueber Luft – Illusion oder schon bald Realitaet? *Deutsche Gesellschaft fur Zerstoerungsfreie Pruefung Jahrestagung*; Bamberg, 7–9 September 1998: 241–249.

Luukkalla, M., Heikkili, P. and Surakka, J. 1971. Plate wave resonance – a contactless test method. *Ultrasonics* 9: 201–208.

Rogovsky, A.J. Development and application of ultrasonic dry-contact and air-contact C-scan systems for non-destructive evaluation of aerospace composites. *Material Evaluation* 50: 1491–1497.

Solodov, I.Yu. & Maev, R. 2000. Overview of opportunities for nonlinear acoustic applications in material characterisation and NDE. In D.V. Hemelrijck, A. Anastasopoulos & T. Philippidis (eds), *Emerging technologies in NDT*: 137–144. Rotterdam: Balkema.

Zheng, Y., Maev, R. and Solodov, I.Yu. 1999. Nonlinear acoustic applications for material characterization: A review. *Can. J. Phys.* 77: 927–967.

Emerging Technologies in Non Destructive Testing, Van Hemelrijck, Anastasopoulos & Melanitis (eds)
© 2004 Swets & Zeitlinger, Lisse, ISBN 90 5809 645 9

Use of ultrasounds to estimate the composition and textural properties of a meat-based product

S. Simal, A. Femenia & C. Rosselló
Department of Chemistry, Univ. of Illes Balears, Palma de Mallorca, Spain

J. Benedito
Food Technology Department/IAD, Univ. Politécnica of Valencia, Valencia, Spain

ABSTRACT: The composition and textural properties of dry-cured meat products frequently can be considered as good indicators of their maturity. Due to the variability among producers and even within one company, rapid and non-destructive techniques to assess maturity and quality of these products are of great interest. The use of ultrasonics as an analytical technique to estimate the chemical composition and the textural properties of a fermented meat-based product (sobrassada) was assessed. The ultrasonic velocity temperature dependence allows the determination of fat, moisture and protein+others contents by measuring the ultrasonic velocity in the meat-based product at 4 and 12°C using a semiempirical equation. The ultrasonic velocity can be mathematically related with the textural parameters such as hardness and compression work. These results show the feasibility of using ultrasonic velocity measurement to assess, in a rapid and non-destructively way, the chemical composition and textural properties of a meat-based product.

1 INTRODUCTION

Low-intensity ultrasound can be used to provide information about the physicochemical properties of many foods having some advantages over other traditional analytical techniques because measurements are rapid, non destructive, precise, fully automated and might be performed either in a laboratory or on-line (Abouelkaram et al., 2000). Ultrasonic techniques have been used to determine the beef carcass value and quality attributes. The composition of chicken, cod and pork meat mixtures has also been addressed (Ghaedian et al., 1997; Benedito et al., 2001).

The ultrasonic properties of a multicomponent material can be described to a first approximation by the following relationship (eq. 1) (Ghaedian et al., 1997) where Φ_j and v_j are, respectively, the mass percentage and ultrasonic velocity in component j:

$$\frac{100}{v^2} = \sum_{j=1}^{n} \frac{\Phi_j}{v_j^2} \tag{1}$$

In solid materials, ultrasonic velocity (v) is related to the square root of the elastic modulus of the material (E) and its density (ρ) by the relationship (Povey &

McClements, 1988):

$$v = \sqrt{\frac{E}{\rho}} = \sqrt{\frac{K + \frac{4}{3}G}{\rho}} \tag{2}$$

This expression relates velocity with elastic constants for a homogeneous isotropic and elastic medium where K is the bulk modulus and G the shear modulus. As a first approximation, equation 2 could be written (eq. 3) (Benedito, 1998):

$$v = A\sqrt{TP} + B \tag{3}$$

where A and B are constants and TP is a textural parameter.

The chemical composition of meat-based products has a large influence on its nutritional value, functional properties, sensory quality, storage conditions and commercial value. Analytical techniques are therefore needed to grade the quality of raw meat and to characterize meat composition during processing and storage. The final composition of these products constitutes one of the major quality attributes, and its knowledge is of great importance for quality control, not only for producers but also for retailers and consumers. Also textural properties of foods are frequently determinants of their acceptability by consumers. The evaluation of the main textural properties of a meat-based product, such as sobrassada, based on non-destructive methods might be important for manufacturers.

The aim of this work was to review the existing applications for the use of ultrasonic techniques to assess the chemical composition and the textural properties of a fermented meat-based product, Sobrassada of Majorca.

2 MATERIALS AND METHODS

2.1 *Raw material*

All the experiments reported in this paper were conducted using meat-based products (sobrassada) manufactured at the island of Mallorca (Spain) by a local factory following the methodology suggested by the Protected Geographical Indication (PGI) for the "Sobrassada de Mallorca". Sobrassada is composed by a mixture of lean pork meat, white fat, paprika and salt. The meat ingredients are kneaded until particle size of 4 mm is achieved and then mixed with the other ingredients. Afterwards, the mince was filled into artificial casings ($25 \times 10^{-2} \pm 10^{-2}$ m of length, 7.5×10^{-2} m of diameter and 2.68×10^{-12} g/Pa·s·m of permeability at 12°C and 75% of relative humidity) and ripened at 14°C and 70% of relative humidity. In order to obtain samples with different composition, sobrassadas were elaborated by using different proportions of meat and white fat.

2.2 *Chemical and textural analysis*

All samples of sobrassada were analysed in triplicate for moisture, fat and protein contents according to the official methods (ISO R-1442, ISO R-1443 and ISO R-937, respectively).

A Texture Profile Analysis (TPA) was performed on all meat-based product samples. Slices of 2.5×10^{-2} m height and 7.5×10^{-2} m diameter were cut from the samples of this meat product. Then, eight cylinders of 2.9×10^{-2} m diameter and 2.5×10^{-2} m height were cut (two cylinders from each slice). A 100 kN load cell was fitted to a Universal Testing Machine (Model Zwick-Z100). The cross-head speed was 20 mm/min and the cylinders were compressed 1.75×10^{-2} m (70%). From the TPA curves, hardness (H) and the compression work (CW) were determined.

2.3 *Ultrasonic measurements*

Measurements of ultrasonic velocity at different temperatures were carried out by using the experimental set-up described previously by Benedito et al. (2001) which consisted of a couple of narrow-band ultrasonic transducers (1 MHz, 0.75″ crystal diameter,

A314S-SU Model, Panametrics, Waltham, MA), a pulser-receiver (Toneburst Computer Controlled, Model PR5000-HP, Matec Instruments, Northborough, MA) and a digital storage oscilloscope (Tektronix™ TDS 420, Tektronix Inc., Wilsonville, OR) linked to a personal computer using GPIB interface. To compute the time of flight for a sample, five acquisitions were performed and averaged. All analyses were performed in duplicate on two different samples without removing the casing.

3 ULTRASONIC ASSESSMENT OF MEAT-BASED PRODUCTS

3.1 *Estimation of the chemical composition*

The mean values of composition for the assayed samples ranged from 43.1 to 67.1% (wet basis) for fat content, from 19.6 to 37.2% (wet basis) for moisture content and from 5.0 to 13.5% (wet basis) for protein content.

The measurement of ultrasonic velocity at different temperatures showed that velocity decreases linearly when temperature increases. The slope of the linear relationships decreased with the increase in fat content and the decrease of water content from -1.93 m/s°C for a sample with a 43.1% fat content (% wet basis), to -4.21 m/s°C for a sample with a 67.1% of fat content (% wet basis). The ultrasonic velocity temperature dependence allows the determination of fat, moisture and protein+others contents by measuring the ultrasonic velocity in the meat-based product at 4 and 12°C using a semiempirical equation (Simal et al., 2003).

It can be considered that the sobrassada is formed by three different constituents: fat, water and protein+others. Based on equation 1, the ultrasonic velocity of a meat-based product can be calculated by the following equation (eq. 4):

$$\frac{100}{v^2} = \frac{\Phi_f}{v_f^2} + \frac{\Phi_w}{v_w^2} + \frac{\Phi_{p+o}}{v_{p+o}^2} \qquad (4)$$

where v is the ultrasonic velocity and v_f, v_w and v_{p+o} are the ultrasonic velocities in fat, water and protein + others, respectively. Moreover, Φ_f, Φ_w and Φ_{p+o} are the percentages of fat, water and protein+others (wm), respectively.

The relationship between water and temperature (eq. 5) can be taken from the literature (Kinsler et al., 1982). Further, the relationship between the ultrasonic velocity and temperature in fat was obtained by Benedito et al. (2001) from experiments carried out using vacuum dried pork fat (eq. 6).

$$v_w = 1403 + 5T - 0.06T^2 + 0.0003T^3 \qquad (5)$$

$$v_f = -5.6076T + 1651.7 \qquad (6)$$

An empirical equation was proposed to relate the v_{p+o} with temperature and fat content (eq. 7). The Φ_f was included due to the fact that the protein proportion varies in "protein+others" between samples with different chemical composition.

$$v_{p+o} = a_1 + a_2\,\Phi_f + a_3\,T \qquad (7)$$

In order to determine the parameters a_1, a_2 and a_3 of equation 7, calculated values of v_{p+o} at 4 and 12°C through equations 4 to 6 for five samples with different chemical composition can be used. SOLVER, an optimization tool (GRG2 method) included in Microsoft Excel 7.0™ spreadsheet was used, obtaining equation 8, which provided the lowest sum of the square differences between the values calculated using equations 4 to 6 and those estimated using the proposed equation 8 (Simal et al., 2003).

$$v_{p+o} = 3262.67 - 15.48\,\Phi_f - 27.56\,T \qquad (8)$$

The chemical composition as Φ_w, Φ_f and Φ_{p+o} was estimated for each sobrassada sample by using the equations representative of v_w, v_f and v_{p+o} (eqs. 5, 6 and 8) and the results obtained in the measurements of ultrasonic velocity at 4 and 12°C for sobrassada samples. Figure 1 shows the comparison between the estimated composition (fat, moisture and protein+ others contents) and the experimental values. As it can be observed in this figure, there was a close fit between both estimated and experimental data. The explained variances obtained by comparison of the estimated composition and the experimental one were satisfactory, 96.8% for the fat content, 95.7% for the moisture content and 95.8% for the protein+others content.

The results show the feasibility of using ultrasonic measurements to assess sobrassada composition. Ghaedian et al. (1998) also used equation 2 for the composition determination of cod fillets considering only two phases, fat and an aqueous phase. Similar results were found in raw meat mixtures by Benedito et al. (2001) considering three constituents.

3.2 Estimation of textural parameters

H and CW exhibited an increase during ripening which can be related to the state of maturity. The increase of H is probably linked to an increase in the bulk modulus, therefore an increase in velocity is expected. Based on equation 1, the ultrasonic velocity can be fitted to the square root of the textural parameters by using the results obtained in the measurements of ultrasonic velocity and textural parameters hardness (H) and compression work (CW) (Llull et al., 2002b).

$$v = 1556.3 + 15.23\sqrt{H} \qquad r^2 = 0.95 \qquad (9)$$

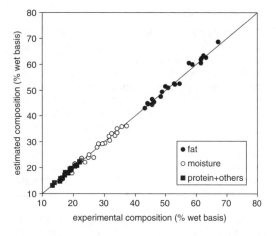

Figure 1. Estimated vs experimental composition of sobrassada samples: fat, moisture and protein+others contents (% wet basis) (Simal et al., 2003).

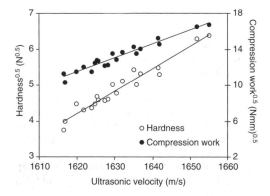

Figure 2. Ultrasonic velocity dependence on the hardness and the compression work.

$$v = 1550.9 + 6.07\sqrt{CW} \qquad r^2 = 0.94 \qquad (10)$$

As it can be seen in figure 2, the ultrasonic velocity could be mathematically related to the textural parameters of sobrassada. The results showed that using this non-destructive method, hardness and compression work could be accurately estimated (%var = 95% in both parameters).

Benedito (1998) also found similar relationships between the ultrasonic velocity in cheese and several textural parameters. The change in the textural parameters is linked to the water losses during ripening as a consequence the ultrasonic velocity is also related to the water content (Simal et al., 2003). As seen in the previous section, velocity is greatly influenced by temperature and therefore this variable must be accurately controlled.

3.3 Changes during ripening

Periodically measuring the ultrasonic velocity, two different periods of ripening can be clearly distinguished, a first period (0–40 days of ripening) and a second period (40 days to the end of the ripening), under the experimental conditions of ripening at 15°C and 75% RH (Llull et al., 2002b). The ultrasonic velocity could be mathematically related with the moisture content and textural parameters such as hardness, compression work, maximum puncture force and puncture work during the first period of ripening. In the second period, some physical modifications were inferred to occur which could be detected through the textural parameters, the ultrasonic velocity and by using optical microscopy. Overall, these results showed that the ultrasonic techniques could be used to accurately estimate the moisture content and the textural parameters of sobrassada during the first period and also as a reliable method to detect changes in textural properties after prolonged ripening (Llull et al., 2002b).

4 CONCLUSIONS

The ultrasonic properties of a meat-based product depend on composition and temperature. Semiempirical equations accurately predict fat, water and protein+others contents of a meat-based product (sobrassada) from ultrasonic velocity measurements at two different temperatures. Furthermore, relationships can be obtained to adequately relate the ultrasonic velocity and the characteristic textural parameters of this meat product. Using non-destructive measurements a quality control of these products can be carried out, resulting into an increase of the uniformity of the batches.

Although these results were found for a typical meat product from Spain, the obtained results pointed out that a non-destructive ultrasound technique is expected to be successfully used for accurately estimating the chemical composition of other meat-based products.

ACKNOWLEDGEMENTS

The authors would like to acknowledge the financial support of CICYT (FD97-1246-C03).

REFERENCES

Abouelkaram, S., Suchorski, K., Buquet, B., Berge, P., Culioli, J., Delachartre, P. & Basset, O. 2000. Efects of muscle texture on ultrasonic measurements. Food Chemistry 69: 447–455.

Benedito, J. 1998. Contribución a la caracterización de quesos mediante el uso de ultrasonidos de señal. Thesis. Universidad Politécnica de Valencia, Spain.

Benedito, J., Carcel, J., Rosselló, C. & Mulet, A. (2001). Composition assessment of raw meat mixtures using ultrasonics. Meat Science 57(4): 365–370.

Ghaedian, R., Decker, E.A. & McClements, D.J. 1997. Use of ultrasound to determine cod fillet composition. Journal of Food Science 62: 500–504.

Kinsler, L.E., Frey, A.R., Coppens, A.B. & Sanders, J.V. 1982. Fundamentals of acoustics. New York: John Wiley and Sons.

Llull, P., Simal, S., Femenia, A., Benedito, J. & Rosselló, C. 2002a. The use of ultrasound velocity measurement to evaluate the textural properties of sobrassada from Mallorca. Journal of Food Engineering 52: 323–330.

Llull, P., Simal, S., Femenia, A., Benedito, J. & Rosselló, C. 2002b. Evaluation of textural properties of a meat-based product (sobrassada) using ultrasonic techniques. Journal of Food Engineering 53: 279–285.

Povey, M.J.W. & McClements, D.J. 1988. Ultrasonics in food engineering. Part I: Introduction and experimental methods. Food Control 54: 7–14.

Simal, S., Benedito, J., Clemente, G., Femenia, A. & Rosselló, C. 2003. Ultrasonic determination of the composition of a meat-based product. Journal of Food Engineering 58: 253–257.

Nondestructive evaluation of multilayered composites using low frequency, oblique incidence acousto-ultrasonics

V.F. Godínez-Azcuaga, R.D. Finlayson, R.K. Miller & S.J. Vahaviolos
Physical Acoustics Corporation, NJ, USA

A. Anastasopoulos
Envirocoustics A.B.E.E., Athens, Greece

Philip T. Cole
Physical Acoustics Limited, Cambridge, UK

ABSTRACT: A portable field-application nondestructive system for the inspection of multi-layer composite sections based on oblique-incidence, low-frequency Acousto-Ultrasonics (AU) has recently been developed by Physical Acoustics Corporation (PAC). This development was the result of an SBIR (Small Business Innovative Research) project sponsored by The Lightweight Structures Team at TARDEC (Tank Automotive Research, Development and Engineering Center). Although developed for the particular purpose of inspecting multi-functional composite armor plates, PAC has used this field inspection system in the evaluation of many critical composite structures. Some of these components include Fiber-Reinforced Polymer "wraps" around concrete structures, graphite-epoxy, foam-core composite centrifugal arms for simulators, and type III Navy diving bottles.

1   INTRODUCTION

Multi-layer composite structural elements have many advantages over comparable metal designs. Among these are a very high stiffness to weight ratio, strength vs. weight and resistance to corrosion. However, in order to realize these benefits, the layers that make up the composite structure must be fully bonded to each other during the original manufacturing or repair process.

To determine the quality (integrity) of a bonded structure, inspection techniques have been developed which include various forms of ultrasonic inspection. Due to the nature of the material – including the bonding process itself – these techniques require sophisticated test set-ups that are typically available only in large laboratories or production facilities.

Once structures are actually put into place, in field applications, the same techniques that can be used in the production environment are often not usable in the field. For instance, the ultrasonic technique most often used for composite panel inspection typically involves a through-transmission setup in an immersion tank to enable ultrasonic waves to penetrate the difficult built-up structure. Thus, once composite structures are

in-service, returning to an "as produced" inspection environment can be very difficult, if not impossible.

As a result, field-based NDT inspection of multi-layer composite structures is a very challenging problem for which the traditional NDT techniques do not offer a solution.

As a successful result of a recent SBIR (Small Business Innovative Research) project sponsored by the Lightweight Structures Team at TARDEC (Tank Automotive Research, Development and Engineering Center) and executed by Physical Acoustics Corporation (PAC), a portable, field-application technique to fully inspect multi-layer composite sections has been developed. This technology, based on oblique incidence angle, low frequency Acousto-Ultrasonics (AU) has been demonstrated to be capable of penetrating through different layers of multi-functional composite and has been used to detect the presence of delaminations in multi-functional composite armor plates (Ji et al. 1996).

In addition to the initially demonstrated application, this field inspection technology will find use in the inspection of many critical composite structures, such as military and commercial aircraft. Typical Air Force and Navy applications might be for inspecting thick-section polymeric composites. The technology will be especially applicable where other technologies cannot be utilized; such as for inspecting thick section (25–50 mm) composites in aircraft fuselage, wings and tail sections, helicopter composite rotor blades and Navy ship hull structures. PAC has already demonstrated the technique for inspecting other composite structures, such as composite wrapped high-pressure bottles, composite structural members, and composite wrapped concrete.

In this paper, a brief discussion on the wave propagation theory used to optimize the inspection parameters of the oblique incidence, low frequency AU technique for different applications will be presented. The development of an imaging technique based on AU, as part of a prototype portable system for field inspection of composite armored vehicles will also be addressed. Finally, results obtained in applications such as the inspection of FRP retrofitted civil structures, graphite-epoxy structures with foam core, and type III FRP Navy diving bottles will be presented.

2 THEORETICAL BACKGROUND

2.1 Theoretical model

The model used in this investigation assumed a generally anisotropic multilayered composite plate immersed in water, with a plane wave of frequency ω impinging upon the surface of the plate at an angle of incidence θ as shown in Figure 1 (Ji et al. 1999). The n layers of the composite are assumed to be rigidly bonded at their interfaces.

It is assumed that the propagation direction of guided waves in the composite plate is in the x_1 direction. Assuming plane wave front propagation, the displacements of the wave are given by

$$u_i = AU_i e^{j(k_i x_i - \omega t)} \tag{1}$$

with $i = 1, 2, 3$, A = amplitude, U_i = displacement vector, k_i = cosine components of the wave vector, and ω = circular frequency.

Based on the Thomson-Haskell transfer matrix method, the relationship between wave displacements, u_i, and stresses, σ_{ij}, at the top and bottom interface of the composite plate can be written as

$$X^t = TX^b \tag{2}$$

with

$$X^t = \begin{bmatrix} u_1^t & u_2^t & u_3^t & \sigma_{33}^t & \sigma_{23}^t & \sigma_{13}^t \end{bmatrix}^T$$
$$X^b = \begin{bmatrix} u_1^b & u_2^b & u_3^b & \sigma_{33}^b & \sigma_{23}^b & \sigma_{13}^b \end{bmatrix}^T \tag{3}$$

The calculation of the reflection and transmission coefficients for the composite plate immersed in water requires that the shear stresses at the upper, t, and lower,

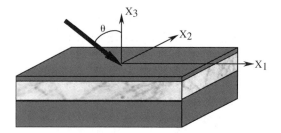

Figure 1. Theoretical model for wave propagation in a multi-functional composite plate. The polarization of the incident wave is longitudinal.

b, interfaces vanish and the normal displacements in the fluid, f, and in the surface of the plate are continuous. These boundary conditions can be written as

$$u_3^{ft} = u_3^t; \quad \sigma_{33}^{ft} = \sigma_{33}^t; \quad \sigma_{13}^t = \sigma_{13}^t = 0;$$
$$u_3^{fb} = u_3^b; \quad \sigma_{33}^{fb} = \sigma_{33}^b; \quad \sigma_{13}^b = \sigma_{13}^b = 0; \tag{4}$$

Considering the boundary conditions, equation (2) can be modified and written as

$$B^I = MA \tag{5}$$

with

$$B^I = \begin{bmatrix} 0 & 0 & 0 & 0 & u_1^I & \sigma_3^I \end{bmatrix}^T$$

$$M = \begin{bmatrix} -1 & 0 & t_{11} & t_{12} & 0 & s_1 \\ 0 & -1 & t_{21} & t_{22} & 0 & s_2 \\ 0 & 0 & t_{51} & t_{52} & 0 & s_5 \\ 0 & 0 & t_{61} & t_{62} & 0 & s_6 \\ 0 & 0 & t_{31} & t_{32} & -a_1 & s_3 \\ 0 & 0 & t_{41} & t_{42} & -a_2 & s_4 \end{bmatrix} \tag{6}$$

$$A = \begin{bmatrix} u_1^t & u_2^t & u_1^b & u_2^b & A_1^r & A_1^t \end{bmatrix}^T$$

Guided waves will propagate in the composite plate without external excitation if B in equation (5) equals zero for a non-zero solution of A, that is

$$\det(M) = 0 \tag{7}$$

When a solution for equation (7) using equations (5) and (6) has been found, the particle velocities can be obtained by taking the derivative of the particle displacement with respect to the time t.

$$v_i^\alpha = -j\omega u_i^\alpha \tag{8}$$

where α is the index for the guided wave modes, and i is the index for the propagation direction.

The stress field in the composite plate can be obtained from the generalized Hooke's law.

$$\sigma_{ij} = C_{ijkl}\varepsilon_{kl} \tag{9}$$

Using equations (8) and (9) the power flow vector for each propagation mode α can be calculated.

$$P_i^\alpha = \frac{1}{2} \sigma_{ij}^\alpha \upsilon_j^{\alpha*} \qquad (10)$$

Finally, the reflection and transmission coefficients of energy for mode α can be calculated using

$$c_e^\alpha = \frac{P_3^\alpha}{P_3^I} \qquad (11)$$

2.2 Calculation of reflection coefficient for composite plates with simulated delaminations

Calculations of the reflection coefficient of a multi-functional composite armor plate were carried out according to the theoretical model described by equations (1)–(11). It was assumed that the composite plate consisted of a 4 mm, 4-layered outside-composite cover of S2 fiber glass plain weave SBA 240 and vinyl ester 411 350, a 0.5 mm ground plane, an 18 mm AD 90 ballistic grade alumina ceramic, a 1.5 mm EPDM rubber, and a 22 mm, 24-layer inner-shell similar to the outside cover. Figure 2 shows a schematic of the simulated multi-functional composite armor.

Using the elastic constants, C_{ij}, of each of the composite layers as input to the theoretical model, the reflection coefficient was calculated for a composite plate without any defects. Similar calculations were carried out for a plate with simulated delaminations located at different depths through the thickness of the composite.

Figure 3(a) shows the reflection coefficient variation in the 10 to 20 degrees incident angle range for a 250 kHz fixed frequency. Figure 3(b) shows the variation of the reflection coefficient in the 240 to 300 kHz frequency range for a fixed incidence angle of 15 degrees. These calculations show that the reflection coefficient changes significantly due to the presence of a delamination. Moreover, the reflection coefficient as a function of the incidence angle and frequency changes significantly with the location of the delamination through the thickness of the plate.

2.3 Contrast Index calculation

As shown in Figures 3(a) and 3(b), the reflection coefficient depends on both the incidence angle and the frequency. That is, in its general form, the reflection coefficient is a surface in which every point has coordinates x = frequency, y = angle of incidence, and z = coefficient amplitude.

The z coordinate can be correlated to a value between 0 and 1 and assigned to a gray scale.

By comparing the reflection coefficient surface for a sample without a delamination with that corresponding to a sample with a delamination, the surface

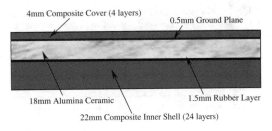

Figure 2. Layout of the multi-functional composite armor plate used in the theoretical simulation.

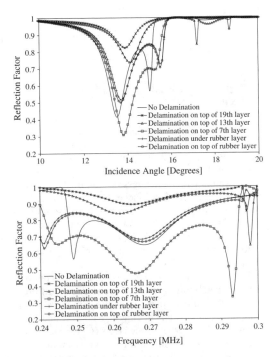

Figure 3. Calculated reflection coefficients for composite armor plate with and without simulated delaminations. (a) Reflection coefficient as a function of angle of incidence. (b) Reflection coefficient as a function of frequency.

corresponding to a parameter called "Contrast Index" (CI) can be generated (Godínez-Azcuaga et al. 2000a).

The CI of a delamination indicates the likelihood of that delamination to be detected by the change in amplitude of the reflected signal, in comparison with the reference signal reflected by the delamination-free surrounding areas of the sample. By examining the values of the CI as a function of incident angle and frequency for delaminations located at different depths in a composite plate, the combination of angle of incidence and frequency that will provide the best contrast, thus the best probability, for detecting the delamination can be determined.

119

Table 1. Optimum incidence angle–frequency combinations theoretically calculated for delaminations located at different depths in the armor plate.

Location of defect	Incidence angle [degrees]	Frequency [kHz]	Contrast Index
Top of rubber layer	15	190–210	0.80
Underneath rubber layer	15	190–210	0.80
Top of layer 7 in the inner shell	5	190–210	0.85
Top of layer 13 in the inner shell	20	260–290	0.93
Top of layer 19 in the inner shell	20	260–280	0.85

Figure 4. Field-portable Acousto-Ultrasonic prototype system for inspection of multi-functional composites. The system was developed under an SBIR Phase II project sponsored by the U.S. Army TACOM, Materials Research Branch.

3 PREPARATION OF SAMPLES FOR EXPERIMENTS

The structure of the samples used in the experimental verification of the theory is similar to the layout described in Figure 2, with exception of the ground plane, which was not present in these samples. Circular teflon wafers with a thickness of 0.04 inch and diameters ranging from 0.5 to 3 inches were used to simulate delaminations. Table 1 summarizes the best combinations of angle of incidence and frequency, theoretically calculated, for delaminations located at five different depths in the composite plate (Godínez-Azcuaga et al. 2000b).

It is important to note that in order to obtain optimal results when applying the AU technique to different composites, information about the structure of the composite (number and thickness of composite layers) and material properties of the layers (C_{ij} for each layer) is necessary. Using this information as input to the theoretical model, the best combination of frequency and angle of incidence for that particular composite can be obtained.

4 PROTOTYPE SYSTEM FOR MULTI-FUNCTIONAL COMPOSITES INSPECTION

Under an SBIR Phase II "Acousto-Ultrasonic Defect Detection in Composite Armor Material" project PAC developed a prototype system for the inspection of multifunctional composite armor plate. At the conclusion of this project, PAC delivered the prototype system shown in Figure 4 to the U.S. Army Tank-Automotive and Armaments Command (TACOM) in Warren, MI (Godínez-Azcuaga et al. 2000b).

This system contains a unique multi-sensor probe (MSP) with four pairs of pulser-receiver sensors embedded in specially designed wedges. This array of sensors provides four different incidence angles and a total of sixteen incident-reflected angle combinations. The wedges where the sensors are mounted, one containing the pulsing sensors and the other containing the receiving sensors, were fabricated of a special dry-coupling elastomer, which eliminates the need for any extra-coupling substance between the sensor and the sample. Only a fine mist of water is required to reduce the friction between the wedges and the composite plates during scanning.

The pulsing sensors are excited with PAC's Arbitrary Waveform Generator (AWG) Board capable of driving the pulsing sensors with either amplitude- or frequency-modulated signals. The signals are recorded with a PAC Acoustic Emission-Digital Signal Processing (AE-DiSP) board and PAC's Ultrawin imaging software is used to generate A-, B-, and C-scan images of the samples.

The multi-sensor probe is mounted on a computer-controlled, high resolution scanning bridge for C-scanning the composite samples. In addition to the high resolution scanning capability, the prototype system offers the option of using the MPS as a hand-held manual scanner in applications where it would not be practical to use the scanning bridge. For manual operation, the position of the MSP is continuously reported by a wireless Position Tracking System (PTS) developed especially for this project. The PTS consists of an electromagnetic transmitter, which sends pulses to a receiver mounted on top of the MSP so the AU signal captured at a particular point is associated with the recorded position. The prototype system then extracts the features of that signal, e.g. amplitude, and it associates them with the recorded position. By repeating this process over an entire section of the vehicle, an image of that section is generated.

5 EXPERIMENTAL RESULTS

5.1 *Simulated delaminations in multi-functional composite armor plates*

Figure 5 presents the results obtained from the inspection of a sample of multi-functional composite armor with simulated delaminations. The inner shell is critical since it provides structural support to the armor and therefore it is important to detect the presence of delaminations in the inner shell.

Figure 5 shows automatic and manual C-scans of three delaminations located inside the inner shell of the composite. Additionally, this figure shows the RF signals reflected from each of the simulated delaminations

The C-scans in Figure 5 were generated using different incidence angles. The sensors used in imaging the delamination located closer to the surface of the composite plate were oriented at 5 degrees, while the pair used for the defects located deeper inside the composite plate was oriented at 20 degrees.

It is important to point out that even after traveling through several interfaces the AU signal is capable of revealing the presence of a delamination and even providing a signal amplitude with contrast enough to produce a good image. This can be clearly seen from the RF signals.

5.2 *Fiber Reinforced Polymer "wraps" in concrete structures*

Figure 6 shows the results obtained when low frequency, oblique-incidence AU technique was applied to the inspection of structures retrofitted with Fiber Reinforced Polymers (FRP).

An image generated using the inspection system is shown in Figure 6. This figure shows the presence of two delaminations in dark gray color, in contrast to the non-damaged areas in light gray. Also, the images reveal the presence of the steel reinforced brackets or "stirrups" as strips in dark gray tones, located at 5 inches interval along the column length.

The column was cut in the locations where the images suggested the possible stirrup locations. All the stirrups were found within 0.25 inch of the cuts.

5.3 *Graphite-epoxy panels with foam core*

Another application in which the AU technique was used was in the inspection of graphite-epoxy composite with foam core. The composite structure was formed by two graphite-epoxy plates, approximately 0.25 inch thick separated by a hard foam insert 0.5 inch thick.

The areas inspected were large flat sections of a centrifugal arm of a racing car simulator and the main concerns were the presence of air bubbles or large resin

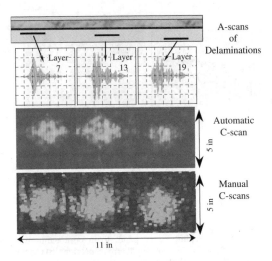

Figure 5. RF signals and C-scan images, automatic and manual, of circular simulated delaminations 2 inches in diameter located inside the inner shell of a multi-functional composite armor plate.

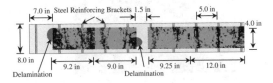

Figure 6. C-scans of a concrete column section reinforced with FRP. The images show the location of the delaminations and stirrups.

pockets in the graphite-epoxy plies, and disbonds between the foam inserts and the composite plates. Figure 7(a) shows a side view of a calibration panel with artificial defects that were prepared for this inspection. Figure 7(b) presents a C-scan of the panel, in which the inserted delaminations between the top composite plate and foam insert and between the foam insert and bottom composite plate are visible in bright gray color.

Also, the dotted box on the C-scan shows the area of the tapered transition between the foam insert and full-thickness graphite-epoxy composite plate.

On the right-hand side of this transition the C-scan shows areas that reflected high amplitude signals, most likely caused by lack of bonding between the composite plies (air bubbles) or areas with excess of resin. Finally, to the left-hand side of the seeded delaminations another area of high amplitude reflected signals can be seen. Due to the very regular pattern presented by this area, it was concluded that these indications were produced by weak bonding between the foam insert and the composite plate.

121

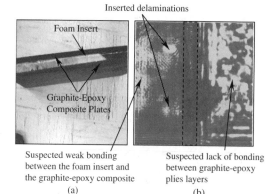

Inserted delaminations

Foam Insert

Graphite-Epoxy
Composite Plates

Suspected weak bonding
between the foam insert and
the graphite-epoxy composite

(a)

Suspected lack of bonding
between graphite-epoxy
plies layers

(b)

Figure 7. Graphite-epoxy panel with foam core. (a) Side view of the graphite-epoxy panel foam core calibration panel prepared for the centrifugal arm inspection. (b) C-scan of the calibration panel.

(a)

Damaged Area

(b)

Figure 8. Type III FRP with aluminum liner Navy diving bottle. (a) Inspection of the bottle using the AU technique. (b) C-scan of a 10 × 2 inches of the diving bottle showing the area damaged by an Instron machine.

5.4 Type III FRP vessels with aluminum liner

AU has also been applied to the inspection of FRP pressure vessels with aluminum liners as shown in Figure 8(a). These bottles were damaged by pressing a steel plunger on the FRP wrap using an Instron machine, which caused cracking of the FRP matrix.

The result of this inspection is presented in Figure 8(b) as a C-scan of the 10 × 2 inches area where the bottle was damaged.

It can be seen that the AU technique detected the damage. The image in Figure 8(b) shows an area that presented almost no reflected signal, dark tones, in contrast with the surroundings where high amplitude reflections are observed in lighter tones. The cause of the low reflectivity in the damaged area is the cracking of the matrix, which disperses the acoustic signals.

6 CONCLUSIONS

The oblique-incidence, low-frequency acousto-ultrasonic technique has been demonstrated to be a useful tool for the inspection of composite materials in general even though it was originally developed for the inspection of multi-functional composite armor. The prototype inspection system has been used to inspect composite structures in a variety of applications of composites. FRP "wraps" on concrete columns, graphite-epoxy composite with foam inserts, and FRP Type III vessels are among these applications. For each of these applications the inspection parameters, incidence angle and frequency, can be optimized. This is achieved by calculating the composite reflection factor according to the structure and material properties of the different layers of the composite.

The prototype inspection system can generate A-, B-, and C-scan images of the samples. These images are color-coded according to the amount of energy reflected by the composite at different points. This presentation of the recorded data provides not only detection but also position and approximate size of the defect area, and can be easily interpreted by the operator.

Currently PAC is commercializing a version of the acousto-ultrasonic inspection system known as T-SCOUT (Thick-Section Composite Oblique Ultrasonic Testing), with all the capabilities of the prototype system and the flexibility to be customized for particular applications.

REFERENCES

Ji, Y., Sullivan, R. & Balasubramanian, K. 1996. Guided Wave Behavior Analysis in Multi-Layered Inhomogeneous Anisotropic Plates. Review Progress in Quantitative Nondestructive Evaluation Vol. 15. Thompson, D.O. & Chimenty, D.E., Eds. Plenum Publishers, New York, pp. 217–222.

Ji, Y., Vahaviolos, S.J., Miller, R.K. & Raju, B.B. 1999. Acousto-Ultrasonic Evaluation of Hybrid Composites Using Oblique Incident Waves. Review Progress in Quantitative Nondestructive Evaluation Vol. 18. Thompson, D.O. & Chimenty, D.E., Eds. Plenum Publishers, New York, pp. 217–222.

Godínez-Azcuaga, V.F., Finlayson, R.D., Miller, R.K. & Raju, B.B. 2001a. Acousto-Ultrasonic Inspection of Hybrid Composite Armor Plate. Review Progress in Quantitative Nondestructive Evaluation Vol. 20. Thompson, D.O. & Chimenty, D.E., Eds. Kluwer Academic/Plenum Publishers, New York, pp. 1898–1905.

Godínez-Azcuaga, V.F., Carlos, M.F., Delamere, M., Hoch, W., Fotopoulos, C., Dai, W. & Raju, B.B. 2001b. Acousto-Ultrasonic System for the Inspection of Composite Armored Vehicles. Proceedings of the Twenty-Seventh Review Progress in QNDE. Vol. 20. Thompson, D.O. & Chimenty, D.E., Eds. Kluwer Academic/Plenum Publishers, New York, pp. 1890–1897.

Emerging Technologies in Non Destructive Testing, Van Hemelrijck, Anastasopoulos & Melanitis (eds)
© 2004 Swets & Zeitlinger, Lisse, ISBN 90 5809 645 9

Acoustic visualization and characterization of thin interface deterioration in adhesively bonded joints: theory and experiment

R.Gr. Maev

Dept. of Physics, University of Windsor, Windsor, Ontario, Canada

ABSTRACT: A typical problem with using standard time domain based C-scan images lies in the acoustic mismatch between the metal and the adhesive, when much of the pulse is reflected from the first interface and "rings" in the upper sheet. Little energy is passed into the adhesive; even less makes its way back from the second interface to the transducer. The resulting time domain signal is a long train of superimposed pulses, consisting, for the greater part of echoes in the upper metal sheet. To make this easier to interpret, this picture can be moved to the frequency domain, where it becomes much clearer. In this case, we will have only a few peaks, corresponding to the resonance peaks of the upper sheet, the adhesive and the lower sheet. The presence, absence or position of these resonance peaks allows one to easily analyse the existence, quality, and thickness of the different adhesives. The results demonstrate the possibility of combining resonance spectroscopy with 2D array technology to form simple and reliable images and are applicable to the evaluation of adhesion quality in real industrial environments.

1 INTRODUCTION

The present work represents theoretical analyses and an experimental study of an imaging technique for the quality control of adhesive bonding. A typical problem with using standard time domain based imaging (C-scans) lies in the acoustic mismatch between the metal and the adhesive. This means that much of the pulse is reflected from the first interface and "rings" in the upper sheet. Little energy is passed into the adhesive; even less makes its way back from the second interface to the transducer.

The resulting time domain signal is a long train of superimposed pulses, consisting, for the greater part, of echoes in the upper metal sheet. To make this easier to interpret, this picture can be moved to the frequency domain, where it becomes much clearer. There, we will have only a few peaks, corresponding to the resonance peaks of the upper sheet, the adhesive and the lower sheet. The presence, absence or position of these resonance peaks allows one to easily analyse the existence, quality, and thickness of different adhesives.

In order for the above resonance spectroscopy technique to be applied, it needs to be processed to allow imaging. This paper demonstrates the possibility of combining resonance spectroscopy with 2D array technology to form simple and reliable acoustic images.

The results given here are applicable to the quantitative estimation of adhesive quality in real industrial environments and the development of specialized ultrasonic equipment.

2 THEORETICAL BACKGROUND – SOUND PULSE PROPAGATION

Let us consider a simple model of the three layered specimen under testing with a flat contact ultrasonic transducer. Assume that the diameter of the transducer is large and the thickness of the layers is small enough to accept one-dimensional formulation of the problem (Fig. 1). Let Z_1, Z_2, Z_3, and Z_0 be the acoustical impedances of the materials of the first, second, third layers and the transducer, respectively. Denote impulse response of the ultrasonic system as $p(t)$, and let $h(t)$ be a response of the sample to the input $\delta(t)$ pulse, where $\delta(t)$ is Dirac Delta-function. In this case, the received waveform $s(t)$ can be expressed as a convolution:

$$s(t) = h(t) * p(t) \tag{1}$$

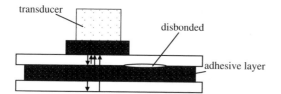

Figure 1.   Experimental set-up for three layer specimen study.

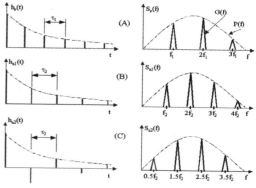

Figure 2.   The pulse sequence and its spectrum for disbonding at the first interface (A), additional sequence for good adhesion on both interfaces (B), and additional sequence for good first and bad second (C).

The impulse response of a single upper layer $h_s(t)$ is a set of exponentially decaying pulses:

$$h_s(t) = g(t) \cdot r_s^{t/\tau_1} \cdot \sum_{n=-\infty}^{\infty} \delta(t - n\tau_1) \qquad (2)$$

Here, $g(t)$ is the data window function: $g(t) = 0$ at $t < 0$ and $g(t) = 0$ at $t > T$, where T is the duration of the recorded waveform, $\tau_1 = 2d_1/c_1$, where d_1 and c_1 are the thickness and sound velocity in the first layer, respectively, and the factor r_s is a product of the reflection coefficients at the boundaries of the layer and the optional attenuation factor η_s:

$$r_s = \frac{Z_0 - Z_1}{Z_0 + Z_1} \cdot \frac{Z_2 - Z_1}{Z_2 + Z_1} \cdot \eta_s \qquad (3)$$

For a good adhesive joint, $Z_0 < Z_1$ and $Z_2 < Z_1$ (O'Neill et al., 2002 and Maeva et al., 2002). In the case of absence of the adhesive layer we should assume $Z_2 = 0$. Thus, in both these cases $r_s > 0$, and the impulse response $h_s(t)$ consists of positive, monotonically decaying spikes (Fig. 2A).

For the wave reverberating in the adhesive layer between two metal sheets, we write a similar formula:

$$h_{a1}(t) = q^2 \cdot g(t) \cdot r_a^{t/\tau_2} \cdot \sum_{k=-\infty}^{\infty} \delta(t - k\tau_2) \qquad (4)$$

Here, $\tau_2 = 2d_2/c_2$, where d_2 and c_2 are the thickness and sound velocity in the adhesive layer, q is the transmission coefficient for the upper metal sheet – adhesive boundary, and:

$$r_a = \frac{Z_1 - Z_2}{Z_1 + Z_2} \cdot \frac{Z_3 - Z_2}{Z_3 + Z_2} \cdot \eta_a \qquad (5)$$

Since $Z_2 < Z_1$ and $Z_2 < Z_3$, the factor $r_a > 0$, and again the response $h_a(t)$ consists of positive, monotonically decaying pulses separated by the time period τ_2 (Fig. 2B). If there is no bond at the second interface, $Z_3 = 0$ and, therefore $r_a < 0$ becomes negative:

$$h_{a2}(t) = q^2 \cdot g(t) \cdot |r_a|^{t/\tau_2} \cdot \sum_{k=-\infty}^{\infty} (-1)^k \delta(t - k\tau_2) \qquad (6)$$

In this case, the pulses of the response h_{a2} have alternating signs (Fig. 2C). Let us now calculate the spectra $S(f)$ of the output waveforms for these cases. According to the convolution theorem (O'Neill et al., 2003), for the first and the second cases we have:

$$S_{s(a1)}(f) = F\{s(t)\} = [G_{s(a1)}(f) * \sum_{m=-\infty}^{\infty} \delta(f - mf_{1(2)})] \cdot P(f) \qquad (7)$$

Here $f_{1(2)} = 1 / \tau_{1(2)}$:

$$P(f) = F\{p(t)\}; \quad G_s(f) = F\{g(t) \cdot r_s^{t/\tau_1}\};$$
$$G_{a1}(f) = q^2 \cdot F\{g(t) \cdot r_{a1}^{t/\tau_2}\}.$$

The characteristic form of the spectra $S_s(f)$ and $S_{a1}(f)$ are presented in Figure 2A, B. Thus, the spectra consist of equidistant peaks at frequencies $f_{1(2)}$, $2f_{1(2)}$, $3f_{1(2)}$, ... whose amplitudes are modulated by the transfer function of the transducer $P(f)$. Due to the monotonic behavior of the function $g(t) \cdot r_{s(2)}^{t/\tau_{(2)}}$ the width of the peaks $G_{s(a1)}$ can be estimated as $\Delta f \propto 1/T$. In the third case:

$$S_{a2}(f) = [G_{a2}(f) * \sum_{m=-\infty}^{\infty} \delta(f - (m+1/2)f_2)] \cdot P(f) \qquad (8)$$

Thus, for poor adhesive bonding at the second interface, the spectral peaks of the response h_{a2} are located at frequencies $0.5f_2$, $1.5f_2$, $2.5f_2$, etc.

We propose the following strategy for the NDE of adhesive joints. The spectral components detected at frequencies different from the harmonics of the upper sheet are evidence of the presence of good bonding at the first interface. Determining the frequencies of the peaks it is possible then to distinguish case (B) from (C) (see Fig. 2) and to characterize the quality of the second interface.

3 EXPERIMENTAL RESULTS – SINGLE TRANSDUCER MEASUREMENTS

For experimental verification of the proposed method, a set of 0.8 mm adhesively bonded steel samples was used. The thickness of the adhesive layer was 1, 0.6 and 0.4 mm.

The experimental waveforms were recorded with a single high-resolution contact probe (6 mm, 5 MHz Xactex transducer). The waveform and its spectrum measured for a steel plate without adhesive is shown in Figure 3.

Only two resonance peaks, the first and the second harmonics, are observed in Figure 2 because of the limited bandwidth of the transducer. The fundamental resonance frequency can be calculated as $f_0 = C_s/2d_s = 5.96 \text{ MHz}/1.6 = 3.73 \text{ MHz}$. The calculated value is in agreement with the experimental frequencies f_1 and f_2. Experimental spectra measured for 2 and 3 layer structures are shown in Figures 4–6. The 2 layer samples "steel–adhesive" represent joints with bad bonding at the second interface, and the 3 layer specimens "steel–adhesive–steel" represent good joints. The thickness of the upper and lower steel plates is 0.8 mm, and the thickness of the adhesive layer is 1, 0.6 or 0.4 mm.

Besides the strong first and second resonance peaks associated with the upper plate, the spectra in Figures 4–6 demonstrate additional weak peaks produced by reflections from the second interface of the adhesive layer. Values of the first four resonance frequencies calculated for the parameters of the specimens are listed in Table 1 and indicated on the graphs in Figures 4–6.

Comparison of the experimental data with the theoretical estimates shows that generally the values of the resonance frequencies represent the correct positions of the spectral peaks. Thus, measuring the vector $f = (f_1, f_2, f_3, \ldots)$ it is possible to classify the object under the test and estimate the thickness of adhesive. On the other hand, a practical implementation of the robust decision algorithm meets several problems.

Some spectral peaks may be small compared to the noise produced mostly by numerous reverberations inside the layered structure. Very often, the resonance peaks overlap and interfere with the powerful peaks produced by the upper sheet.

It appears also that the peak amplitudes are sensitive to the quality of the transducer–sample interface.

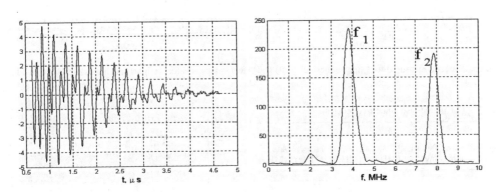

Figure 3. The waveform and its spectrum measured for 0.8 mm steel plate.

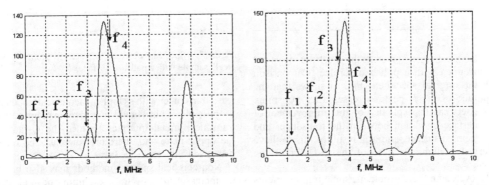

Figure 4. Spectra measured for "steel–adhesive" (left) and "steel–adhesive–steel" (right) specimens with 1 mm adhesive.

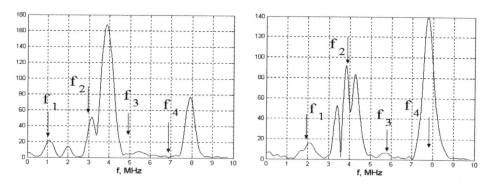

Figure 5. Spectra measured for "steel–adhesive" (left) and "steel–adhesive–steel" (right) specimens with 0.6 mm adhesive.

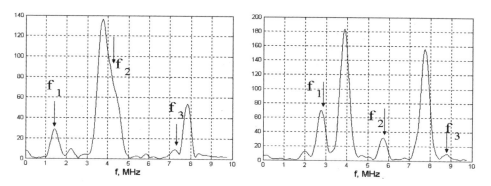

Figure 6. Spectra measured for "steel–adhesive" (left) and "steel–adhesive–steel" (right) specimens with 0.4 mm adhesive.

Table 1. Calculated resonance frequencies.

Adhesive layer thickness (mm)	No. of layers	f_1 (MHz)	f_2 (MHz)	f_3 (MHz)	f_4 (MHz)
1	2	1.15	2.3	3.45	4.6
	3	0.58	1.73	2.88	4.03
0.6	2	1.95	3.9	5.85	7.8
	3	0.98	2.93	4.88	6.83
0.4	2	2.9	5.8	8.7	11.6
	3	1.45	4.35	7.25	10.2

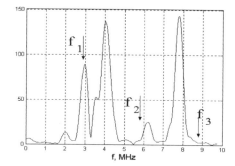

Figure 7. Spectrum obtained for "steel–0.4 mm adhesive–steel" sample with a polished surface.

For example, Figure 7 presents the spectrum measured for specimen with a polished surface. Comparison with the original specimen (Fig. 6) shows that the amplitude of the f_1 peak increases after polishing. This effect can be explained by the improvement of the interface condition between the transducer and the specimen surface, and as a result of this we have observed a slight decrease of the Q-factor of the signal.

To overcome these problems, some improvements of the method may be considered. It is advisable to use a transducer with as high an acoustical impedance as possible. This should improve matching at the transducer–metal interface, reduce the Q-factor of the upper plate and, therefore, increase the relative response of the adhesive layer.

Non-classical spectral methods may also be used to increase the spectral resolution beyond class-ical criteria and separate components with close frequencies. For example, the Prony method allows measurement of the amplitudes and attenuation factors of components with equal frequencies.

4   CONCLUSIONS

The acoustical data, obtained several places on the tested area can be combined into an acoustical image. The two most commonly used kinds of image – B-scan and C-scan represents the amplitude of sound in a fixed time domain. A similar method should be suitable for frequency domain representation as well. Sequenced switching of the measurement locations can be achieved by mechanical scanning of a single probe (as it realized in scanning acoustical microscopy) or by using an array transducer with electronic switching between channels.

The first method gives a good spatial resolution, especially in the case of using focused transducers.

The second method requires a specially designed probe, consist of a set of identical independent transducer elements with corresponding wiring and multi-channel electronics. This probe should overlap the whole testing area and its resolution is limited by the element dimensions. The main advantage in this case is portability of the system and its greatly shorter time of scanning. Successful realization of a 52 element matrix probe for weld quality inspection (Maev et al., 1999) stimulated further development of the technique.

REFERENCES

O'Neill B., Sadler J., Severin F., and Maev R. Gr., Theoretical and experimental study of the acoustical non-linearities at an interface with poor adhesive bonding, Acoustical Imaging, Vol. 26 (ed. by R. Maev), Plenum Publishers, pp. 309–317, 2002.

Maeva E., Chapman G., Severin F., and Maev R. Gr., Development of nondestructive method for ultrasonic evaluation of adhesive bond joint performance, Acoustical Imaging, Vol. 26 (ed. by R. Maev), Plenum Publishers, pp. 345–353, 2002.

O'Neill B., Maev R. Gr., Severin F., and Titov S., An array implementation of the resonance spectroscopy method for adhesive bond imaging, Acoustical Imaging, Vol. 27, Plenum Publishers, 2003 (in press).

Maev R. Gr., Ptchelintsev A., and Denissov A., Acoustic Imaging using matrix of piezoelectric transducers for nondestructive evaluation of plastic composites, Proc. of VI Annual Int. Conf. on Composites Engineering, pp. 665–667, June 1999, Orlando, Florida.

Emerging Technologies in Non Destructive Testing, Van Hemelrijck, Anastasopoulos & Melanitis (eds)
© *2004 Swets & Zeitlinger, Lisse, ISBN 90 5809 645 9*

Impact damage detection and evaluation in graphite epoxy motorcases (GEM)

Richard D. Finlayson, Marco A. Luzio & Valery Godínez-Azcuaga
Physical Acoustics Corporation, Princeton Junction, NJ

Athanasios Anastasopoulos
Envirocoustics A.B.E.E., Athens, Greece

Philip T. Cole
Physical Acoustics Limited, Cambridge, UK

ABSTRACT: There is a sizable need for composite health assessment measures for the detection of damage from impacts and other sources of high mechanical and thermal stresses. Even low-momentum impacts can lead to "barely visible impact damage" (BVID), corresponding to a significant weakening of the composite. This damage can in turn lead to sudden and catastrophic failure when the material is subjected to a normal operating load. The results presented in this paper contribute significantly to the understanding of the acoustic wave propagation properties of a graphite epoxy motorcase (GEM), and present a substantial advancement in the development of a damage source location algorithm. By performing an in-depth analysis of the behavior of travelling acoustic waves in the graphite epoxy structure, it was possible to understand how graphite epoxy properties influence acoustic wave propagation. An analysis of waveforms, attenuation and velocity measurements of different propagation modes, and analysis of Acoustic Emission (AE) features was performed.

1 INTRODUCTION

Physical Acoustics Corporation (PAC) has recently developed a monitoring technique to detect impact type damage in graphite epoxy motor cases for the US Air Force using Acoustic Emission (AE) technology. A special 8-channel digital signal processing (DSP) board was designed and built to accommodate the specific responses obtained from the acoustic properties of graphite epoxy motor cases. Also, software was developed to establish wireless communication between a wireless node (acquisition site) and a base station (controller site).

Acoustic Emission detection offers the potential to monitor damage in real time, to differentiate between micro-damaging mechanisms and to establish their spatial and temporal coordinates. Some of the AE characteristics of composites are:

- Very informative waveforms (advantageous for analysis) and large amounts of AE;
- Large increase in activity before failure;
- At higher stress levels, emission continues during load holds;

- Felicity effect* is a good indicator of prior damage;
- Abundant informative amplitude distributions;
- Friction at damaged surfaces is a major AE source.

As a result of these characteristics, AE is widely considered a very effective Non-Destructive Testing (NDT) method for assuring the structural integrity of composite material fabrications.

2 INVESTIGATION OF AE PROPERTIES

2.1 Introduction to AE properties

As a preliminary study to understand the acoustic properties of the graphite epoxy motorcase (GEM), a series of tests were performed. Due to the generally anisotropic nature of composite materials, these tests included measuring and evaluating the velocity, attenuation, and waveforms of acoustic stress waves as they propagated in different directions. These "AE"

---

*Felicity Effect (1977) – Specimens emit acoustic waves before reaching previous maximum load.

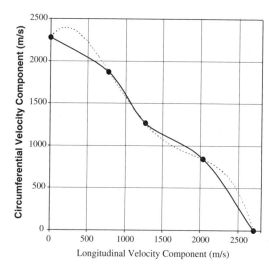

Figure 1. Coordinate plot showing the velocity components for various directions.

signals are collected using PAC's newest AE acquisition software, AEwin, that operates PAC's DiSP boards (AE digital signal processing boards), including the newly developed PCI-8 board that was developed for monitoring Air Force composite GEMs.

2.2 Velocity measurements in the GEM

Figure 1 shows data points representing the change in the velocity of the signal as a function of direction, taken from actual testing on GEMS. Velocity measurements were taken from the axial position on a cylindrical motor case (0°) moving towards the circumferential position of the cylindrical motor case (90°), with equal angle increments. Experimentally, this was achieved by simulating AE waves using the standard method of pencil lead break (PLB), impacting the material, or by sending a pulse from a waveform generator using an AE sensor as the pulser. The main information carrier in a stress wave is the acoustic waveform. Because AE sensors respond to the displacements on the surface of the material, or the "out-of-plane" component of the wave, flexural wave modes are the primary forms detected. The mechanical oscillations detected by the piezoelectric sensors are in turn converted into a voltage output directly related to the amplitude of the peaks. That is, the values of the displacement (mechanical variations) on the surface of the material are converted to voltage outputs, which are then expressed in AE decibels (dB_{AE}). Finally, based on the information from the results, an equation for velocity (speed as a function of direction) can be derived.

2.3 Attenuation measurements in the GEM

Another important factor that was investigated was how much a signal attenuates with respect to distance from the source as most composite materials have high attenuation properties.

The strength of the signal received can be described in terms of the voltage output decaying as a function of distance (i.e. distance from source site to detecting sensor). Attenuation generally follows this decay (theoretically exponential) of pressure created on the sensing surface of the transducer caused by the passing of the acoustic wave. Using this theoretical background, the AE features (e.g. energy, RMS, duration, amplitude, counts, etc.) collected from the tests conducted on the composite materials, were analyzed to establish an attenuation pattern characteristic for this specific material. In Figure 2 (A), this concept is illustrated by plotting the amplitude of the signals versus distance. Each x-axis marking corresponds to the distance from the sensor where the impacts occurred. Figure 2 (B) shows energy levels for impacts on graphite epoxy, taken at distances of 1 meter (black), 3 meters (light grey), and 5 meters (dark grey). The three distinct bin groups, 1, 2, and 3, correspond to impact energies of 3ft-lb., 5ft-lb., and 7ft-lb., respectively.

2.4 Waveform analysis

In addition to gathering and saving AE features, AE waveforms can also be stored for post-test analysis using a special utility of the software.

AE Waveforms give information on the acoustic signal in terms of time and amplitudes.

In order to perform frequency content analysis, the Fast Fourier Transform (FFT) is used to extract the signal's frequency content from the time domain. During the analysis the data from GEMs frequency information was helpful in determining if damage was present on the structure. Figure 3 shows waveforms obtained from GEMs, and the FFTs of the corresponding signals for three different impact energies. It is clear from Figure 3 that there are substantial differences observed in both the amplitude of the waveforms as well as the amplitude and frequency response of the FFTs, making it possible to establish a difference between various impact energy levels. Besides containing the information necessary for source location purposes, that is, time of arrivals, there are several acoustic features that can be extracted from a waveform. These characteristics are intrinsic to the waveform and can reveal important and useful information pertaining to the signal.

2.5 AE features

Some of the most useful AE features to analyze from the GEM data are shown in Figure 4.

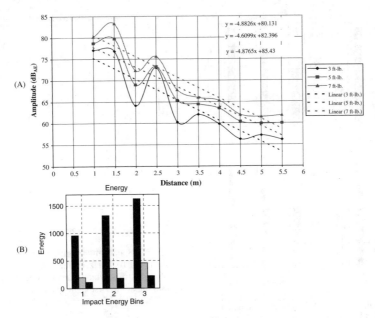

Figure 2. (A) Attenuation plots for three different impact energy levels. The amplitudes are given in AE decibels. From the slope of the line it is possible to obtain a quantitative value for the damping of the signal as a function of distance. (B) Energy histograms. The x-axis represents the impact energy bins, with 1 corresponding to 3 ft-lb., 2 to 5 ft-lb., and 3 to 7 ft-lb.

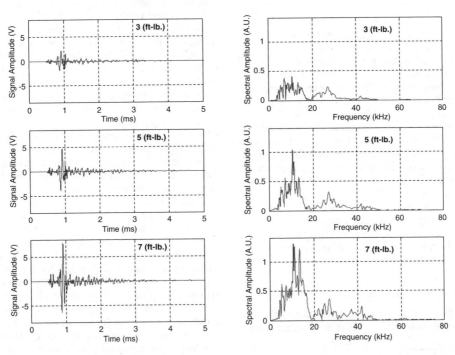

Figure 3. Sample waveforms and FFTs for 3 different impact energies (3, 5 and 7 ft-lb.) taken in the longitudinal direction at a distance of 1 meter from a sensor. The data corresponds to the response of a Commercial Off the Shelf (COTS) sensor (operating range: 30–55 kHz).

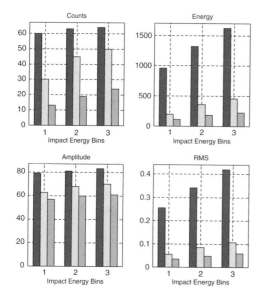

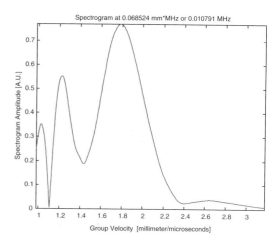

Figure 5. Spectrum amplitude versus group velocity for a frequency of 10.7 kHz. Highest peak corresponds to a velocity of ~1800 m/s, which matches well the velocity calculated by looking at the first significant motion of the waveform.

Figure 4. Histograms for four AE features, counts, energy, amplitude, and RMS. The x-axis represents the impact energy bins, with 1 corresponding to 3 ft-lb., 2 to 5 ft-lb., and 3 to 7 ft-lb. The shades of grey correspond to the distance dark 1 m, light 2 m and medium 3 m.

These data characteristics will be used in developing a method to classify[†] impact energy and to establish a relation to the distance from a source.

Figure 4 illustrates the differences for each AE parameter, for the three impact energy levels, and for three different distances between source and sensor. It is evident that the amplitude levels of the AE features increase as a function of both energy and distance of impact. Using this type of information, simultaneously with waveform analysis, limiting levels of non-damaging impact energies can be established. For the GEMs tested, it was found that attenuation patterns are the same and the slope (attenuation coefficient) variation remains within 10% difference regardless of the impact energy used. This is very useful in determining sensor spacing and maximum separation to achieve reliable detection. Once the separation distance between sensors is known, a calculation of the numbers of sensors to be used can be performed.

2.6 Lamb wave study (modal)

The study of Lamb waves is an effective method for damage detection and to evaluate material conditions.

A Lamb wave is an elastic perturbation that propagates in solid, unbounded plates (or layers of some media). In a plate of a certain thickness and at a specific frequency there can be a finite number of symmetrical and anti-symmetrical Lamb waves which differ amongst themselves by their phase and group velocities, as well as by the displacements and stresses throughout the plate. Figure 6 shows a spectrogram for experimental data taken during the studies of the GEMs in the Air Force program.

One of the approaches to consider in "modal" analysis is to focus on the arrival times of the extensional and/or flexural wave modes. The S_0 wave mode (extensional) speed is faster than the A_0 wave mode (flexural) at lower frequencies, which is the frequency range of interest for composite materials. Using this theory and the measured time differences (Δt) between the wave arrival at the sensors, the distance from source to sensor can be calculated. By repeating this calculation for each of the sensors, a group of Δt's is obtained. By using the velocities measured experimentally for the composite structure, and comparing arrivals at different sensors, the position of the source of emission can be estimated. Analyzing the frequency spectrum will allow the development of an algorithm based solely on the main frequency component and its' corresponding velocity. Figure 5 shows a "cross-section" of the dispersion curve (spectogram shown in Figure 6) at approximately 10.7 kHz. This third dimension, amplitude or energy content, is graphed versus group velocity. It can be seen that the highest peak in Figure 5 (~550 ms) corresponds to a group velocity of just under 1800 m/s (Figure 5), again in agreement with experimental measurements.

[†]Classification of impact energy levels and/or other quantities of interest can be performed using P Neural Network Artificial Intelligence software package (NOESIS).

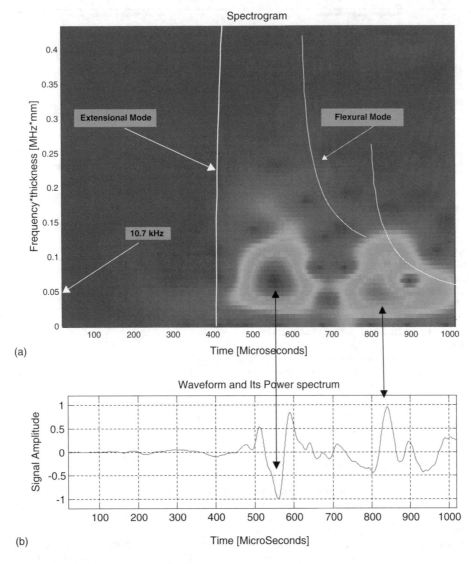

Figure 6. (a) Theoretical dispersion curves for the extensional and flexural modes, superimposed onto experimental data. Dark arrows indicate areas of higher amplitudes. (b) Waveform corresponding to this specific data.

2.7 Active and passive monitoring

For composite motor case damage monitoring, it is possible to use of both passive (no pulser) and active (pulser) monitoring. There are advantages to both of these methods. In the "passive" mode (discussed so far), sensors (removable) are placed on the structure to detect emission from damage, ranging from impact damage to defects caused by stress or overweight. This "*passive*" mode uses the time of arrival to locate an impact/defect, and measures the magnitude of the signals received.

Some of the advantages of using the passive mode include; monitoring of the structure during transportation, sensing impacts and defects during movement, and continuous monitoring during unattended storage periods. During "*active*" mode monitoring, a sensor (pulser) transmits a pulse (Lamb wave) generated by an electronic waveform simulator device. Other sensors will then detect the pulse transmitted and the AE system will analyze the digitized waveforms and resultant FFTs (Figure 7) to determine the severity of the damage compared to a calibrated baseline.

135

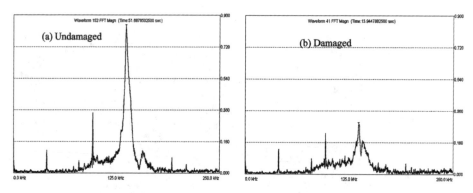

Figure 7. Fast Fourier Transforms (FFT) for (a) a signal passing through a good section on the GEM and (b) a signal passing through a damaged section on the GEM.

Advantages of using the active mode include; ability to inspect a suspect damage site on the structure compared to a previously established base line, periodic checks of motor case integrity, and quick preliminary assessment of type and magnitude of damage. Figure 7 shows some results obtained using active monitoring. The FFTs clearly show differences between an undamaged and a damaged area on a GEM. The FFT on the left was calculated from data taken in a region before introducing damage, and the one on the right shows the FFT for that same exact region after damage was introduced.

3 WIRELESS TRANSMISSION CAPABILITIES

To continuously and wirelessly monitor GEM structures, remote-monitoring capabilities were developed for the US Air Force under the program "Continuous Health Monitoring of Composite Structures" (Finlayson, 1999). The entire monitoring system developed is composed of two main sections, a mobile station with a wireless node, and a base station (see Figure 8). The mobile station is the front end of the system used for data collection, and consists of an acoustic emission system (PAC's 8-channel DSP board) data acquisition software (PAC's AEwin), and other hardware, such as sensors. The wireless node is fitted on the composite structure at all times when monitoring is required. It can collect, process, store and transmit data to the base station. Alarms, due to impact or high intensity acoustic emission, can be set to be triggered at the base station alerting an operator of a potential damaging situation. Also, files containing information about the monitoring tests can be reviewed numerically and/or graphically. Finally, a summary of the conditions of the structure can be uploaded to a monitoring web site which can be accessed from any computer (provided a valid user and password is entered) in order to check on the status of the structure.

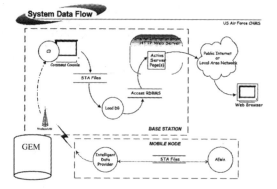

Figure 8. System data flow developed for the Air Force composite monitoring program. The GEM has sensors attached for signal (data) collection.

4 CONCLUSIONS

In this paper we have shown that by performing an in-depth analysis of the behavior of travelling acoustic waves in a GEM structure, it is possible to understand how graphite epoxy properties influence acoustic wave propagation. An analysis of AE features was performed by measuring and evaluating the velocity, attenuation, and waveforms of acoustic stress waves as they propagated in different directions. We also showed that a GEM can be monitored/evaluated with both passive and active sensing systems. From this work we have developed a prototype hardware and software system to detect impact type and classify damage in the GEMs and to transmit the results wirelessly.

REFERENCE

Finlayson R.D. (Major). Continuous Health Monitoring of Composite Structures (CHMCS). Air Force contract F33615-99-C-3800, PAC # R00-704, September 1999.

Emerging Technologies in Non Destructive Testing, Van Hemelrijck, Anastasopoulos & Melanitis (eds)
© 2004 Swets & Zeitlinger, Lisse, ISBN 90 5809 645 9

Gear defect diagnosis using acoustic emission

T. Toutountzakis & D. Mba
School of Engineering, Cranfield University, UK

ABSTRACT: It is widely recognized that Acoustic Emission (AE) is gaining ground as a Non-Destructive Technique (NDT) for health diagnosis on rotating machinery. The source of AE is attributed to the release of energy within and/or on the surface of a material due to micro-structural changes such as the friction process. The released energy manifests itself in the form of AE that propagate in all directions on the surface of the material. This paper presents an experimental investigation on the application of acoustic emission for gear defect diagnosis. This research program was centred on an experimental test rig onto which known defects were seeded. Furthermore, the possibility of monitoring defects from the bearing casing is examined. It is concluded that AE offers a complimentary tool for health monitoring of gears.

1 INTRODUCTION

With increasing demands for high production and high level of quality, condition monitoring techniques for life prediction and failure diagnosis are in great demand for many engineering applications. A common way to increase productivity is to monitor large and expensive machinery in order to reduce equipment down time and reduce, or illuminate, catastrophic failures.

Visual inspection, noise/vibration analysis, lubricant debris analysis and ultrasonic measurements are some widely used methods for condition monitoring of rotating machinery. Over the last few years' attempts have been made to employ Acoustic Emission (AE) technology for condition monitoring of gears with limited success. AE method is widely used in the petrochemical industry, and for condition monitoring large structures such as motorway bridges (Hamstad et al, 1998), pipelines and pressure tanks.

2 REVIEW

The U.S. Army in collaboration with Boeing has funded AE research projects since 1975 for the prediction of bearing failures in the CH-47C Chinook Medium Lift Helicopter. The AE method proved to be advanced to the vibration method for diagnosis and prognosis (Board, 1975). Since then, a few attempts have been undertaken to apply AE technique, to condition monitoring and diagnosis for gears.

Spencer (1988) and Yaghin (1991) both concluded that the AE signal can be used for the detection of faulty gears.

Tandon and Mata (1999) tested spur gears with 15 and 16 teeth in a back-to-back gearbox. The speed of the motor was kept constant at 1000 rpm while the load varied during the test. A jet oil lubrication system was also used.

A simulated pit of constant depth (500 μm) and variable diameter (250/350/450/550/1100 and 2200 μm) was introduced by spark erosion.

The authors concluded that the AE parameters increased as the defect size (diameter of pit) and the load increased. The AE measurements indicated a sharp increase when the defect size was around 500 μm in comparison to the vibration data where a distinct increase occurred when the defect size approach 1000 μm. This indicates earlier detection of gear defects with AE than vibration.

Al-Balushi and Samanta (2002) used a back-to-back gearbox (using spur gears) for comparison reason between statistical techniques and several energy-based methods. The applied load was varied during the course of the test (40 hours), using a set of coupling flanges.

Levels of relative energy were compared with statistical methods (Kurtosis etc) for diagnosis. More specifically, square root of energy index (rEI), cumulative energy index (cEI) and cumulative square root of energy index (crEI) were computed. The results were also compared with vibration data for a helicopter gearbox. The author suggests that AE technique is advantageous to the vibration method as far as identification of the location is concerned.

Singh et al (1996) used two different types of gearboxes in order to investigate early pitting. For both set-ups a resonant type AE sensor with resonant frequency 280 KHz was used. In both set-ups an accelerometer was also used for comparison between AE method and vibration method.

The first set-up was a "Generator Drive Offset Quill" which is used on some UH-1 helicopters. The gear ratio was 41:55 and the speed was maintained constant at 1400 rpm. A simulated pit was introduced at the pitch line (using EDM) in just one tooth, with dimensions 1.25 mm depth and 1.25 mm diameter.

The second set-up was a back-to-back gearbox with gear ratio 1.5 (42/28 teeth). In this set-up no fault was introduced. Pitting started to appear after about half an hour.

Amplitude measurements were used for comparison between AE and vibration technique. The authors concluded that both methods can help the detection of pitting in gears. However AE method has an advantage if early detection is the target.

In most of the above experiment the fault was placed on the pitch line with no justification.

In all the cases reported the AE technique was found to detect gear defects earlier than Vibration analysis.

3 EXPERIMENTAL SET-UP

3.1 *Test rig and gears*

The experimental test rig consisted of a back-to-back gearbox with a set of two identical gears. The gears had a module of 3 mm and were made of 045 M15 steel with no heat treatment or case hardening. The pinion and the wheel had 49 and 65 teeth respectively. Both gears had a standard 20° pressure angle and a surface roughness 2–3 μm. Four identical ball-bearing were used to support the gearboxes shafts. The gearbox, figure 1, was oil bath lubricated and driven by a 2.2 KW DC motor with maximum speed of 2820 rpm. The speed of the motor was controlled by an electronic speed controller.

3.2 *Acquisition system*

A two-channel MISTRAS AEDSP-32/16 board made by Physical Acoustic Corporation was used for data acquisition with a capability to record up-to to 256000 points. A resonant type AE sensor with flat response in the area between 150 KHz to 750 KHz was used. The sensor was attached on the ball-bearing casing of the shaft carrying the driven wheel, see figure 1. The sensor was held in place with a metallic fixture.

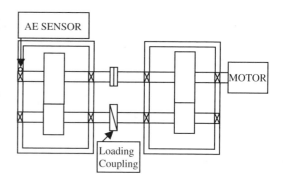

Figure 1.   Diagram of the gearbox.

The preamplifier used was made by Physical Acoustic Corporation (model 2/4/6). It has an integrated filter between 100 KHz to 1200 KHz. The output signal could be amplified by 20 dB, 40 dB or 60 dB. For this experiment a 60 dB amplification level was employed.

The signal from the AE transducer was sampled at 2.0 MHz, which in combination with the capability of the board gave a total recording time of 0.128 sec. This time represented 1.3 revolutions for the driving wheel. The gearbox was run for about one hour prior to each test. For each speed case, the acquisition system was armed three times at one hour intervals. Another hour was allowed in-between the change of the speed for the gears to settle down. So the total time required for the test was approximately 10 hours.

4 ATTENUATION TEST

Attenuation tests were undertaken on the test rig by breaking a lead pencil in accordance to the British European Standards BS-EN 1330-9:2000. The lead pencil had the following specifications: diameter 0.5 mm, length 3 mm, 2 H.

Prior to the testing, the lead pencil was broken at first next to the sensor for reference purpose. Then the pencil was broken again at several positions in order to calculate the degree of attenuation. All calculations were based on the initial reference tests. The signal became weaker passing through several interfaces as it can be clearly being seen in figure 2.

In addition to the standard gears of the gearbox two extra gears were placed on top of the driving wheel gear, in order to observe the attenuation level across more interfaces and to monitor the sensitivity of the sensors (figure 3).

Results of attenuation tests are detailed in figure 2. It can be seen that the attenuation level for the driven wheel is greater than the driving wheel, as the driving wheel is positioned further away from the sensor than

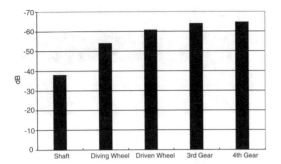

Figure 2.   Attenuation across different interfaces.

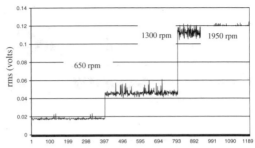

Figure 4.   Background noise for three speeds.

Figure 3.   Attenuation test using extra gears.

the driven wheel. However is worth noting that the levels of attenuation from the driven wheel, 3rd and 4th gears are similar. The attenuation results were very encouraging and vindicated the applicability of AE to gear defect diagnosis.

5   BACKGROUND NOISE

During the test, the AE background noise levels for defect free gears were recorded, at three different speeds, 650 rpm, 1300 rpm and 1950 rpm. The load for all the speed was kept constant during all experiments.

Figure 5.   Pit faults on the driving wheel tooth indicated by arrows.

Figure 4 shows the rms values recorded from the sensor for three speeds. Every point on the x-axis is equivalent to 0.016 sec of data, however, these readings were taken over three 15 minute intervals for every test speed. It can be clearly seen that as the speed increased the AE rms values increased.

6   PITTING

Hatton stated that the most likely area to suffer from pitting is the dedendum of a gear tooth. This is because in that areas of the gear tooth profile, sliding occurs in contrast to the pitch line where pure (theoretical) rolling is taking place during gear meshing.

A defect was introduced into the gear tooth (wheel) in order to simulate pitting, which is one of the major fatigue problems associated with gears and can potentially lead to a fatigue failure (Myers & Shaw, 2002). A small handheld enscriber was employed and the pitting was placed on the addendum of the gear. The addendum area was chosen due to the small module of the gear (3 mm).

Fatigue can be defined as the repeated application and removal of load. By definition, gear tooth are subject to fatigue loads as the gear meshes.

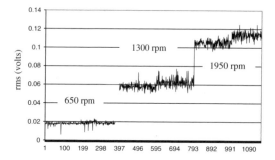

Figure 6.   AE signal after pit generation.

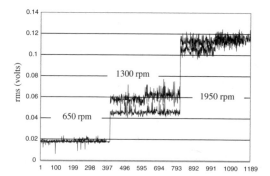

Figure 7.   Background noise and AE signal after pit generation.

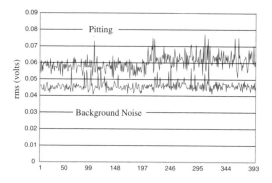

Figure 8.   Background noise and pitting signal for 1300 rpm.

Figure 6 represents the AE signal (rms value) after the simulated pit was introduced on the gear teeth. It is clear that, as with the background noise, the AE rms value increased as the speed increased. The two signals, background noise and detect signal were of very similar levels, see figure 7, with a distinct deference only in the speed of 1300 rpm (figure 8).

As the resonant frequency of the gearbox was around 1200 rpm, it is anticipated that the increased excitation

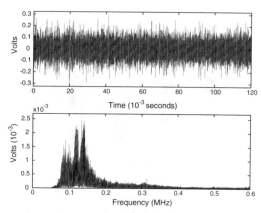

Figure 9.   Noise signature with frequency response.

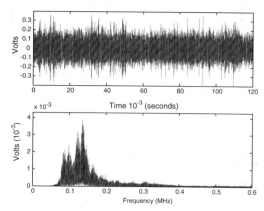

Figure 10.   Defect signature with frequency response.

emphasized the difference between the background and defect AE signatures.

Figures 9 and 10 shows the frequency response for background noise and a defect signature response. The frequency spectra response for both signals are very similar as can clearly be seen, and thus does not provide any additional information.

7   CONCLUSION

The transmissibility of the AE signal and the ability of the sensor to detect the AE signal passing through several interfaces has been established. It is a great achievement that the signal can be detected after travelling through many interfaces. This could have a big impact in condition monitoring for more complex gearboxes. Nevertheless further and more intense investigation is needed in order for the AE method to be well established. Although the results obtained are limited to one speed, it forms the basis for further investigation. This could be the subject of future reports.

REFERENCES

Al-Balushi K.R. & Samantha B. Gear Fault Diagnosis using Energy-Based features of Acoustic Emission Signals, *Systems and Control Engineering*, Vol. 216(I), 2002.

Board B.D. Incipient Failure Detection in CH-47 Helicopter Transmissions, (1975), *Winter Annual Meeting*, Houston, Texas, 30 November–4 December 1975.

European Standards. Non-destructive testing Terminology-Part 9:Terms used in acoustic emission testing, EN-1330-9:2000. January 2000.

Hatton D.R. *Gear Design Course,* Cranfield University and British Gear Association, Cranfield 2003.

Holroyd J. Trevor. *The Acoustic Emission & Ultrasonic Monitoring Hand book,* 1st ed. 2000.

Myers J. & Shaw B. *Gear Wear and Failure Recognition,* British Gear Association, 2002.

Spencer V. The detection of Root Faults in Gear Teeth by Acoustic Emission, MSc Thesis, Cranfield University, 1988.

Tandon N. & Mata S. Detection of Defects in Gears by Acoustic Emission Measurements, *Journal of Acoustic Emission,* Vol. 17, Num. 1–2, 1999.

Yaghin D.L. Condition Monitoring of Gearboxes using Acoustic Emission Techniques, MSc Thesis, Cranfield University, 1991.

Emerging Technologies in Non Destructive Testing, Van Hemelrijck, Anastasopoulos & Melanitis (eds)
© 2004 Swets & Zeitlinger, Lisse, ISBN 90 5809 645 9

On the characterization of acoustic emission activity generated by continuous fibres breakage

Y.Z. Pappas, A. Kontsos, T. Loutas & V. Kostopoulos
Applied Mechanics Lab, Dep. of Mechanical Engineering & Aeronautics, University of Patras, Greece

ABSTRACT: The characterization of the elastic waves generated during mechanical tests of commercially available fibres is the aim of the present work, using the Acoustic Emission (AE) and Acousto-Ultrasonic (AU) methods. To this target, a large number of quasi-static tensile tests were conducted on Organic (Kevlar and Polyethylene), Ceramic (Al_2O_3, SiC and Glass) and Carbon fibre bundles according to DIN 53942. An in-house developed analysis and classification methodology of the captured AE and AU waveforms is proposed for the creation of a useful database, concerning AE features ranges and representative frequency values for fibre failure identification.

1 INTRODUCTION

The use of several non-destructive inspection methods has proved to be a valuable and reliable tool for the structural integrity characterization and the damage evaluation of complex material systems and structures, under different loading conditions. In the case of continuous fibre reinforced composites, the failure mode of fibre fracture (single fibres as well as fibre bundles) is strongly correlated to the final failure of structural component made by composites (ASM 1987). For these materials, it is the integrity of the reinforcing phase that determines the ability of the structure to withstand the external loads. Thus, the ability to identify the damage of fibre bundles during the loading stage could provide valuable information concerning the remaining life of the composite structure.

In general, the genesis of this type of destructive event is accompanied by the release of considerable amount of elastic energy; part of it (AE) propagates through the other intact fibres of the reinforcement and/or through the surrounding matrix of the composite structure. The identification of an electrical signal, captured by a piezoelectric sensor, as the result of reinforcing fibre breakage offers great advantages in the health monitoring process of the overall composite structure. However, this result could not be achieved using a straight-forward approach, since the captured electrical signals (waveforms) contain information not only for the initiation event but also for the medium that propagates the elastic energy, the propagation path (Achenbach 1973), the statistical nature of the physical phenomenon and the used acquisition

system. Thus, any effort to create useful knowledge about the characteristics of the elastic waves that correspond to fibre failure should be referred to well-controlled testing conditions, targeting to the core of the problem, which is the initiation event under a statistical and multi-dimensional analysis scheme.

In the present work, the characteristics of the elastic waves generated as AE activity by the failure of different fibre types are classified, in time and frequency domain. A large number of quasi-static mechanical tests were conducted on commercially available and extensively used fibre bundles, such as Al_2O_3, Carbon, SiC, Kevlar, Glass and Polyethylene, utilizing a special device according to DIN 53942. The applied classification of AE activity leads to the development of a database concerning time and frequency "signature" of each fibre type failure, justifying a new methodology for the determination of the initiation event characteristics. Furthermore, the information gained from each test by the use of the AU method was evaluated in order to identify the propagation medium characteristics, adopting useful approaches of de-convolution methodologies.

2 PROPOSED METHODOLOGY FOR DATABASE DEVELOPMENT

The development of a database containing values of characteristic features of AE events for many damage phenomena is not a new issue, since many researchers provide this kind of information for specific testing conditions, material structure and damage modes.

However, a useful discussion concerning this aspect is whether these "numbers" can be used to evaluate the response of the same material in different testing conditions and structures. Thus, the real question that arises is whether the developed database created by information extracted from a micro-scale approach (fibre, mini-composite tests) is applicable to the macro-scale (plates, big structures, etc.) and vice-versa (Hamstad et al. 1986, Giordano et al. 1998, Pappas 2001, Ni et al. 2002).

Probably, the most effective methodology to solve the stated problem is to work on the captured waveforms, in time and frequency domain, assuming that, in micro-scale, matter vibration relates to eigenmodes, while all the classical frequency domain analysis techniques (Fourier transform, dispersion curves, etc.) are applicable. This approach is called modal AE and is based on the assumption that the most dominant eignemodes of the microstructure are present during AU excitation and AE activation (Prosser 1995).

An important issue in the analysis of the AE waveforms and the construction of a valid and useful database is the ability to characterize efficiently the initial event of each captured waveform. In the case of fibre bundle testing, initial events could be single fibre failures, sliding friction between fibres, etc. However, during the propagation of an emitted elastic wave from the source to the sensor and then to the monitoring system, a number of external factors modify its characteristics, in time and frequency domain. These factors are the propagation medium and the acquisition system. Thus, in order to create a valid database for the evaluation of each capture waveform, it is important to know the contribution of the acquisition system and the propagation medium in the total response. In addition, it is crucial to have a set of features/parameters that could help to identify the "signature" of specific damage mechanisms and thus to assist the evaluation of the captured AE waveforms. For these reasons, de-convolution methodologies are usually applied in order to determine the characteristics of the system and the propagation medium, and to separate this knowledge from the source event contribution (Qi et al. 1997, Giordano et al. 1998, Ageorges et al. 1999). Alternatively, the analysis effort could target at the frequency domain, in order to obtain a representative set of dominant frequencies that corresponds, with a high probability, at specific damage mechanisms (Russell et al. 1977, Clough et al. 1981).

In the present work, a new methodology (Fig. 1) for the development of a reliable database concerning AE characteristics of fibre failures, in time and frequency domain, is proposed, combining both presented approaches. This methodology is based on the assumption that there is a distinct fibre failure "signature". For the evaluation of this "signature" in time domain, various techniques that quantify the information

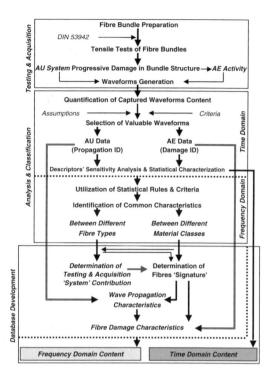

Figure 1. A schematic representation of the applied methodology for the analysis and classification, in time and frequency domain, of the captured AE and AU waveforms, during fibre bundles quasi-static tensile tests.

extracted from AE waveforms, such as AE characteristic parameters histograms and multi-dimensional projections, as well as advanced signal processing methodologies are proposed. For the determination of the filtering effect of the propagation path on the source event, the use of AU method is proposed.

3 TESTING PROCEDURE

3.1 Material

In order to investigate the role of some material parameters, such as fibre structure, anisotropy level and fibre mechanical properties, on the characteristics of the emitted elastic waves during tensile loading, a group of different fibre types were tested. The most important physical and mechanical properties of these fibres are shown in Table 1 (ASM 1987).

3.2 Testing conditions

For the investigation of the failure modes of the selected fibre types, a number of fibre bundles quasi-static tensile tests were performed. All tensile tests were conducted utilizing the special instrumentation shown

144

Table 1. A list of physical and mechanical properties of the tested commercial fibres, together with the used coding.

Fibre type	ρ (g/cc)	d (μm)	E_{11} (GPa)	UTS (GPa)	Fibers #	Code
Nextel 312	2.70	12	150	1.70	400	H
Nextel 440	3.05	12	190	2.00	400	J
Nextel 610	3.88	12	373	2.93	400	K
Nextel 720	3.40	12	260	2.10	400	A
Altex SN-11	3.30	15	210	2.00	1000	B
S-Glass	2.48	20	87	4.89	6000	E
Hi Nicalon	2.74	14	270	2.79	800	T
Carbon M40-B	1.81	7	392	3.90	3000	P
Carbon M40-J	1.77	7	377	4.41	6000	S
Kevlar 29	1.44	12	83	3.60	1000	L
Spectra PE-40	0.97	6	101	3.08	120	C
Spectra PE-30	0.97	10	73	2.57	120	D

ρ: volume density, d: fibre diameter, UTS: Ultimate Tensile Strength.

Figure 2. Experimental set-up for fibre bundle tensile tests (DIN 53942), together with AE and AU sensors positioning.

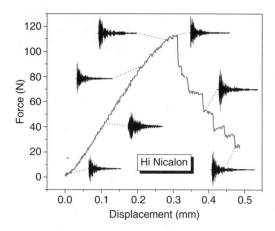

Figure 3. A typical mechanical response plot for fibre bundle tensile test, together with representative AE waveforms corresponding to several loading stages.

To capture waveforms, a 4 MHz acquisition frequency and a 4 k-sample signal size were applied.

4 RESULTS AND DISCUSSION

4.1 Mechanical response and non-destructive evaluation

The applied testing program was based on the dry-bundle approach, which enables the stimulation and identification of single fibre failure events (Hamstad et al. 1986, Cowking et al. 1991, Jihan et al. 1997). However, the use of dry-bundle testing conditions can trigger several other damage mechanisms (Hamstad et al. 1986, Cowking et al. 1991). In Figure 3, a typical mechanical response plot of Hi Nicalon fibre bundle tensile test is presented (0.05 mm/min), together with representative/repeatable captured AE waveforms.

According to the resulting force–displacement plots, it can be stated that the displacement rate is a critical testing parameter that affects the mechanical behavior of different types of fibres. For example, testing of ceramic fibres (like Hi Nicalon) requires low rates in order to achieve well-controlled force–response plots (continuous increase until maximum force and then multiple decrease steps corresponding to multiple fibre failures), instead of sudden total failure after the maximum force. However, in the case of other fibres like Polyethylene, high displacement rates are suggested for reliable results, due to experienced high strain to failure. As long as the non-destructive evaluation procedure is concerned, it can be noted that, in most of the tensile tests, AE activity was continuous (Pappas 2001). However, there was no evidence of dependence between the AE activity and the displacement rate, while the type of the tested

in Figure 2 (DIN 53942) attached on a MTS Testing Machine, under displacement control loading conditions and controlled environmental conditions of 25°C and 70% relative humidity. For each material type, a gauge length of 30 mm was used for every bundle and three different displacement rates were applied (0.05, 0.1 and 0.5 mm/min).

During the testing procedure, AU signals were generated by a pulsing system, utilizing a AD85 sensor, placed on the upper grip. Two NANO 30 transducers, provided by Physical Acoustic Corporation (PAC), were placed permanently on the upper and lower grips of the used instrumentation and detected AE activity. For each one of the AE sensors, 2/4/6-AST pre-amplifiers were used (gain 40 dB and band pass filtering of 20–1200 kHz), while AE activity and AU signals were recorded, as pairs for each sensor, by a Mistras 2001 acquisition system. For the active channels, Threshold and gain was set at 40 dB and 20 dB respectively, while the Peak Definition Time (PDT), Hit Definition Time (HDT) and Hit Lockout Time (HLT) were set at 80 μs, 800 μs and 1200 μs.

Table 2. A representative part of the developed database based on AE features' values, related with fibre failures under quasi-static tensile tests in room temperature (Mistras 2001, NANO 80 and grips based on DIN 53942).

Fiber type	THA	Amp (dB)	RT (μs)	Dur (us)	TE (fJ)
Nextel 312	0.46–0.80	61.8–72.0	13–27	160–370	11–17
Nextel 440	0.54–0.79	58.5–74.3	12–26	180–356	8–29
Nextel 610	0.58–0.77	57.5–68.9	9–39	149–341	1–14
Nextel 720	0.51–0.71	64.6–71.2	23–35	189–455	3–23
Altex SN-11	0.52–0.84	54.4–71.2	8–24	323–447	4–24
S-Glass	0.61–0.71	60.0–63.4	12–18	186–332	4–20
Hi Nicalon	0.51–0.83	58.0–71.8	23–71	166–298	5–30
Carbon M40-B	0.45–0.83	64.6–81.2	14–36	200–324	22–148
Carbon M40-J	0.63–0.75	57.8–65.1	20–24	171–275	22–126
Kevlar 29	0.61–0.74	58.1–77.9	8–27	229–331	38–69
Spectra PE-40	0.47–0.65	70.9–73.8	21–43	352–412	19–42
Spectra PE-30	0.55–0.87	55.0–73.6	39–51	254–410	6–39

Amp: Amplitude, RT: Rise Time, Dur: Duration, TE: True Energy.

material affected significantly the monitored AE and AU activity.

4.2 Quantification of waveforms

4.2.1 Selection of valuable waveforms

In order to achieve the original target of the present study, a representative set of the captured waveforms was created for the characterization of the fibre failure event. However, the existence of such a representative set of waveforms is obviously an assumption that proved to be valid in most types of fibres and for the larger part of the loading procedure.

In the case of dry-bundle tensile tests, the main types of source events that produce significant elastic waves are fibre failures (single or multiple) and sliding friction between the fibres. Waveform selection criteria for single fibre failure were established, taking into account the available knowledge from the literature (Hamstad et al. 1986, Cowking et al. 1991, Prosser 1995, Jihan et al. 1997, Ageorges et al. 1999, Pappas 2001). These criteria concern waveform voltage levels, force–displacement plots, location events and noise reduction methods, sensitivity analysis results, Fast Fourier Transform (FFT) analysis and literature review findings. The selection of each candidate AE waveform is accompanied by the proper AU signal, identified as the first signal prior to the selected AE waveform. The success of the proposed overall selection procedure has to be evaluated. To this aim, the presented methodology for the development of a database can assist this evaluation but the used assumptions have to be justified by the practical use of such an extracted knowledge.

4.2.2 Time domain analysis for database development

The development of a time domain database will be presented in this section. The aim of this approach is the identification of specific events by the use of AE features, following procedures established in the literature (Curtis 1972, Russell et al. 1977, Okada et al. 1987, Anastasopoulos 1995, Groot et al. 1995).

In order to construct the time domain database, the selection of each representative set of captured AE/AU waveforms pair is justified for each single event, following the proposed feature extraction procedure (see Fig. 1). However, it has been proved (Pappas 2001) that a parametric sensitivity analysis should be applied in order to identify the proper Threshold value for reliable feature extraction and to decrease its effect on the resulted values. Moreover, it has been proposed that the Threshold/Amplitude ratio, THA, is a useful tool for AE features comparison between different materials and damage mechanisms. For the present study, feature extraction was performed only for signals with THA <0.8 (mean value <0.6), in order to increase the reliability and the efficiency of the developed database. In Table 2, ranges of some basic AE features are presented concerning different types of fibres.

Based on these values, a general comment could be that a single fibre failure generates elastic waves with medium Amplitude (50–80 dB), low Rise Time (<40 μs) and medium Duration (150–400 μs). In addition, the calculated distributions of AE features values are not Uniform or Gaussian, but most of them are non-symmetrical. Therefore, the use of feature mean values is not a reliable tool for AE event characterization.

4.2.3 Frequency domain analysis for database development

The procedure that was followed for the classification of fibre failure frequency response is given in the present section, justifying the proposed methodology. In order to develop the frequency domain database, it is necessary to identify all the possible "sources" of the dominant frequencies in the Fourier Transform (FT) of the captured waveforms. These "sources" characterize the overall system and the goal is to "extract" these frequencies from the FT of the waveforms. Under the term "system" is included the effect on the captured waveforms that comes from: the acquisition system (calibrated A/D board, sensors and pre-amplifiers), the pulsing system (pulser and sensor), the gripping/supporting system for fibre testing (vibration eigenvalues) and the propagation path at gripping. It is assumed that the "system" is constant for every test, since the settings were kept constant, but for different fibre types AE activity and AU response should be different due to the contribution of different initiation events and propagation path characteristics. The determination of propagation medium characteristics was accomplished in the present study by the use of AU method instead of the pencil break method. However, the whole scheme is based on the aspect of frequency "signature" existence in AE waveforms for failure modes identification (Groot et al. 1995, Qi et al. 1997, Giordano et al. 1998, Mizutani et al. 2000, Ni et al. 2002, Park et al. 2002, Mattei 2003).

In order to perform a statistical analysis in the frequency domain content of the valuable waveforms, a general procedure scheme was utilized to assure that the same rules and criteria were used for all the input data that produces the extracted information. Using the information mentioned above about the overall frequency content of captured AE waveforms, the following summarizing iterative scheme was applied on the AE and AU waveforms for frequency domain database development:

- Identification of the common frequencies in each fibre type (by using statistical rules and criteria) and subtraction of these frequencies from the main set.
- Repetition of the same procedure among the different fibre types for the same material family and among all material families.
- Identification of system frequencies, by comparison of the common frequencies between the extracted from the FFT of both the AE and the AU waveforms.
- Characterization of the remaining set of frequencies as the one that contains information about the fibres. Any common frequency between the FFT of AE and AU waveforms corresponds to the propagation frequencies while the remaining are representative of fibre breakage damage mechanism.

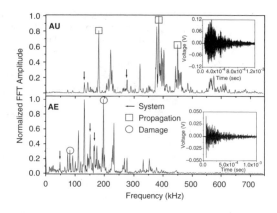

Figure 4. Captured pair of AE and AU waveforms and their frequency domain analysis, monitored during a quasi-static tensile test of a Nextel 720 fibre bundle at 27 N (80% of maximum force) and using a 0.1 mm/min displacement rate.

A representative example of the above-described procedure is given graphically in Figure 4. In that graph, the identification of dominant frequencies in FFT plot is presented, for a selected pair of AE/AU waveforms, nominating frequency "signature" of propagation and fibre failure for Nextel 720 fibres.

In Table 3, a list of frequencies identified by the developed database is given. For comparison reasons, the estimated mean values of the Reverberation Frequency (propagation characteristics) and Rise Frequency (source properties) are given, for every presented fibre type. According to these results, frequency "signatures" as FFT discrete peaks exist for many fibre types, associated with propagation "filtering" and damage mechanisms. However, the individual frequency peaks that characterize the propagation via each fibre type vary in a range of 100–400 kHz (middle frequencies), while the corresponding peak values for fibre failure varies in a range of 20–400 kHz.

Moreover, it is obvious that values of Reverberation Frequency (AE feature in time domain) can be used in a reliable way as a mean approximation of the list of propagation frequency peaks for the presented fibre types. Unfortunately, this is not the case for the Rise Frequency, the values of which overestimate the frequency response of the failure events.

5 CONCLUSIONS

In the present work, a characterization of the elastic waves generated as AE activity by the failure of different types of fibres is proposed, in time and frequency domain, conducting tensile tests on commercially available and extensively used fibre bundles. The applied classification of AE activity

Table 3. List of frequencies (in kHz), including system, propagation and damage (fibre failure) representative values, for different fibre types (criterion 10 kHz), in comparison with the corresponding values of relative AE features.

Material	Applied methodology						AE feature[*]
	Major frequencies						
			System				
Sensors	48	102	144	–	280	320	
Others[+]	–	–	154	169	213	328	
			Propagation				*Reve. Freq.*
Nextel	–	94	–	185	–	382	129–165
S-Glass	–	–	136	176	–	360	132
Kevlar	–	–	–	–	229	352	151
			Damage				*Rise. Freq.*
Nextel	26	84	–	204	–	–	400–593
S-Glass	59	–	115	–	223	–	411
Kevlar	31	–	–	196	–	–	639
Hi Nical	–	–	–	–	230	343	410
C M40-B	32	67	–	–	–	395	573

*Extracted from the captured waveforms.
+Gripping apparatus and acquisition system.

leads to the development of a database, which can identify the time and frequency "signature" of each fibre type failure, justifying a new methodology for the determination of the initiation event characteristics. However, proofs for the feasibility of the AE database in a micro-to-macro scale approach should be given using different testing conditions.

ACKNOWLEDGMENT

The authors are grateful to Professor G. Grathwohl and Dr. M. Kuntz (University of Bremen, Germany) for collaboration and technical support for utilization of DIN 53942, under the framework of Greek–German collaboration Project (1999–2000). Also, the authors want to acknowledge the technical assistance received by ICE-HT/FORTH for the execution of the experimental work.

REFERENCES

Achenbach, J.D. 1973. *Wave propagation in elastic solids.* North-Holland Publishing Company.
Ageorges, C., K. Friedrich & L. Ye 1999. Experiments to relate carbon-fibre surface treatments to composite mechanical properties. *Comp.Scie.Tech.* 59: 2101–2113.
Anastasopoulos, A.A. 1995. Failure mechanism identification in composite materials by means of acoustic emission: is it possible? In Proceedings of Non Destructive Testing, Patras.
ASM International, 1987. *Composites.* Engin. Mat. Handbook.
Clough, R.B., J.C. Chang & J.P. Travis 1981. Acoustic emission signatures & source microstructure using indentation fatigue & stress corrosion cracking in aluminum alloys. *Scri.Met.* 15: 417–422.
Cowking, A., A. Attou, A.M. Siddiqui & M.A.S. Sweet 1991. Testing E-glass fibre boundless using acoustic emission. *J.Mat.Scie.* 26: 1301–1310.
Curtis, G.J. 1972. The characterization of failure mechanisms by means of acoustic emission. In Proceedings of Symposium on Acoustic Emission, Imperial College.
DIN 53942, 1984. *Testing of textiles-flat bundle method.*
Giordano, M., A. Calabro, C. Esposito, A.D. Amore & L. Nicolais 1998. An acoustic emission characterization of the failure modes in polymer-composite materials. *Comp.Scie.Tech.* 58: 1923–1928.
Groot, P.J., P.A.M. Wijnen & B.F. Janseen 1995. Real-time frequency determination of acoustic emission for different fracture mechanisms in carbon/epoxy composites. *Comp.Scie.Tech.* 55: 405–412.
Hamstad, M.A. & R.L. Moore 1986. AE from single & multiple Kevlar 49 filament breaks. *J.Comp.Mat.* 20: 46–66.
Jihan, S., A.M. Siddiqui & M.A.S. Sweet 1997. Fracture strength of E-glass fibre strands using acoustic emission. *NDT.E.Int.* 30(6): 383–388.
Mattei, C. 2003. Identification of delamination onset in CFRP laminates under fatigue testing using modal analysis of AE signals. Presentation by CSM Materialteknik, Sweden.
Mizutani, Y., K. Nagashima, M. Takemoto & K. Ono 2000. Fracture mechanism characterization of cross-ply

carbon-fiber composites using AE analysis. *NDT.E.Int.* 33: 101–110.

Ni, Q. & M. Iwamoto 2002. Wavelets transform of AE signals in failure of model composites. *Eng.Frac.Mech.* 69: 717–728.

Okada, A., T. Yasujima & T. Tazawa 1987. Fracture behavior and AE signals in carbon fiber reinforced aluminium. *Trans.Jap.Inst.Mat.* 28: 1004–1011.

Pappas, Y.Z. 2001. Stochastic modeling of composite materials fatigue response. PhD Dissertation, University of Patras.

Park, J.M., J.W. Kim & D.J. Yoon 2002. Interfacial evaluation & microfailure mechanisms of single carbon fiber/ bismaleimide composites by tensile and comprehensive fragmentation tests and acoustic emission. *Comp. Scie. Tech.* 62: 743–756.

Prosser, W.H. 1995. Advanced waveform-based acoustic emission detection of matrix cracking in composites. *Mat.Eval.* 53: 1052–1058.

Qi, G., A. Barhorst, J. Hashemi & G. Kamala 1997. Discrete wavelet decomposition of acoustic emission signals from CFR composites. *Comp.Scie.Tech.* 57: 389–403.

Russell, S.S. & E.G. Henneke 1977. Signature of acoustic emission from graphite/epoxy composites. *NASA Grant NSG 1238*, Rep.No. VPI-E-77-22.

Emerging Technologies in Non Destructive Testing, Van Hemelrijck, Anastasopoulos & Melanitis (eds)
© 2004 Swets & Zeitlinger, Lisse, ISBN 90 5809 645 9

Fracture behaviour and damage mechanisms identification of SiC/Glass ceramic composites using DEN specimens and AE monitoring

T. Loutas, G. Sotiriadis, K. Dassios, I. Kalaitzoglou & V. Kostopoulos
Applied Mechanics Laboratory, Dept. of Mechanical Engineering & Aeronautics, University of Patras, Greece

ABSTRACT: The phenomenon of Large Scale Bridging in SiC/MAS-L composites was studied by tensile testing of DEN specimens accompanied by in situ continuous Acoustic Emission monitoring. The AE data were successfully classified using Unsupervised Pattern Recognition Algorithms and the resulting clusters were correlated to the dominant damage mechanisms of the material. Their evolution in time is feasible after the pattern recognition classification. Microscopic data are also used to enhance confidence for the failure mode identification.

1   INTRODUCTION

It is now well established that, of all types of reinforcement, continuous ceramic fibre reinforcements results to a ceramic composite having high fracture toughness and damage resistance. The enhanced damage resistance and increased fracture toughness of continuous fibre reinforced ceramic matrix composites (CFCCs) is due to their inherent ability to effectively redistribute stresses around holes, notches and cracks, a phenomenon which stems from the development of shielding forces in the process zone around the crack tip. Fracture in the vast majority of CFCCs is associated with the formation and propagation of macrocracks (Class I and Class II fracture) [Evans et al. 1994]. The corresponding process zone consists of two parts: the so called bridging zone with fiber bridging and pull-out developing within the macrocrack and a matrix cracking process zone ahead the macrocrack. The increase in fracture resistance is the result of several energy-dissipating mechanisms acting in the two zones. In the matrix process zone a complex set of phenomenae such as matrix microcracking, fibre/matrix interfacial debonding and transformation toughening may take place concurrently. In the bridging zone the cracked matrix is bridged by intact and/or failed fibres, which debond, slip and pull-out. The role of the bridging zone in the fracture resistance of the composite is of particular importance as the bridging fibres carry a significant portion of the applied load, hence resisting to further crack opening. When the size of the bridging zone is comparable to the specimen dimension, a phenomenon known as Large Scale Bridging (LSB) encountered in most CFCCs. Then,

the well established rising crack growth resistance of the ceramic composite (R-curve) is a material-extrinsic property which depends upon specimen geometry and dimensions [Cox et al. 1991, Fett et al. 2000]. On the other hand, AE is a powerful non-destructive technique for real-time monitoring of damage development in materials and structures, which has been used successfully for the identification of the damage mechanisms in composite structures under quasi-static and dynamic-cyclic loading. The aim of the present work is to investigate in the case of SiC/MAS-L Ceramic Composites subjected to Double Edge Notche (DEN) fracture measurement experiments the activated damage mechanisms and their evolution in time by using an in-house unsupervised, Pattern Recognition (PR) technique. The applied methodology proved to be capable to classify the AE data, recorded during the DEN tests and to illuminate the different failure modes. Up to now, the majority of the proposed applications of AE for the characterization of composite materials are based on the 'Conventional' AE analysis, which usually incorporates investigation of the activity in diagrams of cumulative hits versus load and the correlation of some AE features, such as Amplitude and/or Duration, to basic damage mechanisms. Nevertheless, in the case of ceramic composites, these techniques are not sufficient due to the large number of damage and stress redistribution mechanisms, such as matrix cracking, fibre/matrix debonding and sliding (fibre pullout), as well as stochastic fibre failure, which are active continuously during loading. In contrast, the unsupervised PR technique proposed in this work takes into consideration a large number of AE descriptors and thus, it is a more powerful tool since its multiparameter approach provides detailed information for the activated damage mechanisms within the material structure and their evolution during loading.

2 EXPERIMENTAL PROCEDURE

The material used in this study is a laminated cross-ply SiC/MAS-L composite processed by EADS (ex-Aérospatiale, France). The reinforcing SiC fibres are grade Nicalon NL202 with a chemical composition in weight concentration terms of 56.6% Si, 31.7% C and 11.7% O. The glass-ceramic matrix contains MgO, Al_2O_3, SiO_2 and LiO_2 and is made via the solgel route. Plates of 2.0 and 3.0 mm thickness with 8 and 12 plies respectively were produced via hot pressing. The laminae were stacked together in a symmetric $[0–90°]_{2s}$ and $[0–90°]_{3s}$ lay up for the 2.0 and 3.0 mm thick plate respectively. The effective volume fraction of the fibres in the loading direction is 0.17 [Brenet et al. 1996] whereas the matrix stiffness (75 GPa) and failure strain are lower than the corresponding values for the fibre and hence cracks appear first in the matrix. Typical Double Edge Notch specimens were used in this study and their geometry is presented schematically in Figure 1.

All DEN fracture characterization tests under tension were performed using a MTS Universal Testing Machine equipped with hydraulic gripping system, under displacement mode, at controlled environmental conditions of 25°C and 70% relative humidity. AE activity was continuously recorded during the testing of the materials. The data acquisition system used was a Mistras 2001 of Physical Acoustics Corporation. The AE signals were monitored by using two resonant transducers (NANO 30), which were attached to each specimen by means of a suitable coupling agent. Pre-amplification of 40 dB and bandbass filtering of 20–1200 kHz was performed by 2/4/6-AST pre-amplifiers.

3 AE MONITORING

Acoustic emission was monitored during a number of tensile tests. The acquisition parameters for the two active channels were: threshold and gain equal to 30 and 20 dB respectively, while the Peak Definition Time (PDT), Hit Definition Time (HDT) and Hit Lockout Time (HLT), were set at $50 \mu s$, $100 \mu s$ and $500 \mu s$. Pencil break tests were used for the calibration of the applied set-up.

Figures 2 and 3 are two representative AE data plots of AE energy versus time and AE hits versus time respectively. The background continuous line represents the applied load versus time. Both Figures clearly indicate the association of load drop to AE increased activity and the load tail due to bridging effect which preserves significant AE activity too.

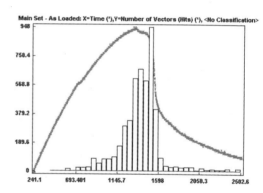

Figure 2.   AE energy vs time.

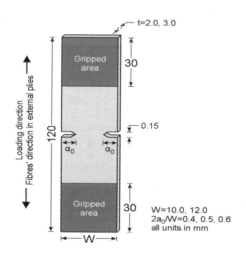

Figure 1.   DEN specimens configuration.

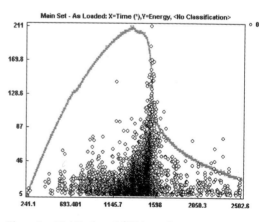

Figure 3.   Distribution of AE hits vs time.

4 PATTERN RECOGNITION

A schematic representation of the proposed pattern recognition scheme which was used for the analysis of AE signals monitored during the DEN tests, is presented in Figure 4. Noesis Professional 3.1, software by Physical Acoustics Corporation, offered a variety of algorithms for unsupervised PR such as Maxmin Distance, Cluster Seeking, Forgy, k-means and Isodata. With the exception of Cluster Seeking (a generalized radial seeking algorithm), all the others are widely known and used in the literature [Pappas et al. 1998,

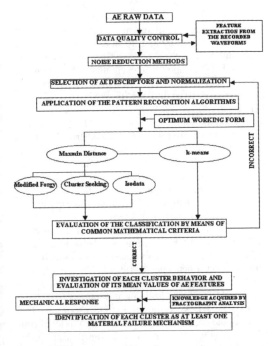

Figure 4. Pattern recognition scheme.

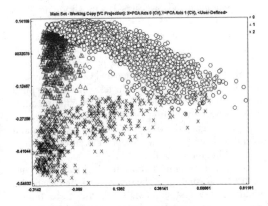

Figure 5. Principal components projection.

Jain et al. 1987]. The determination of the initial conditions is critical for the performance of the above algorithms. A most interesting approach [Loutas et al. 2002] utilizes the Maxmin Distance algorithm for a primary classification of the data and use the resulting cluster centers as initial conditions for the application of a PR algorithm.

Moreover, it was decided to test extensively all the available algorithms and check their performance for the same problem. Consequently the unsupervised PR schemes, which were used in this work, are: (1) Maxmin Distance-Forgy, (2) Maxmin Distance-Cluster Seeking, (3) Main Distance-Isodata, (4) k-means. Only k-means was used alone, since it does not demand initial conditions but the definition of the number of the desired classes. A complete procedure for performing unsupervised PR was of primary importance for the authors. The main problem was the fact that every algorithm involves a number of predefined parameters which must be set by the user. These parameters determine the internal operation of the PR scheme and differ in each case. After systematic 'trial and error', a parametric analysis was conducted and the range of interest of those parameters was located as well as the step increase of their values. The clustering results are evaluated using cluster validity criteria. From a plethora of validity criteria, R criterion and τ criterion [Tou et al. 1974, Fukunaga et al. 1990] were chosen as they have the advantage of being independent with the number of classes. These two criteria give an indication of the compactness and the separation among the resulting classes. Low values for R and high values for τ reveal a successful classification and the formation of well-defined compact clusters. The application of the above described procedure led to a 3-class clustering. The results of the pattern recognition algorithms application is shown in Figure 5 in principal axes projection.

The classification is successful and compact clusters in the n-th dimensional space were created. In our study n equals 4 because this is the number of the chosen AE descriptors where the pattern recognition scheme is based.

5 FAILURE MODE IDENTIFICATION

After a successful, according to R and τ criteria, classification of AE data, the resulting clusters should be correlated with the material's damage mechanisms. The most demanding point of the PR approach is to identify the damage mechanisms [Barre et al. 1994, Curtis et al. 1972, Mizutani et al. 2000, Ohtsu et al. 1987, Loutas et al. 2002] that these clusters correspond to, in order to understand the damage evolution in SiC/MAS-L ceramic composites under fracture loading conditions. This is due to the fact that the classification process does not lead to a unique solution and there do not exist any solid and indisputable criteria to determine which classification result is more appropriate and representative for the actual damage mechanisms. The first goal of a PR algorithm is to result in compact and well-separated classes. In the present work, this is accomplished and this is proved by the values of R and τ validity indices, discussed in the previous paragraph.

Towards a successful damage mechanisms identification, the mechanical behaviour of a fibre-bridged crack must be well understood. The Class I fracture characteristics (formation and development of a single macrocrack) of a brittle-matrix fibre-reinforced composite are presented in the load–displacement (F–d) plot of Figure 6 together with a schematic depiction of the mechanical processes occurring during testing.

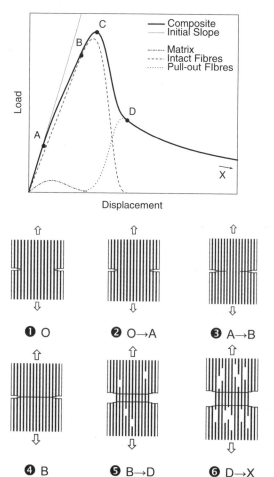

Figure 6.   Load–displacement (F–d) curve typical of Class I composite fracture and schematic overview of the main fracture processes in the composite.

During the initial loading stages, reversible mechanical phenomenae occur within the composite (region O→A, stage 2 in Figure 6).

The first matrix crack (point A in Figure 6) triggers the appearance of the bridging zone while cracking evolves (region A→B, stage 3 in Figure 6) until the macrocrack has fully developed, spanning the total width of the specimen (point B, stage 4 in Figure 6). Beyond this stage, fibres start failing within the volume of the composite and the load-carrying capacity of the remaining fibres decreases until a critical number of fibres have failed (point C in Figure 6). Failed fibres undergo pull-out and an additional contribution to the recorded load arises, owing to friction at the fibre/matrix interface (region B→D, stage 5 in Figure 6). Under the global load-sharing principle, each fibre failure is followed by a uniform redistribution of the remaining load to the surviving fibres. As the portion of load that corresponds to each intact fibre is greater after the redistribution, more failures are induced and the process evolves until all fibres have failed (point D in Figure 6).

Beyond this stage, the load carried by the composite corresponds entirely to interfacial friction due to pull-out of failed fibres (region D→X, stage 6 in Figure 6). With increasing displacement, fibre ends that were originally located at various statistical locations inside the composite are sequentially disengaged from the matrix until, eventually, the composite separates in two parts.

In summary, the main mechanisms [Dassios et al. 2003] typical of Class I composite behaviour are, in order of appearance: Linear elastic composite behaviour (O→A), linear elastic fibre behaviour (A→B), crack bridging by intact and pulled-out fibres (B→D) and purely frictional bridging due pull-out of failed fibres (D→X). Accordingly, the corresponding F–d curve of the composite is the sum of 3 individual contributions (Figure 6): A limited contribution corresponding to the load carried during the early loading stages by the fibres and the brittle matrix, a contribution corresponding to the load carried by the surviving fibres which can be assumed identical to that of a fibre bundle where load is carried by a large number of fibres acting independently and a contribution corresponding to friction by fibre pull-out. Beyond the physical understanding of the bridging phenomenon and Class I behaviour, the activation of the different clusters versus the applied load, as it is presented in Figure 7, provide an additional extremely useful piece of information.

In this figure the different clusters activation and evolution is very clear. Matrix cracking is dominant in the very beginning of the load rise.

Since matrix cracks come to a critical point and before the formation of the macrocrack that spans the width of the test coupon, stochastic fibre failure becomes the prevalent failure mechanism and is responsible for significant AE activity. In the final stage interfacial sliding due to fibre pullout is clearly shown.

During the testing of double edge notched SiC/MAS-L specimens (see Figure 8), the outer surface of the specimen (0° ply) was constantly monitored through optical microscopy by using a low magnification (×4) lens focusing along the ligament between the notches. Shortly after the application of load, matrix microcracks form and develop at both notch roots at random vertical positions.

The microcracks are oriented normal to the direction of loading and increasing load, propagate within the matrix towards the facing notch. The phenomenon evolves until a critical crack density is established

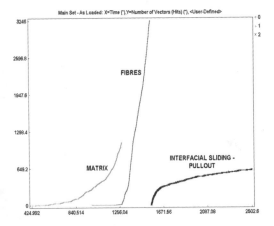

Figure 7. Cluster evolution in time.

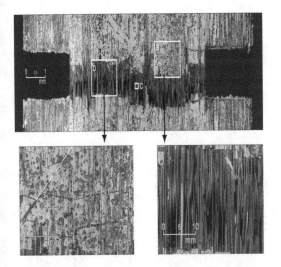

Figure 8. Large scale bridging in SiC/MAS-L composite: in situ microscopical damage observation during testing. Magnifications: (a) Matrix microcracking during the early fracture stages, (b) Fibre bridging and pull-out.

where the different paths of adjacent microcracks merge to form a dominant, fully developed macro-crack that spans the width of the ligament.

Upon formation of a fully developed macrocrack, the remaining microcracks do not propagate further. In the bridging zone bridging fibres stretch, fail into the matrix and pull-out. The bridging phenomenon is particularly prominent for the material tested in this study. The pull-out contribution was also extensive owing to a weak fibre/matrix interphase.

6 CONCLUSIONS

Failure mode identification for SiC/MAS-L composites, during quasi-static tensile loading was accomplished using an unsupervised pattern recognition analysis to AE data monitored during the tests. A number of algorithms were used and the results were evaluated based on R and τ validity criteria. The cluster activation plotted against the normalized applied load, was proved very useful in the identification of the damage mechanisms of the materials. This knowledge was supported by the assistance of extensive microscopy.

Unsupervised pattern recognition proved a powerful tool for the classification of AE hits, in the case of ceramic composites, and the procedure established is repeatable and reliable.

ACKNOWLEDGMENTS

The authors wish to thank Mr. Patrick Peres of EADS (France) for providing the composite material used in this study. This work is part of an AGARD/AVT support program (G-100).

REFERENCES

Evans A.G. and Zok F.W., *The physics and mechanics of fibre-reinforced brittle matrix composite*, Journal of Materials Science, V.29, pp. 3857–3896, 1994.
Cox B.N., *Extrinsic factors in the mechanics of bridged cracks*, Acta Metall. Mater, V.37, pp. 1189–1201, 1991.
Fett T., Munz D., Geraghty R.D. and White K.W., *Influence of specimen geometry and relative crack size on the R-curve,*Eng. Fract. Mech.,V.66, pp. 375–386, 2000.
Brenet P., Conchin F., Fantozzi G., Reynaud P., Rouby D. and Tallaron C., *Direct measurement of the bridging constraint as a function of crack displacement in ceramic-ceramic composites*, Compos. Sci. Technol., V.84, pp. 817–823, 1996.
Noesis© 3.1 Reference Manual, *Pattern Recognition & Neural Networks Software for Acoustic Emission Applications*, Ver. 3.1, *Envirocoustics ABEE*, 1999.
Loutas T.H., Kontsos A., Sotiriadis G., Pappas Y.Z. and Kostopoulos V., *On the identification of the damage mechanisms in oxide/oxide composites using Acoustic Emission*,

Proceedings of the 25th European Conference on Acoustic Emission Testing, Prague, Czech Republic, 2002.

Pappas Y.Z., Markopoulos Y.P. and Kostopoulos V., *Failure mechanisms analysis of 2D carbon/carbon using acoustic emission monitoring*, NDT&E International, V.31(3), pp. 157–163, 1998.

Jain A.K., *Advances in statistical Pattern Recognition*, NATO ASI Series, V.F30, pp. 230–249, 1987.

Fukunaga K., *Introduction to statistical pattern recognition*, 2nd English edn, Academic Press, San Antonio, CA, USA, 1990.

Tou J.T. and Gonzales R.C., *Pattern Recognition principles*, Addison-Wesley, Reading, MA, 1974.

Barre S., *On the use of Acoustic Emission to investigate damage mechanisms in Glass-fibre-reinforced polypropylene*, Composite Science and Technology, V.52, pp. 369–376, 1994.

Curtis G.J., *The characterization of failure mechanisms by means of Acoustic Emission*, Symposium on Acoustic Emission, Imperial College, 1972.

Mizutani Y., Nagashima K., Takemoto M. and Ono K., *Fracture mechanism characterization of cross-ply carbon fibre composites using Acoustic Emission analysis*, NDT&E International, V.33, pp. 101–110, 2000.

Ohtsu M. and Ono K., *Pattern Recognition analysis of Acoustic Emission from unidirectional carbon-fibre epoxy composites by using Autoregressive modeling*, Journal of Acoustic Emission, V.6(1), pp. 61–71, 1987.

Dassis K.G., Galiotis C., Kostopoulos V. and Steen M., 2003 (to be published).

156

Emerging Technologies in Non Destructive Testing, Van Hemelrijck, Anastasopoulos & Melanitis (eds)
© 2004 Swets & Zeitlinger, Lisse, ISBN 90 5809 645 9

Valve condition monitoring of reciprocating compressors using Acoustic Emission

S.C. Kerkyras
Hellenic Petroleum, SA

ABSTRACT: This paper briefly summarizes a part of the work that was done under the European Union funded *BRITE III RECIP-AE* ("The Monitoring of Reciprocating Plant & Machinery for Improved Efficiency and Reduced Breakdown") Research Project focusing on on-line condition monitoring of reciprocating compressors. It contains a technical overview and a short discussion of how this system might be used as a stand-alone application or as a part of an integrated system.

1 ACOUSTIC EMISSION OVERVIEW

Acoustic Emission (AE) can be defined as the transient elastic energy spontaneously released in materials undergoing, for instance, deformation, fracture, or both and is dependant on basic deformation mechanisms such as dislocation motion, grain boundary sliding, cleavage, twinning and vacancy coalescence. The AE signal occurs as one of two distinctive types:

(a) Burst; which is a high amplitude low frequency signal generally associated to surface events such as slip-line formation and surface micro cracks.
(b) Continuous; which is a lower amplitude high frequency signal generally associated to internal mechanical activity.

AE has been widely used as a condition monitoring technique for the evaluation of structural integrity of pressure vessels and storage tank bottoms. Recently, this method was used as a PdM method for rotating machinery under the European Union funded *BRITE II PC-CONMON* Project ("The Development of a PC-based Artificial Intelligence System for the Diagnosis and Prognosis of Machine Condition using Acoustic Emission and Acceleration Monitoring") producing very promising results. Then, the same method was applied in reciprocating machinery under *RECIP-AE* Project, the results of which are briefly summarized in what follows. Furthermore, several other authors have used AE as a diagnostic tool of rotating and reciprocating machinery like *Brown et al* (1996) & *Kerkyras et al* (1999). Moreover, AE has been used for the monitoring of turning operations by *Kerkyras et al* (1991).

2 HISTORICAL BACKGROUND

The methodology and the algorithms that will be described here were mainly the output, as stated above, of the European Union funded *BRITE III RECIP-AE* ("The Monitoring of Reciprocating Plant & Machinery for Improved Efficiency and Reduced Breakdown") Research Project. This project commenced on 1996 and had aimed at the development of a non-destructive condition monitoring tool for compressors and diesel engines. Under this Project, several condition monitoring techniques were investigated including, apart from AE, vibration, temperature, pressure, sound, etc. data.

It must be stated that members of the consortium that undertook the Project were some of the leading European companies in their field, including A P Moller (the biggest Danish shipping company), MAN B&W (the well known engines manufacturer), HEL.PE. (the biggest Greek refining company) and Perkins Engines (another well established British diesel engines manufacturer). These companies planned to use the outcome of the Project to their benefit and some of them are currently using some kind of prototype to some extent. Our company mainly focused on reciprocating compressors. The Project is completed.

3 SYSTEM HARDWARE DESCRIPTION

System consists of a suitable AE transducer placed on pre-selected positions on the machine under consideration. The sensor that is typically used is PAC D95203B. The signal is driven into a suitable pre-amplifier. This pre-amplifier also incorporates a band-pass filter (0.1 to 1.2 MHz) which accommodates further sampling and processing. The preamplifier that is typically used is the PAC 1220A model. The next step is a decoupling network that is also responsible for the power supply of the preamplifier. The signal is then fed into a circuit that can provide a 2nd stage of amplification and filtering depending on the data acquisition method followed. Within this stage several other features can be incorporated (such as integration, if it is decided that RMS measurements are going to be used). The last two stages are implemented in a single piece of hardware. The last part is a PC where a suitable data acquisition board is installed in. A wide range of National Instruments ADCs have been used throughout the development of the project.

4 BRIEF ALGORITHM DESCRIPTION

The data acquired the way described above were analyzed using MatLab. A systematic study of a wide range of signals obtained at compressors at the sites of several of the project partners shows that when a fault is actually present the corresponding signal can be generally categorized into two major categories. The first category of signals has a low amplitude noise (grass noise) at certain parts of the compressor cycle. This phenomenon is associated with gas leakage that is present in the compressor. The second category of signals exhibits a considerably different type of main bursts and is associated with wet phase and/or broken valve plates. Figure 1 shows three typical raw AE signals from a single stroke of a reciprocating compressor. The first appears to be perfectly OK (a) whilst in the second the gas leakage is obvious (b) and in the third one can very clearly note how the main bursts differ from the typical good condition (c). In these figures the dominant events are the two main bursts that can be observed in all of them. The first one is associated with the gas revealed after the valve has lifted and the second with the valve impact on the guard.

Time domain signal is acquired over an entire stroke and then normalized over the mean energy of each signal. Then the kurtosis vs time signal is calculated and max. kurtosis value which corresponds to the strike of the valve plate on the guard and is characteristic of the cycle is regarded as a timing signal. Then signal analysis proceed by regarding two types of valve faults:

(i) First type of fault (gas leakage). Signal is divided into two parts: one part that contains the main burst and a second part that contains the rest of the signal. Since the kind of noise that is present in these two parts have amplitude less than 10% of the maximum, focus is placed on this range of amplitudes via a selection making routine. The

parameter that is finally calculated is the standard deviation of the remaining symmetric signal. This parameter when calculated for both parts proved to have a strong relationship with the gas leakage condition.

(ii) Second type of fault (wet phase and/or broken valve plates). Several parameters have been

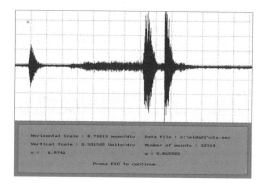

(a)

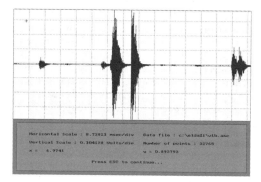

(b)

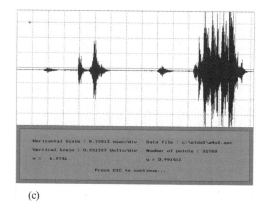

(c)

Figure 1.   Different kinds of AE signals; the difference is obvious.

158

calculated. Signal is again divided into two parts (not exactly in the same way though), one part containing the main bursts and another containing the rest of it. An RMS version of both parts is obtained because some of these parameters are calculated in the RMS domain. Both types of signals (raw & RMS) are fed into suitable modification routines some of which resemble the amplitude selection routine of the previous paragraph. These routines modify the input signals so that, when certain statistical parameters are calculated over them, they provide a result that can be interpreted in condition monitoring terms. Among the parameters that exhibited a straight forward relationship with the actual condition of the compressor are the area under the RMS curve in both parts of the signal, the duration of the bursts the number of bursts in both parts of the signal.

Both types of signals (raw & RMS) are fed into suitable modification routines. The calculated parameters have been found to exhibit a straightforward relationship with the actual condition of the compressor.

The data have been synchronized by a process that is based on a Kurtosis calculation algorithm. This makes unnecessary the use of an external timing signal. A rough block diagram of the steps taken is presented on Figure 2.

5   A CASE STUDY

The system developed for compressors was based on a data ensemble consisting of more than 20 large compressors, measured during a 3-year period. The following is a typical case study recorded at the HEL.PE site during the project. It has to do with V-4001S, a Hydrogen compressor which had its discharge valves replaced due to the indications of the above AE-based condition monitoring method. This method was the only one to suggest that there was a valve problem, although the valves had been recently replaced. Figure 3 shows a schematic representation of V-4001S.

In the figures that follow, the results of the prototype software is presented in chronological order. The output of the software consists of two parts; the first one is the raw and RMS AE data that were acquired and the second is the corresponding calculated parameters mapped in a three color background where green stands for OK, red stands for not OK and the narrow yellow area stands for merely acceptable condition. Figures 4 to 9 show the output of the system. Tracing of signals is obvious.

In these figures the position of the small cross represents the condition of the valve (green, yellow or red). As it can be seen, although the valves had been recently replaced, their condition deteriorated dramatically, suggesting that there was another problem cause. Figures 10 & 11 show photographs of the replaced valves where the deposits and the broken springs are obvious.

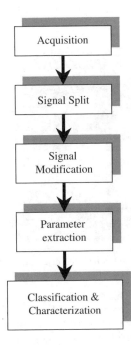

Figure 2.   Block diagram of basic algorithm.

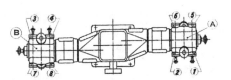

Figure 3.   Hydrogen compressor V-4001S, manufactured by BORSIG GHH.

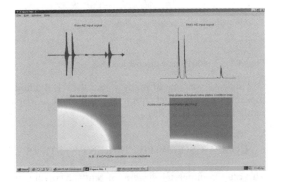

Figure 4.   V-4001S new suction valve.

159

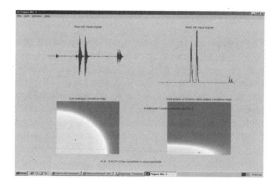

Figure 5.   V-4001S suction valve after 4 days operating.

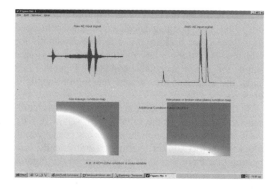

Figure 6.   V-4001S suction valve leaking after 18 days operating.

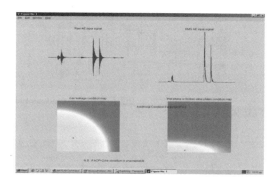

Figure 7.   V-4001S new discharge valve.

   This case was one of many where the AE method presented results satisfactory. The prototype is currently in use by HEL.PE.

6   COMPARISON WITH CURRENT SYSTEMS

The only commercial system currently available in the market for monitoring reciprocating parts is the

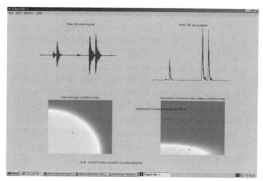

Figure 8.   V-4001S discharge valve after 4 days operating.

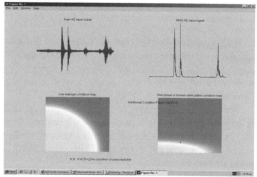

Figure 9.   V-4001S discharge valve leaking after 18 days operating.

Figure 10.   V-4001S replaced discharge valve; the arrows indicate the broken springs.

RECIP-Trap. This is a widely accepted system developed by a US company named Liberty. This system is using pressure cylinder data obtained in an intrusive way together with other non-intrusive data as vibration, temperature, ultrasound and process data. When

Figure 11. V-4001S replaced discharge valve; the arrows indicate deposits.

comparing this system with the system developed under the Project *RECIP-AE* it can be said that:

1. The AE-based system is non-intrusive, whilst *RECIP-Trap* is mainly based on pressure data obtained from in-cylinder measurements.
2. The AE-based system is mainly oriented for valve problems in compressors, which is the dominating problem (valve faults are over 70% responsible for the deterioration of reciprocating machinery) but also gives valuable information for cycle efficiency, particularly in diesels.
3. The AE-based system is a low cost non-intrusive predictive maintenance tool.
4. The AE-based system can be used as a side system for the evaluation of condition of reciprocating components, including valves, with other traditional systems based on vibration and being suitable for monitoring rotating parts as shafts and bearings.

7   CONCLUSIONS

This system has proved very reliable under very hard industrial conditions although it has not yet been finalised. Since there is a lack of similar products in the market, it is strongly believed that it can be a success. It can be used as a stand-alone system or as a part of a more general methodology. It is not intrusive, therefore it is possible to use it without the need of shutdowns.

The system in its current version *is not intrinsically safe*. In a future version a larger scale of integration may be used, including all the hardware parts into a single unity capable of performing the analysis on site, thus making it possible to have an immediate and reliable condition evaluation.

ACKNOWLEDGEMENT

This work was funded by the European Union under the BRITE-EURAM European Union Program RECIP-AE BE96-3491.

REFERENCES

PC-CONMON BE 7405, "The development of a PC-based artificial intelligence system for the diagnosis and prognosis of machine condition using acoustic emission and acceleration monitoring"

RECIP-AE BE96-3491, "The monitoring of reciprocating plant and machinery for improved efficiency and reduced breakdown"

RECIP-Trap, Liberty Corporation, USA

E R Brown, S C Kerkyras, G D Neil, R L Reuben, J A Steel, "A Polyvinylidene fluoride based acoustic emission sensor for industrial applications", pp. 499–508, COMADEM 1996

S C Kerkyras, "The use of Acoustic Emission as a condition monitoring tool of rotating and reciprocating machinery", 2nd International Conference on Emerging Technologies in NDT, Athens, 24–26 May 1999, proceedings ISBN 90 5809 1279

S C Kerkyras, S J Wilcox, W K D Borthwick, R L Reuben, "Acoustic Emission Monitoring of Turning Operations using PVdF film and PZT Sensors", Proceedings of 4th, World Meeting on Acoustic Emission, Boston, Sept. 15–18, 1991

On-stream monitoring of process plant

P.T. Cole & S.N. Gautrey
Physical Acoustics Limited, Cambridge, England

ABSTRACT: Conventional practice for establishing the integrity of process plant is to inspect during shut-down. Financial pressures to keep plant on-line limit the extent to which this is possible, as a result it is impor-tant to make optimum use of maintenance resources to minimize down-time, whilst still avoiding in-service failures. The ability to identify developing problems on-line helps directs inspection quickly to problem areas, increasing efficiency. In addition to this, the identification of areas that do not require inspection avoids wast-ing inspection effort inspecting areas that are free of propagating damage. This paper discusses the use of advanced NDT methods in particular acoustic emission (AE) as a condition assessment and monitoring tool for on-stream process plant.

1 INTRODUCTION

Conventional practice for establishing the integrity of process plant is to inspect during shutdown. Financial pressures to keep plant on-line limits the extent to which this is possible, as a result it is important to make optimum use of maintenance resources to minimize down-time, whilst still avoiding in-service failures. The ability to identify developing problems on-line helps direct inspection quickly to problem areas, increasing efficiency, the identification of areas that do not require inspection avoids wasting time inspecting areas that are not suffering in-service deterioration.

Acoustic emission (AE, ref. [1]), which is comple-mentary to NDT methods, is able to test/monitor 100% of a structure using relatively few stationary sensors. In addition it is primarily sensitive to active defects, and it is often possible to use the method whilst plant is on-line, maximising production time.

2 AE CONDITION MONITORING

Although classified as an NDT (non-destructive test-ing) method, AE does not give information on defect size, and many consider it more akin to condition mon-itoring. The capabilities and limitations of the method when applied to in-service process plant are sum-marised briefly as follows:

Capabilities:

- Sensitive to propagating defect activity and damage to microstructure (cracking, corrosion, overstress).
- Tests/monitors 100% of volume.
- Not affected by defect orientation.
- Identifies when damage occurs (monitoring use).
- Good to direct local NDT.
- Relatively straightforward to apply to high temper-ature equipment.

Limitations:

- Does not measure physical defect size, which requires follow-up NDT.
- Does not detect manufacturing defects, unless they are propagating in-service.
- Not possible to apply on-line to all plant.

It is clear that AE and NDT methods are complemen-tary since NDT methods are applied locally, indicate static presence of defects rather than dynamic growth, and give sizing information.

2.1 *In-service testing or monitoring – the options*

There are a number of ways in which AE may be applied during plant operation; these are a matter of both tech-nical and cost considerations. The options are:

A *In-service "test"*

This involves increasing the stress on the item under test whilst monitoring for AE from any active defects. This usually requires the ability to increase the stress (i.e. pressure or level) to typically 10% above the maximum stress that has occurred during the past 12 months operation. The main limitation of this approach is the requirement to change the pressure, which is not always possible. The test may be carried out on-line or during a temporary process pause, depending upon circumstances.

B Continuous monitoring

Damage to plant may be the result of unpredictable process temperature or pressure transients, in this case continuous monitoring may be the only way to identify propagating damage and the conditions under which it occurs. This approach is also used for alarm monitoring systems such as those used on methanol converters.

C Semi-continuous monitoring

Certain types of plant experience conditions of highest stress during specific conditions, such as during start-up or shut-down, or during thermal cycling. In this case AE "semi-continuous" monitoring should be applied during these periods to identify any defect growth. Deliberate increase in stressing, for example by increasing cooling rates, is not recommended, as this may actually cause defects.

D Corrosion monitoring

Active corrosion may be detected by monitoring for a short period of time, this approach is used for storage tank floors, and, by using higher frequency sensors to minimize process noise, on process plant, particularly stainless steel.

E Particle and leak detection

Not AE by the strict definition, but AE sensors are used to detect leakage in pressure systems and valves, and to monitor burners and other continuous processes, and monitor sand particles in oil production.

2.2 Applicable plant

The types of process plant to which these methods are applied fall into clear categories:

(i) Pressure storage vessels such as spheres and bullets; these are tested using method "A" above, also tested using this approach is the wall and knuckle area of storage tanks, the stress in this case is increased by raising the level. According to the International Process Safety working Group, (IPSG, ref. [2]) the most commonly used procedure for this test is the MONPAC™ procedure (ref. [3]). This also formed the basis for ASME VIII article 12 as applied to new vessels. In addition to the AE test and any required follow-up NDT, complementary NDT should be applied to check for early corrosion damage; a "C-Scan" strip covering all conditions seen, plus IRT or similar method to check for corrosion underneath the leg fire protection, which has led to collapse in the past.

(ii) Thick-walled high temperature plant, such as methanol converters, platformers, and ammonia converters, use method "B" or "C", the choice depending upon process conditions (ref. [4]). Due to the very high temperatures there is not much in the way of complementary NDT that can be applied on-stream, with the exception of thermography, to identify insulation problems. However, the on-stream monitoring allows NDT to be directed to any problem areas immediately the plant is shut-down, helping shut-down planning and minimising down time.

(iii) Most storage tank floors are tested in-service using method "D". The TANKPAC™ procedure accounts for virtually all testing (ref. [5]), and was developed together with an oil industry working group. It has been applied to more than 5000 tanks as a primary maintenance planning tool, and its reliability widely reported (refs. [6, 7]). This method addresses general floor corrosion, corrosion of the shell is easily addressed in-service by ultrasonic C-scan magnetic crawlers, and specialist methods such as TALRUT provide limited information on the critical shell-to-annular area by UT scanning from the chime, provided this is in good condition.

(iv) On-stream process plant may use a similar method to identify active corrosion and stress corrosion cracking (SCC), provided the noise levels are not excessive. Developed by a consortium with EC support, primarily to detect corrosion pitting and stress–corrosion cracking in stainless steel process equipment, the COR-PAC™ procedure (ref. [8]) uses higher frequency sensors to reduce the effect of process noise. The use of higher frequency sensors limits the detection distance to the local area ($\sim + -1$ m).

3 CASE HISTORIES

Examples from each type of test are presented, naturally, for interest, these are cases where defects have been identified and confirmed, however, the majority of tests will not indicate significant problems, allowing valuable inspection resources to be focused elsewhere.

3.1 Stainless steel reactor chain

This set of five vessels could suffer stress corrosion cracking if any moisture got into the process. The internal steam heating coils were a possible source of moisture should they leak, and the cost of shut-down was in excess of £100,000 per day, added to this, the contents were toxic.

The vessels were tested annually using the MONPAC™ procedure, 64 sensors provided 100% coverage of the five vessels, and the test was conducted by increasing the pressure using the plant nitrogen supply, from the operating pressure of 8 psi, to 9 psi, in stages, over a period of ~1 hour. Design pressure of the vessels was 50 psi. The test was always carried out ~16 weeks before shut-down was scheduled, this provided sufficient time to have new top-heads manufactured should a problem be identified by the on-line test. The results of the in-service test usually indicated no identifiable defects, with all sensors graded "insignificant" on the MONPAC™ severity scale, on two occasions however the the in-service test identified "E" grades (the most severe) on a top head. On both occasions the appropriate top-head was manufactured ready for replacement at shut-down. Subsequent inspection and sectioning on both occasions confirmed stress-corrosion cracking up to 80% through wall, the advanced warning saved £m's in lost production. Even had five replacement heads been carried it would have been 2-weeks into the outage before all five reactors could have been decontaminated and fully inspected. On a third occasion "C" sources were identified, however NDT was unable to confirm these at the subsequent outage, 9 months later "E" sources were present in the same locations, and SCC confirmed at the outage.

3.2 Platformers

Platformers comprise three carbon steel vessels stacked on top of one another, plus a lot of stainless internals. During start-up and shut-down the thermal stresses are high where the vessels are joined together. Cracks have been found on platformers where the top heads join the shell, so this critical area is an ideal candidate for monitoring during temperature changes. The temperature of the vessels is up to 550°C so "waveguides" (welded to the shell) are used to allow monitoring, sensors are then outside the insulation on the end of the guide. During one planned shutdown intense AE sources were identified on two lower top-head-to-shell welds, one extending one third of the way around the circumference, the other two-thirds of the way around. NDT confirmed cracking around both heads at these locations. It was not known if the cracking was pre-existing, extending during the shutdown, or caused by the shutdown process itself. The plant would regularly "trip" during operation, and concern was expressed that this may be leading to crack growth. Following repair the decision was taken to permanently monitor the vessels. The following year of operation provided may plant trips, but no evidence of cracking during these. The next shutdown however provided more evidence of cracking. Analysis of the difference between plant trips and shutdown showed that during a plant trip the compressor would stop, leaving the hot gases inside the vessels. Shutdown however left the compressors running circulating

cooler gas, the internals shrouded the shell keeping the shell hot, but the gas cooled the heads quickly, resulting in them "shrinking" away from the shell, resulting in the cracking. A much slower and better controlled cool-down prevents this.

3.3 Methanol converter

The methanol converter has internal refractory and a burner providing internal temperatures of 1300°C at a pressure of 40 bar with hydrogen.

Should the refractory fail, overstress of the transition region could occur within 30 seconds, and within two minutes could burn through the 100 mm shell releasing 200 bar hydrogen. Clearly any "safety" monitoring system must give instantaneous warning and provide this information direct to the plant DCS, enabling rapid shutdown. Although thermal methods are used to identify any long-term refractory problems, the slow heat transfer means that if a rapid refractory failure occurs the shell has almost burned through before the alarm is given, making shutdown before blow-out impossible. Acoustic monitoring of the burner and shell gives instantaneous warning, the monitoring system provides 48 status and alarm signals direct to the plant distributed control system (DCS), plus an interface giving source position information. Operating experience over five years gave no false alarms or down-time, and showed continued effectiveness under simulated alarm conditions.

3.4 Storage tanks

This is one of the largest application areas for in-service testing using AE, special sensors are mounted on the shell above the bottom knuckle, and these detect the sound of corrosion product formation and break-up from the floor. Disbelief is the usual response to this procedure, however an industry user group working

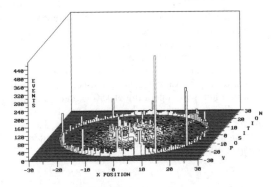

3D LOCATION FOR "ALL DATA"

Figure 1.   TANKPAC™ 3D view of tank corrosion damage.

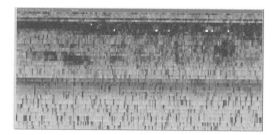

Figure 2. TALRUT C-scan image of annular ring (taken from chine) showing indications of severe corrosion pitting.

with Physical Acoustics Limited developed and proved the method in the field over a 6-year period and more than 600 in-field tests. The results of comparison between the TANKPAC™ test result and subsequent internal inspection and repairs were reported by the user group chairman at the European NDT Conference (ref. [6]). One of the statistics from this showed that up to 50% of tanks were removed from service unnecessarily, wasting considerable maintenance resources and cleaning costs. A look at the statistics more carefully however, showed that some sites had over-maintained their tanks, whereas others had neglected them to the point where the majority was in need of urgent maintenance, such as the "E" grade tank (below) which had lost more than 50% of the annular ring thickness and was immediately retired from service. A similar tank subsequently collapsed when the annular failed during filling, resulting in rapid loss of contents and pulling a vacuum.

A new method called TALRUT allows some indication of annular damage to be obtained by UT from the chine, it is not particularly sensitive, but will identify severe pitting close to the shell and up to ~30 cm in from the chine.

3.5 *Distillation column*

In-service detection of corrosion and stress-corrosion cracking using CORPAC™ technology is ideally illustrated by this example on an operating distillation column. It was impractical to change the stress during operation, and due to flow noise the use of any other standard AE procedure was not possible.

The column was operating at 150°C and the result of the diagnostic, which only required passive listening for a period of one hour, indicated localised active corrosion. Subsequent inspection during shutdown found stress-corrosion cracking in that location.

ACKNOWLEDGEMENTS

The authors would like to acknowledge and thank all members of the UK and Netherlands based user groups who helped to develop the advanced field applications of AE for in-service plant, and EPA Paris and its partners for the CORPAC™ development.

REFERENCES

1. British Standard, 2000. Non-destructive testing Terminology Part 9: Terms used in acoustic emission testing, BS EN 1330–9:2000.
2. Hewerdine, S. Plant Integrity Assessment by Acoustic Emission Testing, Institution of Chemical Engineers International Process safety Working Group, ISBN 0 85295 316 X.
3. Fowler, T.J. Chemical Industry Applications of Acoustic Emission. (1992) – Chemical Engineering Progress.
4. Cole, P.T. On-Line Monitoring of Continuous Process Plant pp 27–39 ISBN 0-85312-683-6.
5. Gautrey, S.N., Cole, P.T. (1997) Proceedings of 22nd European Working Group Conference on Acoustic Emission. Aberdeen.
6. Van De Loo, P.J., Herrmann, B. (1998) Proceedings of 7th European Conference on Non-Destructive Testing.
7. Petroleum Institute of France. Guide for the use of Acoustic Emission on tank Floors.
8. J.C. Lenain, A. Proust, P. Labeeuw, L. Renaud, Y. Cetre. Détection de la corrosion localisée par la technique d'Emission Acoustique – CEFRACOR – *3ème Colloque Européen Corrosion dans les Usines Chimiques et Parachimiques (1997).*

Emerging Technologies in Non Destructive Testing, Van Hemelrijck, Anastasopoulos & Melanitis (eds)
© 2004 Swets & Zeitlinger, Lisse, ISBN 90 5809 645 9

Advanced acoustic emission for on-stream inspection of petrochemical vessels

M.F. Carlos
Physical Acoustics Corporation, Princeton, New Jersey, USA

D. Wang
Shell Oil Products Company, Deer Park, Texas, USA

S.J. Vahaviolos
Mistras Holdings Corporation, Princeton, New Jersey, USA

Athanasios Anastasopoulos
Envirocoustics S.A., Athens, Greece

ABSTRACT: The "Advanced Acoustic Emission (AE) for on-stream inspection" project is a Petroleum Environmental Research Forum (PERF) sponsored initiative. The objective of the project is to develop reliable AE technology for on-stream inspection of pressure equipment in the petroleum refining and chemical industries, instead of internal inspection. To achieve this objective, the project has focused on four key areas of development, including:

• accurate AE source location algorithms,
• quantitative AE-fracture mechanics correlations for fitness for service assessment,
• reliable AE source discrimination techniques, and
• testing guidelines, procedures and inspector qualification.

The completed work items have been integrated into a high-speed, parallel-processing, cost-effective AE system and are currently being evaluated in lab and field AE tests. Many unique and exciting capabilities have been implemented in the "pre-test" AE test planning functions, and "post-test" automated AE data analysis process. Included in the pre-test planning is the capability to predict the AE response in terms of the source moment tensor using Green's functions and integration of the actual sensor response with these predictions. Velocity prediction calculations are aided by three-dimensional dispersion curves (includes signal amplitude) that take into account vessel material as well as AE source type, orientation and depth in the material. In addition, Fitness for Service (FFS) analysis is integrated into the AE software to provide API 579 Level II analysis to help prioritize AE indications for follow-up inspection. Analysis and results of the various software functions, will be presented to show the unique "marriage" of past theoretical models to AE and Fracture mechanics FFS assessment.

1 INTRODUCTION

The overall objective of this effort is to deliver reliable AE methods for global, on-stream inspection of pressure vessels instead of internal inspection. Nine companies, including seven oil companies and two, AE field test companies have joined together in this PERF sponsored initiative. The development work has included investigating and implementing new and improved AE waveform-based techniques for source location, source determination, noise identification and discrimination, correlation of AE indications to severity and ultimately, fitness-for-service assessment.

Much theoretical, laboratory and field test work has been conducted to improve the knowledge and tools necessary to achieve the objectives of this project. These include:

• Developing advanced three-dimensional dispersion curves (wavemode analysis) that take into account, not only the materials associated with the vessel construction, but also vessel source type, orientation

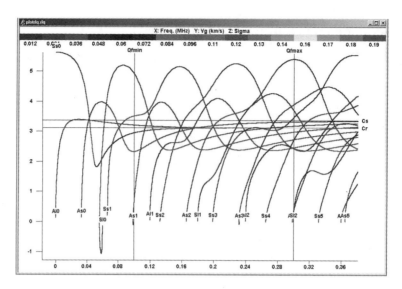

Figure 1. Coloured dispersion curves show velocity as a function of frequency for a specified thickness. Varying wavemode line colors quantify strength.

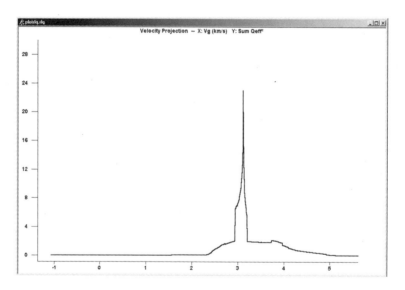

Figure 2. Velocity projection shows the group projected velocity within the Qfmin and Qfmax of frequency limits of the dispersion curve (100 kHz–300 kHz) of figure 1.

and depth in the material (Weaver, 1998 Phase I). An example of calculated 3-D dispersion curves is shown in figure 1.

- Velocity determination (necessary for accurate source location) calculations based on selected frequency ranges of the dispersion curves (figure 2).
- Source waveform prediction (Weaver & Pao 1982, Weaver 1998, Phase II) at the sensor face and sensor output, taking into account source distance, source

height, source moment tensor (which corresponds to source type), wave modes, fluid loading, and the sensor transfer function (figure 3).

- Finite element analysis of pressure vessels taking into account, the vessel construction (including size, mounting, nozzles and welds), its material properties (including strength parameters, physical constants, fracture toughness and material data for crack growth calculations), in order to carry out stress

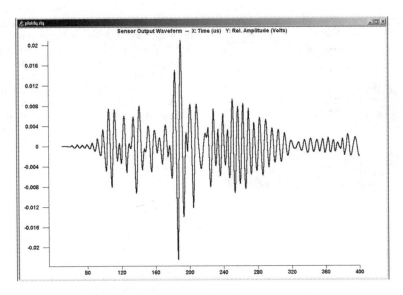

Figure 3.   Source wavefrom prediction at sensor output for a 20 mm thick steel material with surface defect.

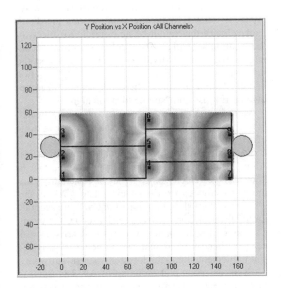

Figure 4.   Unwrapped vessel with color based minimum detectable crack size mapping.

predictions and ultimately fitness for service determination (Tsai et al 1999).

- Level II, Fitness for Service (FFS) analysis calculations within API-579 guidelines as a pre-inspection tool to aid in determining the minimum detectable defect size and to adjust the system in terms of sensor spacing and test pressure to detect defects of interest before the AE test (figure 4).

- Laboratory characterization of petroleum industry steels, to develop an understanding of AE activities due to fatigue cracking and environmentally induced cracking, specifically to determine the load level at which detectable AE signals appear with a given level of damage as well as to determine the relationship of the AE signal to crack size and crack type determination (Rokhlin 1999).
- Calculations to assess crack size from the AE data.
- Level II, FFS analysis during and after the AE test to evaluate the vessel's fitness for service as well as safety of the welded structure (Tsai et al 1999, Wang 1999), using the AE response, source location of AE signal and taking into account the Stress Concentration Factors (SCF) of the various parts of the vessel (welds, nozzles, etc.) as shown in figure 5.
- Location characterization tests carried out to analyze and improve location algorithms.
- Development of advanced event detection and data qualification techniques (Carlos 1999 Event Detection).
- Development of advanced location determination calculations using waveform based AE analysis, frequency filtering and wavemode extraction.
- Development of Time-Frequency Analysis techniques to differentiate embedded AE sources from external noise (Weaver 1999 Phase III).
- Performing Field test work on petroleum vessels to and improve the procedures and software.

As can be seen from the above work, this undertaking represents one of the largest and most comprehensive AE development efforts to date, bringing together these aspects into the AE system and software. Although

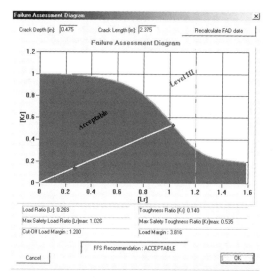

Figure 5. Failure Assessment Diagram (FAD) showing Level II, API Analysis of crack-like flaws.

the project is not yet completed, there are substantial accomplishments and developments to be reported. These are described further below in the following sections.

2 ACCURATE SOURCE LOCATION

Accurate source location requires the application of a series of very important processes, including; accurate detection and processing of AE signals, extraction of critical timing features and assignment of their "correct" propagation velocity, event detection and grouping of AE signals to form an accurate "event record", and application of the appropriate location algorithm. This project has focused on each of these components in terms of pre-test analysis and setup to assure the best up-front test conditions, as well as carrying out the in-test (and post-test) location determination processes and verifications.

PERFPAC software has been developed, based on the theoretical efforts described above with dispersion relations and source waveform prediction. The software provides the user insight into the propagation of AE signals within the structure being analyzed (e.g. vessel or piping) and aid with the selection of AE sensors, frequency filtering and wave velocity determination, for optimum signal detection and accurate location of AE sources. Many variables are acting on AE signals that affect their propagation from the source. This theoretical approach has been integrated to aid in understanding these effects and guide the practitioner into the proper AE system configuration for optimum AE performance.

There are several important tools available within the dispersion relations and source waveform prediction utilities to assist in analyzing and visualizing the propagation of AE signals within the structure being analyzed. First, there is a very versatile dispersion curve generation capability (figure 1) that allows for a 3-dimensional presentation of data.

Second, there is a "best velocity" analysis (figure 2), which provides the user with a theoretical velocity value to use based on a selected frequency range of analysis. This helps in determining the appropriate velocity to use for accurate source location calculations in the presence of multiple wave modes in a structure, each propagating at a different frequency. Third, there is a source waveform prediction capability (figure 3) which shows the user the resultant waveform based on defect type, source height in the structure, source distance and sensor selection. This is useful in determining the best AE sensor to select for the detection of a specific type of defect.

As can be seen by using these integrated utilities, many aspects regarding setting up of the AE system and signal processing prior to a test can be carried out by the program. The user can determine the most appropriate AE sensor, frequency response, and the dominant velocity that might be appropriate to select for the location algorithm. Three other aspects of "accurate source location" have also been investigated and improved in this project. These include, "Extraction of critical timing features", "Event detection and grouping algorithms" and "source location algorithms".

Various timing features are being investigated in this work. A "Timing feature" (also known as "time of arrival") is the determination of the accurate time of arrival of that waveform (AE signal) from the source. Along with a timing feature, it is also important to know a wave velocity, since it may be related to one or more wavemodes or frequencies. The goal of this effort is to allow the extraction of one or more accurate timing features from each AE sensor for a given event. Various timing features including; first arrival, peak analysis, band-limited analysis are being investigated and implemented into the location analysis software.

Event detection and grouping of AE data (hits) to form an event is also a very important consideration in this work. In actual testing on a vessel, there are many noise sources which may be emitting at the same time as a defect event and there is also the possibility of receiving overlapping events. It is important to carefully apply event detection and grouping algorithms to make sure that each received AE "hit" is part of a given event. This is being accomplished by studying various aspects of AE sensor setup, their distance and timing relationships, attenuation and propagation characteristics, and implementing one or more event qualification checks to eliminate errant AE sensor arrivals, in order that the data presented to the location algorithm is correct.

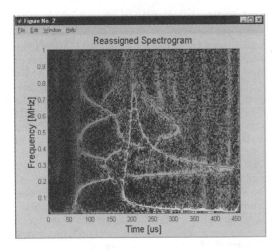

Figure 6. Spectrogram (Time-frequency) graph or a received AE waveform with original wavemode overlaid.

The last important aspect of accurate source location involves the selection and implementation of the source location algorithm. Various source location algorithms have been investigated in this project with an "over-determination" location method being selected, allowing more than the minimum number of AE arrivals to be analyzed in order to improve location accuracy. This is important since it allows an averaging and minimizing of any error due to one or more erroneous data points.

3   RELIABLE SOURCE DISCRIMINATION

The purpose of integrating source discrimination and identification techniques is twofold. First, it provides for the identification of defects as opposed to noise sources and secondly, it allows the identification of different types of defect sources. Removing noise sources and identifying defect sources provides a signal "filtering" function that eliminates processing and analysis of unimportant signals. This can result is "faster" system processing, but more importantly, it provides more accurate results to the Fitness for Service (FFS) algorithms since only data from flaws and known flaw types is analyzed. By performing "Time-frequency" analysis on each received AE waveform, it is possible to extract the original wave modes as shown in figure 6. Since material properties and vessel thickness is already known, and source location provides the information needed to determine distance from the source, then the original moment tensors can be derived, leading to the description of the source. Crack sources can then be distinguished from other sources.

4   ACOUSTIC EMISSION BASED FITNESS-FOR-SERVICE ANALYSIS

With the above AE work being carried out to detect, process, locate and classify AE signals, the next step in the analysis process is to determine the severity of these received AE sources and determine how they affect the overall quality and structural integrity of the vessel. This is the overall purpose of the "Fitness-for-Service" (FFS) part of the project work. It effectively closes the loop and provides a quantitative AE inspection result.

Utilizing the theoretical and laboratory work referenced above in the introduction, AE based Fitness for Service software has been integrated into the PERF-PAC software. The FFS program intention is to provide both "pre-test" analysis and setup support as well as "post-test" FFS analysis results based on Level II, American Petroleum Institute (API), analysis of "crack-like flaws".

In pre-test assessment, the operator is able to view a map of the "minimum detectable defect size" (figure 4), based on the placement and separation of AE sensors on the vessel and the loading (pressurization) schedule planned. This allows some flexibility in sensor and test setup to plan for the detection of a desirable minimum defect size. A minimum detectable defect size map is provided to show the sensitivity of the AE system at different locations on the vessel (as shown in figure 4).

Most of the information needed for the FFS analysis is already entered into the software as the user is setting up the AE system for the test or using the dispersion relation utilities. The vessel geometry is setup (dimensions and thickness) as part of the AE sensor setup process, weld lines and nozzles are setup and dimensioned providing information on the appropriate "Stress Concentration Factors" (SCF) in case an AE source falls along a nozzle weld or weldline. In addition, important material properties are entered simply by selecting the appropriate material from the material database.

The AE-FFS post-test function, analyzes each detected defect based on its AE source location and the AE signal levels (which are related to crack size). The API-579 Level II, crack-like flaw analysis is integrated into the PERFPAC software to evaluate the vessel's "Fitness for Service", based on the detected flaw(s). If the vessel is fit for service, its safety margin is calculated and displayed using the Failure Assessment Diagram (FAD) technique (figure 5).

5   AE INSTRUMENTATION AND GUIDELINES

The efforts, previously described above, are improving on the science and technique of the AE inspection process by providing more accurate and quantitative AE results. In addition to this however, another key part of the project has been to improve further on AE instrumentation and software with a new class of multi-channel, high performance, waveform based, collection and real-time, digital signal processing AE system. This is needed in order to perform the detailed analyses described in the previous sections. This goal has been met with the development of the DiSP, AE system. The DiSP uses multiple PCI/DSP-4's, "AE System on a PCI card", to provide up to 56 AE, high performance channels inside one system enclosure. The features of the system that make it conducive for this project includes, its fully digital design, with 16 bit, 10 Mega-Samples/second A/D waveform acquisition rate, its on-board digital signal processor for processing data at very high speeds, 4 high pass and 4 low pass "real time" filters for selecting the most desirable bandwidth for AE signal processing, its waveform module with separate DSP processor for high speed waveform, all built upon PCI bus, for high speed data transfer to the analysis computer. This multiple digital signal processing system provides the necessary data acquisition and analysis performance which is needed for the success of the project.

The software for the system has been named, "PERFPAC" to denote its origins with the PERF sponsors and the manufacturer's software base. The "user-friendly" and powerful WINDOWS software is based on PAC's AE win software. PERFPAC integrates all the AE system setup, pre-test analysis, acquisition, and post-test analysis functions into one program, incorporating all the features that have been described above. The system and software will expect the user to be Level II certified in AE testing.

In addition to the system and software, a comprehensive set of AE test guidelines are being developed. These guidelines will provide all the necessary background in order to help the user understand all aspects of the test and to guide the user through all parts of the test, including instrumentation description and setup, software description, setup and operation, pre-test planning and analysis, physical setup of the test, data acquisition procedure and guidance, dismantling, data analysis and reporting requirements.

6 SUMMARY

In summary, much work has been completed in each of the four area's identified, including; development of more accurate source location techniques, determination and use of reliable source discrimination techniques, implementation of quantitative AE-fracture mechanics and fitness for service, and integration of these developments into a high speed, cost effective AE system with experienced based, user friendly software and full testing guidelines and procedures. The effort involves, theoretical and laboratory work, field testing as well as system hardware and software development work. The project team is comprised of vessel owners and AE field test companies who are seeking a better, more efficient and accurate NDT test method, AE vendors who are advancing the state-of-the-art of AE, and university professors who are turning theoretical knowledge into practical useful technology. Although not complete, this project is well on its way to a very successful completion, offering a new "Advanced AE for on-stream inspection" technology.

REFERENCES

Carlos, M.F., 1999. Advanced Acoustic Emission For On-Stream Inspection, Progress Report. PERF 95-11 report.

Carlos, M.F., 1999. Event Detection and Grouping for Accurate Source Location. PERF 95-11 report.

Carlos, M.F., Vahaviolos, S.J, Wang, W.D., 2000. Advanced Acoustic Emission For On-Stream Inspection. Acoustic Emission – Beyond the Millennium (Elsevier) pages 159–168.

Rokhlin, S.I., 1999. Assessment of Acoustic Emission from Cracking in A-516 Grade 70 Steel. PERF 95-11 report.

Tsai, Chon, L., Zhao, Yufei, 1999. Fitness-for-Service Assessment for Operating Pressure Vessels and Pipelines In Refinery and Chemical Service by Acoustic Emission Tests. PERF 95-11 report.

Wang, W.D., 1999. Fitness-for-Service/Acoustic Emission correlation program for Pre-Test and Post-Test Analysis. PERF 95-11 report.

Weaver, R., Pao, Y-H., 1982. Axisymmetric Elastic Waves Excited by a Point Source in a Plate. Journal of Applied Mechanics.

Weaver, R., 1998. Theoretical Modeling of Acoustic Emission Wave Propagation, Phase I: Analytic Formalism. PERF 95-11 report.

Weaver, R., 1998. Theoretical Modeling of Acoustic Emission Wave Propagation, Phase II: Numerical Evaluations. PERF 95-11 report.

Weaver, R., 1999. Theoretical Modeling of Acoustic Emission Wave Propagation. Phase III: Preliminary Study of the Inverse Problem. PERF 95-11 report.

Emerging Technologies in Non Destructive Testing, Van Hemelrijck, Anastasopoulos & Melanitis (eds)
© 2004 Swets & Zeitlinger, Lisse, ISBN 90 5809 645 9

Robustness of acoustic emission technique for bearing diagnosis

D. Mba
School of Mechanical Engineering, Cranfield University, Cranfield, Beds, 0AL

ABSTRACT: Acoustic emission (AE) was originally developed for non-destructive testing of static structures, however, over the years its application has been extended to health monitoring of rotating machines and bearings. It offers the advantage of earlier defect detection in comparison to vibration analysis. The investigation reported in this paper was centered on the application of a standard acoustic emission (AE) characteristic parameter on a radially loaded bearing. An experimental test-rig was designed to allow seeded defects on the inner and outer race. It is concluded that irrespective of the radial load and rotational speed, a simple AE parameter such as r.m.s provided an indications of bearing defect.

1 INTRODUCTION

Acoustic emissions (AE) are defined as transient elastic waves generated from a rapid release of strain energy (Pao et al, 1979, Pollock, 1989, Mathews, 1983). In this particular investigation, AE's are defined as the transient elastic waves generated by the interaction of two surfaces in relative motion. The interaction of surface asperities and impingement of the bearing rollers over the seeded defect on the outer and inner races will result in the generation of acoustic emission. Acoustic emission testing on bearings has the added advantage of earlier defect detection in comparison with standard vibration diagnostic techniques.

2 ACOUSTIC EMISSION AND BEARING DEFECT DIAGNOSIS

The formation of subsurface cracks due to the Hertzian contact stress induced by the rolling action of the bearing elements in contact with the inner and outer races, and, the rubbing between damaged mating surfaces within the bearing will generate acoustic emission activity. Catlin (1983) reported AE activity from bearing defects were attributed to four main factors. Roger (1979) utilised the AE technique for monitoring slow rotating anti-friction slew bearings whilst Yoshioka and Fujiwara (1984) have shown that the AE parameters identified bearing defects before they appeared in the vibration acceleration range. Hawman et al (1988) reinforced Yoshioka's observation. The modulation of AE signatures at bearing defect frequencies has also been observed by other researchers (Holroyd, 1993, 2001).

The most commonly measured AE parameters for diagnosis are amplitude, r.m.s, energy, counts and events (Mathews, 1983). Tandon & Nakra (1990) investigated AE counts and peak amplitudes for an outer race defect and concluded that AE counts increased with increasing load and rotational speed. Choundhury et al (2000) employed AE for bearing defect identification and observed that AE counts were low for undamaged bearings. In addition, it was observed that AE counts increased with increasing load and speed for damaged and undamaged bearings. Tan (1990) used a variation of the standard AE count parameter for diagnosis of different sized ball bearings. The acoustic emission technique has also been employed by Miettinen et al (2000) to monitor the lubricant condition in rolling element bearing. And successful applications of AE to bearing diagnosis for extremely slow rotational speeds have been reported (Mba et al, 1999, Jamaludin et al, 2001).

It must be noted that the transmission path between the defect and the receiving transducer affects the observed acoustic emission. The investigation presented in this paper intends to validate the use of AE r.m.s for bearing diagnosis.

3 EXPERIMENTAL EQUIPMENT

A test rig was designed to simulate early stage of bearing defects, see Figure 1. The rig consisted of a motor/gear box unit that providing a rotational speed range of between 10 to 4000 rpm. Two aligning support bearings, a rubber coupling and a larger support bearing was employed. The test bearing investigated

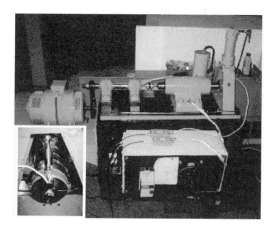

Figure 1.   Bearing test-rig.

Figure 2.   Schematic diagram of acquisition system.

Figure 3.   Seeded 'small' and 'large' defects on the inner race.

was a split Cooper spherical roller, type 01C/40GR. This type of bearing was chosen owing to its ability to be disassembled without removing slave bearings, thereby allowing the test bearing to be regularly inspected throughout the test programme.

Furthermore, it allowed assembly of the defective components with minimal disruption to the test-rig. A radial load was applied to the top of the bearing via a hydraulic cylinder ram supported by an 'H' frame. It must be noted that for all tests, the receiving transducer was cemented onto the test bearing housing, see insert in Figure 1. Characteristics of the test bearing included an internal (bore) diameter of 40 mm, roller diameter of 11.9 mm and the bearing had 10 rollers.

4   DATA ACQUISITION SYSTEM

A piezoelectric type sensor (Physical Acoustic Corporation type WD) with an operating frequency range of 100 kHz–1000 kHz was employed. A schematic diagram of the acquisition system is illustrated in Figure 2.

Pre-amplification ranged from 40 to 60 dB and the acquisition card provided up to 8 MHz sampling rate and incorporated 16-bit precision giving a dynamic range of more than 85 dB. A total of 33,000 data points were recorded per acquisition (data file) at a sampling rate of 4 MHz. One hundred (100) data files were recorded for each simulated case, providing over 0.8 seconds of data per fault simulation. This was equivalent to 8-revolutions of data at 600 rpm; 20-revolutions at 1500 rpm and 40-revolutions at 3000 rpm. A trigger level of 31 mV was employed, this was set above the electronic background noise. The procedure for recording data simply involved arming the acquisition system at random intervals over a 15-minute period for each simulation.

It was thought this would provide a good test on the robustness of specific AE characteristic parameters to diagnosis of operational bearings. The AE parameter measured for diagnosis in this particular investigation was r.m.s.

5   EXPERIMENTAL PROCEDURE

Two types of defects were seeded on the inner and outer races; a 'large' and 'small' defect, see Figure 3. The nominal width, depth and length of the defect on the outer and inner race was measured at 1 mm, 75 μm and 5 mm for a 'small defect', while the 'large defect' had a length of approximately 15 mm, see Figure 3. The test-rig was operated at three different rotational speeds; 600 rpm, 1500 rpm and 3000 rpm. For each rotational speed three load cases were considered: 0 kN, 2.4 kN and 4.8 kN, and for every test condition a total of 100 data files were recorded. Prior to seeding defects the test-rig was operated to provide an indication of background noise levels.

6   EXPERIMENTAL RESULTS

It must be noted that at the higher speed of 3000 rpm, pre-amplification was reduced to 40 dB. All the results presented are comparative at 60 dB, which implied that a multiplication factor of 10 was applied to all data captured at 40 dB. For all outer race defects an increase in r.m.s was observed for increasing speed and defect size. This trend also applied to inner race defects, see Figures 4 and 5. The following format was employed for labeling all AE data presented:

L0; L2; L4; L – load; **0** load value – 0 KN; **2** – 2.4 KN; **4** – 4.8 KN

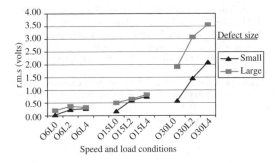

Figure 4.   r.m.s values for outer race defects.

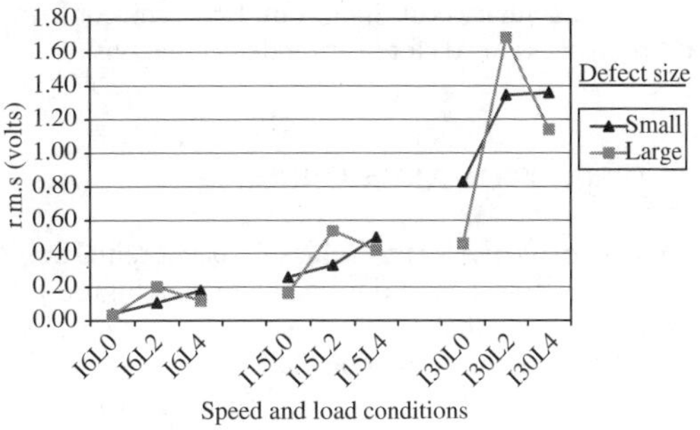

Figure 5.   r.m.s values for inner race defects.

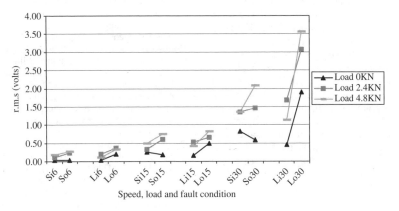

Figure 6.   r.m.s values for inner and outer race defects as a function of load, speed and defect size.

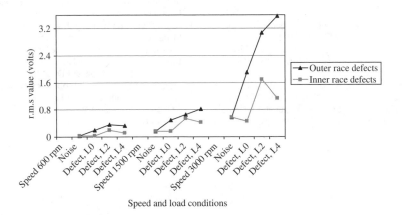

Figures 7.   Background noise and large defect condition for varying loads; speed 1500 rpm.

'Si30'; S – Small defect; i – inner race; 30 speed – 3000 rpm.
'O6L4'; Outer race defect at 600 rpm and 4.8 KN
'I15L2'; Inner race defect at 1500 rpm and 2.4 KN
'Li15L2'; Large inner race defect at 1500 rpm and 2.4 KN.

Comparisons of AE r.m.s values showed an increase for outer race defects in comparison to inner race defects, see Figure 6. This can be attributed to increased attenuation experienced by signatures from the inner race. The results presented thus far are in agreement with several researchers (Yoshioka et al, 1984, Tandon et al,

1990, Choundhury et al, 2000, Tan, 1990) and confirms the robustness of AE for bearing defect diagnosis.

7 DISCUSSION AND CONCLUSION

A comparison of AE r.m.s values for background noise and a large defect condition can be seen in Figure 7. It must be noted that background noise r.m.s values could mask defect conditions at lower speeds. A rise in AE r.m.s was observed for increasing rotational speed. In addition, at fixed rotational speeds there was evidence to suggest that increasing the load on the defect also resulted in an increase in AE r.m.s. The use of AE r.m.s values has been validated as a robust technique for detecting bearing damage and has been shown to correlate with increasing speed, load and defect size. In application of AE to bearing diagnosis, particularly on machinery operated over a range of speed conditions, it would be advisable to investigate background noise r.m.s levels at all process operational speeds. Whilst numerous exotic diagnostic techniques such as wavelets, higher order statistics, neural networks, etc, could be employed to aid diagnosis, all attempts must be made to keep the method of diagnosis simple and robust as this is the only way to encourage the adoption of this invaluable technique.

REFERENCES

Pao, Y-H., Gajewski, R.R and Ceranoglu, A.N. (1979), Acoustic emission and transient waves in an elastic plate, J. Acoust. Soc. Am., 1979, 65(1), 96–102.

Pollock, A.A. Acoustic Emission Inspection, Physical Acoustics Corporation, Technical Report, 1989, TR-103-96-12/89.

Mathews, J.R. Acoustic emission, Gordon and Breach Science Publishers Inc., New York. 1983, ISSN 0730–7152.

Catlin Jr., J.B. The Use of ultrasonic diagnostic technique to detect rolling element bearing defects. Proceeding of Machinery and Vibration Monitoring and Analysis Meeting, Vibration Institute, USA, April 1983, 123–130.

Roger, L.M. The application of vibration analysis and acoustic emission source location to on-line condition monitoring of anti-friction bearings. Tribology International, 1979, 51–59.

Yoshioka T and Fujiwara T. New acoustic emission source locating system for the study of rolling contact fatigue, Wear, 81(1), 183–186.

Yoshioka T and Fujiwara T. Application of acoustic emission technique to detection of rolling bearing failure, American Society of Mechanical Engineers, Production Engineering Division publication PED, 1984, 14, 55–76.

Hawman, M.W. and Galinaitis, W.S, Acoustic emission monitoring of rolling element bearings, Proceedings of the IEEE, Ultrasonics symposium, 1988, 885–889.

Holroyd, T.J and Randall, N., (1993), Use of acoustic emission for machine condition monitoring, British Journal of Non-Destructive Testing, 1993, 35(2), 75–78.

Holroyd, T. Condition monitoring of very slowly rotating machinery using AE techniques. 14th International congress on Condition monitoring and Diagnostic engineering management (COMADEM'2001), Manchester, UK, 4–6 September 2001, 29, ISBN 0080440363.

Tandon, N and Nakra, B.C, Defect Detection of rolling element bearings by acoustic emission method, Journal of Acoustic Emission, 1990, 9(1) 25–28.

Choundhury, A and Tandon, N. Application of acoustic emission technique for the detection of defects in rolling element bearings, Tribology International, 2000, 33, 39–45.

Tan, C.C. Application of acoustic emission to the detection of bearing failures. The Institution of Engineers Australia, Tribology conference, Brisbane, 3–5 December 1990, 110–114.

Miettinen, J and Andersson, P. Acoustic emission of rolling bearings lubricated with contaminated grease, Tribology International, 2000, 33(11), 743–802.

Mba, D., Bannister, R.H and Findlay, G.E. Condition monitoring of Low-speed rotating machinery using stress waves: Part's I and II. Proceedings of the Instn Mech Engr 1999, 213(E): 153–185.

Jamaludin N., Dr. D. Mba and Dr. R.H. Bannister condition monitoring of slow-speed rolling element bearings using stress waves. Journal of Process Mechanical Engineering, I Mech E. Pro. Inst. Mech Eng., 2001, 215(E): Issue E4, 245–271.

Infrared methods

Ultrasound Burst Phase Thermography (UBP) – advances by ultrasound frequency modulation

T. Zweschper, G. Riegert, A. Dillenz & G. Busse

Institute of Polymer Testing and Polymer Science (IKP), Dept. of Non-destructive Testing, University of Stuttgart, Germany

ABSTRACT: Ultrasound excited thermography allows for defect selective imaging using thermal waves that are generated by elastic waves. The mechanism involved is local friction or hysteresis which turns a dynamically loaded defect into a heat source which is identified by a thermography system. If the excitation frequency matches to a resonance of the vibrating system, temperature patterns can occur that are caused by standing elastic waves. These undesirable patterns can affect the detection of damages in a negative way. We describe a technique how the defect detectability of ultrasound activated thermography can be improved. With the objective of a preferably diffuse distributed sonic field we applied frequency modulated ultrasound to the material. That way the standing waves can be eliminated or reduced and the detectability is improved.

1 EXTERNAL EXCITATION: OPTICAL LOCKIN THERMOGRAPHY (OLT)

Thermal waves have been used very early for remote monitoring of thermal features, e.g. cracks, delamination (Wong et al. 1979), and other kinds of boundaries. After the advantage of signal phase had been discovered (Busse 1979, Thomas et al. 1980, Lehto et al. 1981), phase angle imaging using photothermal techniques (Nordal & Kanstad 1980) became a powerful tool for imaging of hidden structures due to the enhanced depth range and its independence on optical (Rosencwaig & Busse 1980) or infrared surface patterns. As the thermal diffusion length is the important parameter for depth range (Rosencwaig 1979), it turned out very soon that imaging of features deep underneath the surface requires very low modulation frequencies and a correspondingly long time to obtain a photothermal image. Unfortunately many industrial questions are related to samples with defects at about a millimeter depth. An image obtained pixel after pixel at a modulation frequency in the 1 Hz range could easily require several hours.

One approach allowing for a reduction of inspection time is lockin thermography where the low frequency thermal wave is generated simultaneously on the whole surface of the inspected component and monitored everywhere several times per modulation cycle in order to obtain an image of amplitude and phase of temperature modulation (Carlomagno & Berardi 1976, Beaudoin et al. 1985, Kuo et al. 1987, Busse et al.

1992). In this case the inspection time is given by a few modulation cycles. As one can image square meters of airplanes within a few minutes (Wu et al. 1997), one has a powerful method for fast inspection of safety relevant structures with a depth range of several millimetres in CFRP.

Figure 1 shows the principle and the set-up of optical excited lockin thermography. Absorption of intensity modulated radiation generated on the whole surface a thermal wave. It propagates into the interior where it is reflected at boundaries and defects so that it moves back to the surface where it is superposed to the initial wave. This way a defect is revealed by the local change of phase angle. Therefore both defects

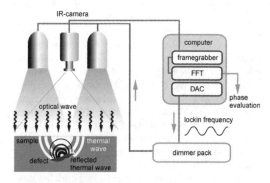

Figure 1.   Principle and set-up of lockin thermography with optical excitation.

and intact structures are imaged at the same time. Defects can be revealed only by comparing the observed features with expected patterns by reference samples or by design drawings. Even for an experienced inspector it is difficult to distinguish defect areas from these thermal features.

2 INTERNAL EXCITATION METHODS

Further investigations aimed at a method where a defect responds selectively so that the image would display only the defect and not the confusing background of the intact structure. Defects differ from their surroundings by their mechanical weakness. They may cause stress concentrations, and under periodical load there may be hysteresis effects or friction in cracks and delamination. As defects may be areas where mechanical damping is enhanced, the ultrasound is converted into heat mainly in defects (Mignogna et al. 1981, Stärk 1982).

2.1 Ultrasound lockin thermography (ULT)

Modulation of the elastic wave amplitude results in periodical heat generation so that the defect is turned into a local thermal wave transmitter. Its emission is detected via the temperature modulation at the surface which is analysed by lockin thermography tuned to the frequency of amplitude modulation (Rantala et al. 1996, Salerno et al. 1998). The ultrasonic transducer is attached at a fixed spot from where the acoustic waves are launched into the whole volume where they are reflected several times until they disappear preferably in a defect and generate heat. These high frequencies are very efficient in heating since many hysteresis cycles are performed per second.

2.2 Ultrasound burst phase thermography (UBP)

Another established method is the use of short sonic bursts for sample excitation. The spectral components of that signal and the following cooling down period provide information about defects in almost the same way as Lockin technique but with reduced measuring duration. As the characteristic defect signal is contained in a limited spectral range while the noise typically is distributed over the whole spectrum, one can reduce noise as well. That kind of evaluation technique using Fourier or Wavelet transformations is also applicable to flash light excited thermography (Maldague et al. 1996, Maldague & Marinetti 1996, Galmiche et al. 2000). The signal to noise ratio of ULT and UBP images (and hence defect detectability) is significantly better than just one temperature snapshot image in a sequence (Thomas et al. 1995, Favro et al. 2000).

The following example confirms that high frequencies probe only near-surface areas while low frequencies with their larger depth range provide information

Table 1. Depths of delaminations in the CFRP sample shown in Figure 2.

Marker	Depth of delamination
1	0.2 mm
2	0.9 mm
3	1.5 mm
4	2.2 mm
5	Ultrasonic transducer

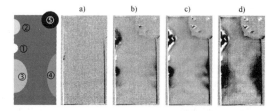

Figure 2. UBP images of a CFRP sample with delamination in different depths. Excitation: 1.1 kW, ultrasonic frequency 20 kHz, burst duration: 200 ms.

about defects deeper inside the component. Thus depth resolved mapping of defects can be performed with only one measurement. For this demonstration we manufactured a CFRP sample (80 mm × 38 mm × 5 mm) which had four delaminations in various depths.

The transducer was attached in the right upper corner (see marker 5 in Figure 2). After an ultrasonic burst of 200 ms duration (1.1 kW) had been applied the resulting temperature sequence was evaluated at several frequencies. In the phase image at 4 Hz (Fig. 2a) only the defect next to the surface causes a change in the phase angle. At 1 Hz (Fig. 2b) already three delaminations in depths from 0.2 mm to 1.5 mm become visible. At 0.1 Hz (Fig. 2c) all defects appear. At an even lower frequency (0.025 Hz, see Figure 2d) the image contrast is reduced due to lateral diffusion effects.

3 EXPERIMENTAL ARRANGEMENT

The experimental set-up is shown in Figure 3. We used a non-standard digital 2.2 kW power supply, driving an ultrasonic welding transducer whose excitation frequency can be modulated with a frequency up to 25 Hz in a range from 15 to 25 kHz. The excitation duration typically ranges from 100 ms up to a few seconds for burst phase thermography and up to some minutes for lockin thermography. A digital to analogue converter triggers the modulation frequency and if required the lockin frequency for amplitude modulation. The temperature sequences were acquired with a focal plane array IR camera (CEDIP Jade MWIR). The software

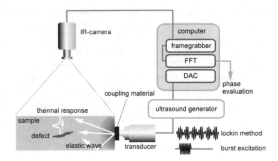

Figure 3. Principle and set-up of ultrasonic excited thermography.

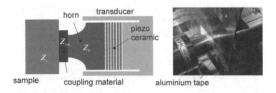

Figure 4. Impedance matching for efficient ultrasound coupling.

(e/de/vis DisplayIMG) installed on a common personal computer performs an online Fourier transformation and subsequent image processing.

The tip of the welding horn is pressed against the component to be tested (Fig. 4). For an efficient ultrasound coupling tight contact as well as impedance matching is important. The acoustic impedance Z is given by the product of the density ρ and the elastic wave velocity v in the material. In close analogy to optical techniques it is advantageous if the acoustic impedance of the coupling material (Z_{cm}) matches with the geometric mean of the acoustic impedance of the material to be tested (Z_s) and the material of the welding horn (Z_h), in our case a titanium base alloy

$$Z_{cm} = \sqrt{Z_s Z_h} = \sqrt{\rho_s v_s \cdot \rho_h v_h} \qquad (1)$$

In terms of optimised ultrasound coupling in CFRP, good results were obtained on a thin sheet of adhesive aluminium tape whose acoustic impedance is close to the square root mentioned above. In comparison to the direct coupling without any coupling material an increase of the efficiency of 35 per cent was achieved. Such materials do not only improve the ultrasound coupling but also the component surface was protected against external damage caused by mechanical and thermal load during the ultrasonic excitation.

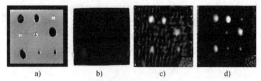

Figure 5. Detection of delamination caused by thermal overload in a CFRP plate (d = 5 mm) using sonic excitation. Comparison of techniques: (a) air coupled ultrasound C-scan (450 kHz, transmission), (b) phase evaluation of optically excited lockin thermography at a frequency of 0.01 Hz, (c) phase signature of ultrasound excited lockin thermography at 0.05 Hz, (d) reduced standing wave pattern and increased defect detectability using frequency modulated ultrasound lockin thermography (f_R = 15–25 kHz, f_{mod} = 20 Hz), phase signature at 0.05 Hz.

4 ENHANCED DEFECT DETECTABILITY USING FREQUENCY MODULATED ULTRASOUND

By applying a monofrequent excitation to a sample it is not unlikely that this frequency matches to a resonance of the vibrating system. The result is a standing wave pattern. Due to hysteretic losses in the elongation maximum, these standing elastic waves can appear as temperature patterns causing misinterpretations: In the worst case the defect could be hidden in a node ("blind spot") while the standing wave maximum might appear as a defect. This can be avoided by using two or more ultrasound converters with several frequencies simultaneously or, even better, by frequency modulation of a sinusoidal signal. In these cases the standing wave pattern is superimposed by a field of propagating waves giving sensitivity also where only nodes existed before.

An example for the advances of frequency modulated sonic excitation is the inspection of a CFRP plate with nine areas of heat damage at the rear side of the sample (Fig. 5). The initial sample with a thickness of 2 mm was damaged by thermal overload and subsequently enlarged to a thickness of 5 mm. To localize the nine damages and to characterize their real size, an air coupled ultrasound C-scan was performed at 450 kHz (Fig. 5a). The limited depth range of optically excited lockin thermography prevents a successful application of this remote technique on the laminate: All damages remain undetected (Fig. 5b). The use of sonic excitation improves the situation. Using the lockin technique with its enlarged depth range five of the nine damages could be found (Fig. 5c). But the monofrequent excitation (f_R = 20 kHz) generates a temperature pattern caused by standing elastic waves with all the drawbacks mentioned before. Especially the smaller defects remain still undetected. A significant enhancement was achieved by wobbling the excitation frequency from 15 to 25 kHz with a modulation frequency of 20 Hz. The pattern was reduced mostly, now eight of the thermal damage are visible (Fig. 5d). In this case the modulation of the excitation frequency enhances defect detectability.

5 DETECTION OF FRACTURES, DELAMINATION, AND CRACKS IN CFRP STRINGER STRUCTURES

Fractures of stringers in aerospace components are a serious problem because this kind of damage is almost invisible from the outer surface of an aircraft for most non-destructive testing methods, since stringers are hidden behind a panel. The access from inside the aircraft is difficult and time consuming.

A stringer reinforced structure (Fig. 6a) where we compared several excitation techniques is the cutout of a flap (850 mm × 240 mm) with a crack and a delamination in the stringer underneath the intact CFRP-skin

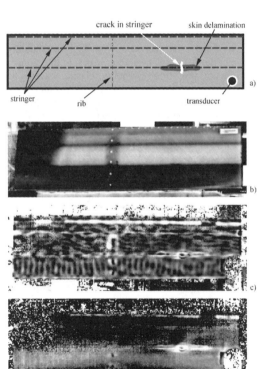

Figure 6. Crack detection in a stringer of a CFRP landing flap (cutout) using different techniques: (a) map of damage location, (b) phase evaluation of optically excited lockin thermography at a frequency of 0.05 Hz, (c) phase signature of ultrasound excited lockin thermography at 0.05 Hz, (d) elimination of the standing wave pattern using frequency modulated ultrasound lockin thermography, phase signature at 0.05 Hz.

(thickness 4.5 mm). The OLT phase image (Fig. 6b) suppresses the optical sample characteristics and displays the whole thermal structure of the component (stringer, rib, and variation of material thickness), but the small effect of the defect easily escapes attention. The temperature pattern in the ULT phase signature (Fig. 6c) caused by standing waves makes the detection of the crack very difficult. When frequency modulated ultrasound (f_R = 15–25 kHz, f_{mod} = 20 Hz) was applied the pattern disappeared and defect detectability increased significantly (Fig. 6d).

6 CONCLUSION

Ultrasound excited thermography is an efficient non-destructive tool. The phase evaluation of the recorded temperature sequences has clear advantages as compared to conventional pulse thermography: Sensitivity variations within the detector array as well as optical sample characteristics such as inhomogeneous temperature distribution and varying emission coefficients on the sample surface are suppressed. Furthermore the signal to noise ratio is improved significantly. The use of short sonic bursts (e.g. for sample excitation in conjunction with phase evaluation) reduce the measuring time without abandoning the advantages mentioned before. Automation is possible, a cycle time less than one second per measurement is achieved. Coupling materials for impedance matching do not only improve the ultrasound coupling but also the component surface is protected against external damage caused by mechanical and thermal load during the ultrasonic excitation. By applying a frequency modulated ultrasonic excitation the disturbing temperature patterns caused by standing elastic waves were eliminated. This results in homogenous power density in the material and in an increased defect detectability.

The simultaneous use of several ultrasound converters provides a more homogenous low power density in the material to be inspected. This results in an efficient and non-destructive excitation so that even larger components can be examined with this technique.

ACKNOWLEDGEMENTS

The authors are grateful to Airbus Deutschland GmbH (Bremen, Germany) for providing samples. Technical support of Cedip Infrared Systems (Croissy-Beaubourg, France) and Branson (Dietzenbach, Germany) is highly appreciated as well.

REFERENCES

Beaudoin, J.L., Merienne, E., Danjoux, R., Egee, M. 1985. Numerical system for infrared scanners and application to the subsurface control of materials by photothermal radiometry. In: Infrared Technology and Applications, SPIE Vol. 590, p. 287.

Busse, G. 1979. Optoacoustic phase angle measurement for probing a metal. In: Appl. Phys. Lett. Vol. 35, pp. 759–760.

Busse, G., Wu, D., Karpen, W. 1992. Thermal wave imaging with phase sensitive modulated thermography. In: J. Appl. Phys. Vol. 71, pp. 3962–3965.

Carlomagno, G.M., Berardi, P.G. 1976. Unsteady thermo-topography in non-destructive testing. In: Proc. 3rd Biannual Exchange, St. Louis/USA, pp. 33–39.

Dillenz, D., Zweschper, Th., Busse, G. Progress in ultra-sound phase angle thermography, Thermosense 2001, Orlando, USA.

Favro, L.D., Xiaoyan Han., Zhong Ouyang., Gang Sun., Hua Sui., Thomas, R.L. 2000. Infrared imaging of defects heated by a sonic pulse, Rev. Sci. Inst. 71, 6, pp. 2418–2421.

Galmiche, F., Vallerand, S., Maldague, X. 2000. Pulsed Phase Thermography with the Wavelet Transform, Review of Progress in Quantitative NDE, D.O. Thompson et D.E. Chimenti eds, Am. Institute of Physics, 19A: pp. 609–615.

Kuo, P.K., Feng, Z.J., Ahmed, T., Favro, L.D., Thomas, R.L., Hartikainen, J. 1987. Parallel thermal wave imaging using a vector lock-in video technique. In: Photoacoustic and Photothermal Phenomena, ed. P. Hess and J. Pelzl. Heidelberg: Springer-Verlag, pp. 415–418.

Lehto, A., Jaarinen, J., Tiusanen, T., Jokinen, M., Luukkala, M. 1981. Amplitude and phase in thermal wave imaging. In: Electr. Lett. Vol. 17, pp. 364–365.

Maldague, X., Marinetti, S. 1996. Pulse Phase Infrared Thermography, J. Appl. Phys, 79(5), pp. 2694–2698.

Maldague, X., Marinetti, S., Busse, G., Couturier, J.-P. 1996. Possible applications of pulse phase thermography. Progress in Natural Science. Supplment to Vol. 6, pp. 80–82.

Mignogna, R.B., Green, R.E., Jr., Duke, Henneke, E.G., Reifsnider, K.L. 1981. Thermographic investigations of high-power ultrasonic heating in materials. In: Ultrasonics 7, pp. 159–163.

Nordal, P.-E., Kanstad, S.O. 1979. Photothermal radiometry. In: Physica Scripta Vol. 20, pp. 659–662.

Rantala, J., Wu, D., Busse, G. 1996. Amplitude Modulated Lock-In Vibrothermography for NDE of Polymers and Composites. In: Research in Nondestructive Evaluation, Vol. 7, pp. 215–218.

Rosencwaig, A. 1979. Photoacoustic microscopy. In: American Lab. 11, pp. 39–49.

Rosencwaig, A., Busse, G. 1980. High resolution photo-acoustic thermal wave microscopy. In: Appl. Phys. Lett. Vol. 36, pp. 725–727.

Salerno, A., Dillenz, A., Wu, D., Rantala, J., Busse; G. Progress in ultrasonic lockin thermography. Quantitative infrared thermography, QIRT 98. Akademickie Centrum Graficzno-Marketingowe Lodart S.A., Lodz 1998, S. 154–160. ISBN 83-87202-88-6

Stärk, F. 1982. Temperature measurements on cyclically loaded materials. In: Werkstofftechnik 13, Verlag Chemie GmbH, Weinheim, pp. 333–338.

Thomas, R.L., Pouch, J.J., Wong, Y.H., Favro, L.D., Kuo, P.K., Rosencwaig, A. 1980. Subsurface flaw detection in metals by photoacoustic microscopy. In: J. Appl. Phys. Vol. 51, pp. 1152–1156.

Thomas, R.L., Favro, L.D., Kuo, P.K., Ahmed, T., Xiaoyan Han, Li Wang, Xun Wang, Shepard, S.M. 1995. Pulse-Echo Thermal-Wave Imaging for Non-Destructive Evaluation, Proc. 15th International Congress on Acoustics, Trondheim, Norway, pp. 433–436.

Wong, Y.H., Thomas, R.L., Pouch, J.J. 1979. Subsurface structures of solids by scanning photoacoustic micro-scopy. In: Appl. Phys. Lett. 355, pp. 368–369.

Wu, D., Salerno, A., Malter, U., Aoki, R., Kochendörfer, R., Kächele, P.K., Woithe, K., Pfister, K., Busse, G. 1997. Inspection of aircraft structural components using lockin-thermography. In: Quantitative infrared thermo-graphy, QIRT 96, Stuttgart, ed. D. Balageas, G. Busse, and G.M. Carlomagno. Pisa: Edizione ETS, pp. 251–256. ISBN 88-467-0089-9.

Emerging Technologies in Non Destructive Testing, Van Hemelrijck, Anastasopoulos & Melanitis (eds)
© *2004 Swets & Zeitlinger, Lisse, ISBN 90 5809 645 9*

Contrast enhancement of nonlinear photothermal radiometry in composite materials

G. Kalogiannakis & D. van Hemelrijck
Department of Mechanics of Materials and Constructions, Vrije Universiteit Brussel, Belgium

ABSTRACT: Great potential has been detected in nonlinear photothermal radiometry. Experimental (Rajakarunanayake & Wickramasinghe 1986) and theoretical (Gusev et al. 1992, Gusev et al. 1993, Mandelis et al. 1994) studies have demonstrated the higher defect contrast, when detecting the second harmonic of the temperature signal with respect to the modulation frequency of the external heat source. When temperature variations inside the material are large, thermal properties can not be considered constant. This fact is simply one of the factors that render the heat diffusion nonlinear. Based on an experimental study (Kalogiannakis et al. 2002), which determined the aforementioned dependencies, a finite element model, simulating the temperature distribution of the one-dimensional transient problem, was developed using modulated thermal excitation. When a defect is introduced, spectral analysis shows that the second harmonic of the surface temperature signal is affected much more than the fundamental frequency component. When treating highly nonlinear materials like composites, the phenomenon is enhanced, demonstrating the potential of the photothermal method. Carbon fiber reinforced epoxy is widely employed in advanced constructions and its thermal properties exhibit quite strong dependence on temperature. Therefore, it was selected to verify the validity of the assertion.

1 INTRODUCTION

The detection of a defect is enhanced in nonlinear thermal imaging by virtue of higher contrast of the second harmonic of the temperature signal. This fact has triggered increasing interest in nonlinear heat diffusion phenomena. Some work has been done both in the experimental (Rajakarunanayake & Wickramasinghe 1986) and the theoretical field (Gusev et al. 1992, Gusev et al. 1993, Mandelis et al. 1999).

The first concern of nonlinear thermal imaging is the study of the thermal properties' dependence on temperature. The influence of this effect on the generation of higher harmonics has to be evaluated before extending the study to strongly nonlinear effects.

Furthermore, nonlinear heat diffusion phenomena appear to have much potential in composite materials. Experimental facts (Kalogiannakis et al. 2002) demonstrate that the dependence of thermal properties on temperature is much stronger for carbon fiber reinforced epoxy matrix than for metals. As the use of composite materials in modern constructions increases, the necessity for efficient nondestructive evaluation grows.

2 PROBLEM FORMULATION

The geometry of a thermal imaging problem is practically confined to one dimension (Maldague 2001). Thus, the study is focused on a numerical nonlinear treatment of the one-dimensional heat diffusion problem in composite materials using the finite element method.

Three-dimensional heat diffusion in solids is mathematically formulated by the following partial differential equation.

$$C\frac{\partial T}{\partial t} - \nabla(k\nabla T) = Q \tag{1}$$

where C = heat capacity; and k = thermal conductivity (both are characteristic thermal properties of the solid material); Q = internal heat source to account for the generation of heat inside the material.

In order to detect a defect inside a solid component using photothermal radiometry, one has to investigate the temperature response at the surface of the solid to a thermal excitation.

A thermal wave, generated by depositing heat on the surface, flows, by means of conduction, into the substance.

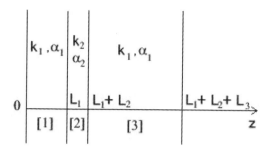

Figure 1. A three-layer solid represents a defective component. The thermal properties of the first and the third layer are identical as shown.

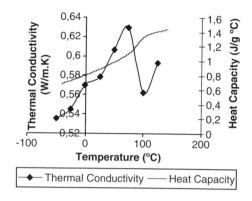

Figure 2. The thermal properties of carbon/epoxy as a function of temperature.

These waves are highly attenuated and the depth of the penetration depends on the thermophysical properties of the substance and the modulation frequency of the heat source. The key variables in this investigation are the amplitude and the phase of the resulting oscillating temperature field in the stationary regime at the surface. When the wave reaches the interface of a defect (Fig. 1), it is 'reflected' back to the surface altering the surface response.

In many cases, the thickness of the defect layer is negligible. Therefore there is contact between layers 1 and 3 without an actual bonding force and with a marginal interface thermal resistance (Maldague 2001). Cielo et al. (1985) used a laser beam to lift one side through local heating. Furthermore, Gusev et al. (1993) presented a theory for a strong nonlinear photothermal signal considering the modulation of the defect thickness originated from the periodical thermal stimulation. For the present study, a delamination of constant thickness was considered having appeared at approximately three layers below the surface of the composite laminate. For the particular material, this corresponds to the penetration depth of a thermal wave having a frequency of 1 Hz. Supposing this delamination is retained opened, it is modeled having a thickness of approximately one fourth of the layer thickness. The properties of the delaminated area were those of dry air. The power density was considered to have an amplitude of 15 kW/m². This value is adequately high to cause significant temperature variations.

3 PROPERTIES DETERMINATION

The thermal properties, as aforementioned, depend in general on temperature T rendering (1) nonlinear. In practice, though, one can simplify the problem, by considering small temperature variations and taking, thus, constant thermophysical parameters. However, when temperature variations are adequately large, a nearer approximation to the actual state may be necessary.

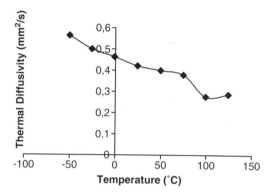

Figure 3. The thermal diffusivity of carbon/epoxy as a function of temperature.

This approach employs the thermal properties as a linear function of temperature (Mandelis et al. 1999).

$$k = k_0(1 + \beta.T), \text{ and } C = C_0(1 + \gamma.T) \qquad (2)$$

For most substances, the dependence factor β is negative. However, for carbon epoxy treated in the problem, a foregoing experimental work (Kalogiannakis et al. 2002) showed that β is positive. The proportionality factor γ is also typically positive for composite materials. The method of Modulated Temperature Differential Scanning Calorimetry (MTDSC) (Annual Book of ASTM Standards, 1998) was used to measure the thermal conductivity and the specific heat of the composite material in a temperature range from -50 to $120°C$.

The variation of the thermal properties with temperature is illustrated in Figures 2–3. The experimental results clearly depict the three regions, which characterize the material behavior. The sudden increase of the gradient for the specific capacity and the existence of a maximum for thermal conductivity indicate the

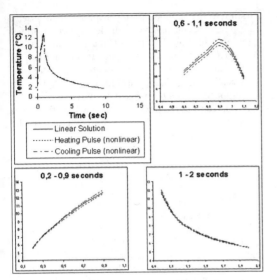

Figure 4. The temperature surface response for a pulse in three different cases: (i) the linear case (the absolute value of the temperature response is identical for a heating and a cooling pulse), (ii) a heating pulse for variable thermal properties (10 kW/m², 1sec), (iii) a cooling pulse for variable thermal properties (− 10 kW/m², 1sec).

microstructural changes taking place during the glass transition stage.

Another plain conclusion is the high dependence of the composite material thermal behavior on temperature, a precursor of significant nonlinear effects. For instance, within a few tens of degrees above the room temperature, the material experiences changes in the heat capacity on the order of 35% whereas the changes in thermal conductivity do not exceed 10%. Due to this fact of smaller dependency of the thermal conductivity, the thermal diffusivity is dropping with temperature rise (Fig. 3).

At high temperatures, conductivity grows and heat is transmitted faster to the interior. At the same time, heat capacity grows and makes temperature increase more difficult. On the other hand, cold temperatures are favored. From a phenomenological point of view, it is easier to descend the slopes of Figure 2 than to ascend. Heat capacity and thermal conductivity drops while thermal diffusivity rises to support this phenomenon. So, less energy needs to be absorbed for the temperature to decrease. To demonstrate this, a comparative diagram for the temperature evolution at the surface for a heating and a cooling pulse (negative power density) is illustrated in Figure 4. The absolute value of the power density of the pulse is 10 kW/m² and its duration is 1 sec. As already discussed, the linear solution, which is of course identical for a heating or a cooling pulse (absolute value), is predictably between the two other

solutions. While the solutions diverge until the maximum absolute value, they quickly converge afterwards.

4 NUMERICAL MODEL

The problem of nonlinear heat diffusion has been analytically treated by Mandelis et al. (1999) and Gusev et al. (1993). The theoretical solution is inevitably based upon premises. In this study the problem is treated numerically based on the finite element method.

In order to solve this problem, the boundary conditions need to be given. In the case under study, the diffusion is considered one-dimensional with a *Neumann* boundary condition of a cosine form at the front surface, whereas a homogeneous *Dirichlet* condition is applied at the rear surface. These boundary conditions are typically corresponding to a test using modulated thermography. In particular:

$$k \frac{\partial T}{\partial z}\Big|_{z=0} = Q_0 (1 - \cos \omega t) \tag{2a}$$

$$T\big|_{z=l.} = 0 \tag{2b}$$

In the equations above, L = thickness of the model under consideration; Q_0 = power per unit surface of the heating source; and ω = modulation frequency.

In order to solve the equation numerically in the time domain, semi-discretization of the spatial domain in finite elements leads to sets of ordinary differential equations (in the case of linear problems).

$$C\dot{a} + Ka + f = 0 \tag{3}$$

In equation (3), C = the assembled heat capacity matrix from the elementary matrices; K = the assembled conductivity matrix from the elementary matrices; f = the forcing vector encompassing external and internal heat sources; and a = temperature. The finite element method (Zienkiewicz & Taylor 1991) (including finite difference) is widely applicable in discrete time approximation yielding the so-called recurrence schemes.

Introducing a single step method for the solution of the first order problem, it is evident that only the value of initial temperature is necessarily specified. A linear expansion in the time interval is, thus, natural and most widely used. At each time step t_{n+1}, the discrete problem in the weighted residual form is described by equation (4a).

$$\psi(a_{n+1}) = Ca_{n+1} + P(\tilde{a}_{n+1} + \theta\Delta t\alpha_{n+1}) + \bar{f}_{n+1} = 0 \tag{4a}$$

$$\tilde{a}_{n+1} = a_n \tag{4b}$$

$$a_{n+1} = a_n + \Delta t\alpha_{n+1} \tag{4c}$$

187

In equation (4a), P = vector of nonlinear internal forces; α = the first time derivative of temperature; θ = the weighting parameter; and $\bar{f}_{n+1}$ = the weighed force vector.

The solution of the nonlinear problem reduces in solving the set of equations in (5a, b, and c).

$$\left(C + \theta \Delta t K_T^{(i)}\right)\delta\alpha_{n+1}^{(i)} = -\psi\left(\alpha_{n+1}^{(i)}\right) \tag{5a}$$

$$\alpha_{n+1}^{(i+1)} = \alpha_{n+1}^{(i)} + \delta\alpha_{n+1}^{(i)} \tag{5b}$$

$$K_T^{(i)} = \left.\frac{\partial P}{\partial a}\right|_{a^{(i)}} \tag{5c}$$

where $\alpha^{(i)}$ = the time derivative at the ith iteration; and the greek δ = difference designation.

For the convergence of any finite element approximation it is necessary and sufficient that it be consistent and stable. The time-stepping scheme of Galerkin ($\theta = 2/3$) was selected to serve this goal.

The modified Newton-Raphson method was employed to solve the nonlinear problem defined in equations (5a, b, and c). It essentially replaces the variable Jacobean stiffness matrix by a constant approximation corresponding to the first iteration.

5 SPECTRAL ANALYSIS

The solution of finite element model was analyzed with the National Instruments Signal Processing Toolkit. (Super-Resolution Spectral Analysis and Joint Time and Frequency Analysis Toolkit).

The latter is using the model-based methods, the auto-regressive model (AR) in particular, to estimate parameters such as the amplitude, phase and frequency of signal components. The method uses a short sample data length to predict future data with error $w[n]$, based on the model of equation (6).

$$x[n] = -\sum_{k=1}^{p} a_k x[n-k] + w[n] \quad , \quad 0 \leq n \leq N \tag{6}$$

Once the coefficients a_k are determined, it is possible to extract the missing data based on the given finite data set. The most significant advantage of the model-based method is the high frequency resolution without the necessity for a large data set. On the other hand, the method is quite sensitive to noise. Therefore, the local error of the finite element method could have a substantial role in generating noisy false signal components detected in the spectral analysis. Besides, the numerical procedure inevitably induces a pulse-like diffusion wave at the beginning of the simulation. Even if a cosine excitation form is used, the pulse has the

duration of the simulation time step, attenuating with time. The method, however, can also detect the signal damping behaviour. This important aspect can accredit the signal components belonging to the domain of interest and exclude the components generated, for numerical reasons, in the short term.

The algorithm, which was used for parameter estimation was the *matrix pencil* method while the super resolution power spectra have been computed by the *covariance* method.

6 RESULTS – DISCUSSION

In order to establish a link between abnormal isotherms recorded and the presence of a subsurface defect, it is necessary to study the thermal front propagation inside the material and the alteration of the temperature response at the surface. Therefore, graphs depicted in the Figures 5–8 show the in-depth temperature signal components. The values of amplitude and phase are normalized with respect to their values in a sound material component as shown in equation (7).

$$R = \frac{R_D - R_S}{R_S} \tag{7}$$

In the above, the subscripts D and S correspond to defective and sound component respectively and R is either the amplitude or the phase response.

For comparison, the analysis is first performed for the fundamental frequency component in the linear case. The existence of the defect is revealed with the peaks of the normalized signal components. These originate from the thermal resistance of the defect interface met by the thermal front. One important fact at this point is that lateral diffusion of the front around the defect (flow effect) is neglected due to the one-dimensional consideration of the problem. This effect reduces the visibility of the defect in real life testing.

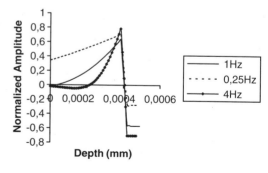

Figure 5. The normalized amplitude of the temperature (linear solution) as a function of the depth inside the material.

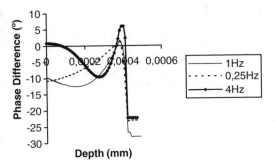

Figure 6. The normalized phase of the temperature (linear solution) as a function of the depth inside the material.

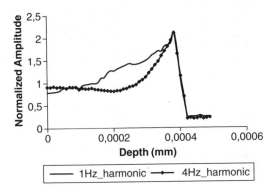

Figure 7. The normalized amplitude of the harmonic temperature signal (nonlinear solution) as a function of the depth inside the material.

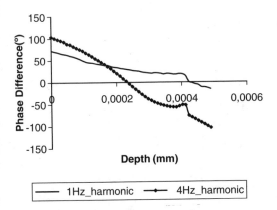

Figure 8. The normalized phase of the harmonic temperature signal (nonlinear solution) as a function of the depth inside the material.

The temperature amplitude exhibits exponential decrease as depth increases when the material is intact. When a defect is present at a depth equivalent to the penetration depth of the thermal wave, in the linear case the amplitude rises to reach a peak of about 50% higher near the defect.

The temperature field along the thickness and, more specifically, the surface response depend on the excitation frequency (Fig. 5). It appears to obtain much higher values if the frequency is lower (0,25 Hz) or even be unaffected by the presence of the defect if the frequency is much higher because of the strong attenuation and the smaller diffusion length.

The inaccuracies (derivative singularities) for the second harmonic are explained by the augmentation of the relative numerical error with respect to the real values, which are much smaller for the second harmonic. Nevertheless, the amplitude of the second harmonic presents the same pattern as for the fundamental component. A thorough observation of the graph in Figure 7, however, would reveal that the normalized

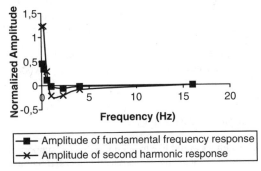

Figure 9. The normalized amplitude of temperature as a function of the excitation frequency for the fundamental and the second harmonic response.

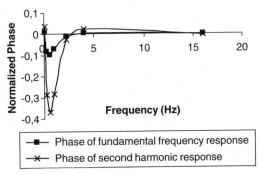

Figure 10. The normalized phase of the temperature as a function of the excitation frequency for the fundamental and the second harmonic response.

189

values are much higher in this case, jumping to 200% near the defect.

The phase difference exhibits the same discontinuity approaching the defect interface. In the linear case the defect interference causes a negative change (Fig. 6) while a great positive change is observed for the second harmonic (Fig. 7), especially at the surface where the response is measured. This is an important feature additional to the fact, that phase remains unaffected by non uniform heating or emissivity over a large area under photothermal radiometry testing.

Several excitation frequencies were used to acquire the graphs depicted in Figures 9 and 10. These graphs correspond to the temperature response characteristics at the surface. As aforementioned, the normalized values converge to zero as the frequency increases due to the heavier damping and consequently the smaller influence of the defect on the surface signal. It is clearly shown that although the contrast for both signal components follows the same pattern for the fundamental and the second harmonic, the relative values have great difference. This fact suggests in practice a better visibility of the defect. Therefore, drawing the conclusions of this study, it has been confirmed and numerically demonstrated that the nonlinear effects of the heat diffusion have great potential in photothermal radiometry. This argument becomes actually stronger for composite materials where nonlinear effects are 'loudly' present.

ACKNOWLEDGEMENTS

The authors would like to acknowledge the support of the Fund for Scientific Research – Flanders (Belgium), which made this work possible.

REFERENCES

Annual Book of ASTM Standards, E1952-98 1998.

Arpaci, V. 1966. *Conduction Heat Transfer*. Massachussets: Addison-Wesley.

Carslaw, H.S. & Jaeger, J.C., (second edition) 1959. *Conduction of Heat in Solids*. New York: Oxford Science Publications.

Cielo, P., Maldague, X., Rousset, G. & Jen, C.K. 1985. Thermoelastic inspection of layered materials: dynamic analysis. *Materials Evaluation* 43(9): 1111–1116.

Gusev, V., Mandelis, A. & Bleiss R. 1992. Theory of Second Harmonic Thermal-Wave Generation: One-Dimensional Geometry. *International Journal Of Thermophysics* 14(2): 321–337.

Gusev, V., Mandelis, A. & Bleiss R. 1992. Theory of Strong Photothermal Nonlinearity from Sub-Surface Non-Stationary ('Breathing') Cracks in Solids. *Journal of Applied Physics* A57: 229–233.

Kalogiannakis, G., Van Hemelrijck, D. & Van Assche, G. 2002. Experimental Study on the Nonlinear Effects of Heat Diffusion in Composites and its Potential in NDT. *Proceedings of European Conference of Composite Materials*, Brugge, 3–7 June 2002.

Maldague, X. 2001. *Theory and Practice of Infrared Technology for Nondestructive Testing*. New York: Wiley Series in Microwave and Optical Engineering.

Mandelis, A., Salnick, A., Opsal, J. & Rosencwaig, A. 1999. Nonlinear Fundamental Thermal Response in Three-Dimensional Geometry: Theoretical model. *Journal of Applied Physics* 85(3): 1811–1821.

Mandelis, A. 2001. *Diffusion-Wave Fields*, New York: Springer-Verlag.

Rajakarunanayake, Y.N. & Wickramasinghe, H.K. 1986. *Applied Physics Letters* 48:218

Zienkiewicz, O.C. & Taylor, R.L. 1991. *The Finite Element Method*, Malta: McGraw – Hill.

Vibrational methods

Emerging Technologies in Non Destructive Testing, Van Hemelrijck, Anastasopoulos & Melanitis (eds)
© *2004 Swets & Zeitlinger, Lisse, ISBN 90 5809 645 9*

Crack identification in beam and plate structures using wavelet analysis

E. Douka
Aristotle University, School of Technology, Mechanics Division, Greece

S. Loutridis & A. Trochidis
Aristotle University, School of Technology, Physics Division, Greece

ABSTRACT: A method for crack identification in beam and plate structures based on wavelet analysis is presented. The existence of a crack in a structural member causes changes in the structural response at the crack site. Such local changes are usually not apparent from the measured response data, they are however detectable by utilizing wavelet analysis due to its high resolution properties. The viability of the proposed method is demonstrated by wavelet analyzing vibrational modes of beams and plates containing open transverse cracks. The cracks are located due to the sudden changes in the spatial variation of the transformed response.

1 INTRODUCTION

Many studies have been carried out in the last two decades in an attempt to find methods for non-destructive crack detection in structural members. A review of the state of the art of vibration based methods for testing cracked structures has been published by Dimarogonas (1996).

A crack in a structure induces a local flexibility, which causes changes in natural frequencies and mode shapes of vibration. Classical methods utilize natural frequency shifts (Chondros & Dimarogonas 1980, Cawley & Adams 1979, Narkis 1993) and mode changes (Rizos et al. 1990, Stubbs & Kim 1996, Kim et al. 2002) to identify cracks. Natural frequencies are rather easy to measure with a desired degree of accuracy but they give a proper estimation only in cases of moderate cracks. Using mode shapes on the other hand has two main drawbacks: the crack may not be influences significantly the lower modes usually measured and furthermore, noise and sensors used can considerably affect the accuracy of the detection procedure.

It appears that wavelet analysis can provide an alternative to classical methods of crack detection (Deng & Wang 1998, Wang & Deng 1999, Quek et al. 2001, Hong et al. 2002). The advantage of wavelet analysis is that allows an accurate identification of local features in the structural response signal and can be easily implemented as fast as the Fourier Transform. In the present work a method of crack identification in beams and plates structures based on wavelet analysis is proposed. The simulated vibration modes of the cracked beams and plates are wavelet transformed and the presence of a crack is easily detected due to sudden changes in the variation of the response. The obtained results provide a foundation for using wavelet analysis as efficient crack detection tool.

2 WAVELET ANALYSIS

In this section the relevant wavelet theory is briefly introduced. A detailed analysis is given by Mallat & Hwang (1992).

Let $\psi(x)$ be a function localized in both space and frequency domains. The function $\psi(x)$ is called the mother wavelet and is used to generate wavelets by translation and dilation defined by

$$\psi_{u,s}(x) = \frac{1}{\sqrt{s}} \psi(\frac{x-u}{s}) \tag{1}$$

where u is the real-valued translation parameter and s is the real-valued dilation (scale) parameter.

For a given function $f(x)$ the wavelet transform (wavelet coefficients) is defined as

$$Wf(u,s) = \int_{-\infty}^{+\infty} f(x)\, \overline{\psi}_{u,s}(x)\, dx \tag{2}$$

where, the overbar denotes the complex conjugate of the function under it.

An important property of the wavelet transform is its ability to react at subtle changes of the analyzing signal. Hence, singularities of the signal $f(x)$ can be detected by performing the wavelet transform to obtain the wavelet coefficients and by examining their variations with position. In particular, Mallat & Hwang (1992) related the singular points of a function to the local maximum of the wavelet transform modulus. This relation indicates the presence of a maximum of the wavelet transform modulus at the finer scales (small values of s), where a singularity occurs. In the general case, a sequence of modulus maxima is detected which converges to the singularity. The singularities can be characterized by examining the evolution of the amplitude of the modulus maxima across scales. Singularities created by the signal have a modulus maxima amplitude which decrease with decreasing scale. This allows their discrimination from the ones produced by noise, since their amplitude increases strongly with decreasing scale. Within these results the wavelet transform can be applied for detecting and characterizing singularities in a signal. The detection of the wavelet maxima can be extended in signals distributed in two dimensions by using the two-dimensional wavelet transform.

In the subsequent section the wavelet transform is applied for identifying singularities caused by the presence of cracks in structures based on their displacement response signals. The effectiveness of the proposed method in identifying cracks is investigated for beam and plate structures.

3 WAVELET ANALYSIS OF CRACKED STRUCTURES

3.1 Cracked beams

As a first example we consider a cantilever beam of length l and rectangular cross section $w \times w$. The beam contains a crack located at l_c from the fixed end as shown in Figure 1a. The crack is assumed to be open and have a uniform depth a.

Due to the localized effect the cracked beam can be simulated by two segments connected by a massless spring (Fig. 1b). The bending spring constant K_T in the vicinity of the cracked section can be evaluated as a function of the crack size using fracture mechanics considerations (Paipetis & Dimarogonas 1986). It can be expressed as

$$K_T = \frac{1}{C}, \quad C = (5.346w/EI)J(a/w) \qquad (3)$$

where E is the modulus of elasticity of the beam, I is the area moment of the beam cross-section and $J(a/w)$ is the dimensional local compliance function.

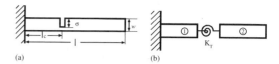

Figure 1. a) Cantilever beam under study, (b) Cracked cantilever beam model.

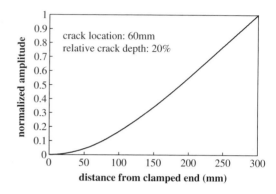

Figure 2. Calculated fundamental vibration mode of the cracked cantilever beam.

For the connection between the tow segments, conditions can be introduced which impose continuity of the displacement, bending moment and shear. Moreover, an additional condition imposes equilibrium between transmitted bending moment and rotation of the spring representing the crack. The resulting system can be solved numerically and the mode shapes of vibration can be obtained.

For numerical simulations a plexiglas cantilever beam of length 300 mm and rectangular cross-section 20×20 mm^2 has been considered. A crack of relative depth 20% is introduced at $x = 60$ mm from the clamped end. The fundamental vibrational mode of the beam was calculated and the results are shown in Figure 2.

To determine the location of the crack the response data of Figure 2 were wavelet transformed. Using "symm 4" as the analyzing wavelet the wavelet transform is implemented for scales 1 to 25. The results of the wavelet analysis are presented in Figure 3 for scales 2, 5, 10 and 15.

It is obvious that the wavelet transform coefficients exhibit a maximum at $x = 60$ mm. This implies the presence of a singularity. To be certain about the presence of a real crack, however, the trend of the wavelet maxima at this point as the scale decreases has been examined. The results are shown in Figure 4. It can be clearly seen that the absolute value of the wavelet maxima decreases in a regular manner with decreasing scale.

194

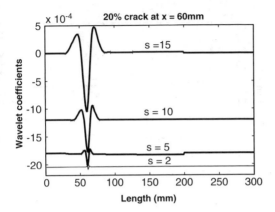

Figure 3. Wavelet analysis of different scales based on the displacement response shown in Figure 2.

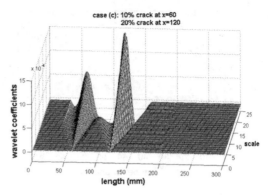

Figure 5. Three-dimensional plot of the wavelet transform showing the trend of the wavelet modulus maxima at crack location.

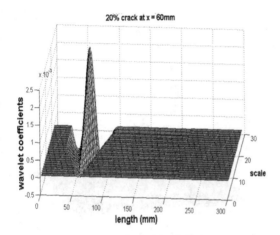

Figure 4. Three-dimensional plot of the wavelet transform showing the trend of the wavelet modulus maxima at crack location.

As a second example we consider a cantilever beam having two cracks located at l_1 and l_2.

For numerical simulations the same beam with dimensions defined as above is considered. Two cracks, at locations $l_1 = 60$ mm and $l_2 = 120$ mm from clamped end, are introduced. Their relative cracked depths are fixed at $a_1 = 10\%$ and $a_2 = 20\%$ respectively. Based on the model described above, the fundamental vibration mode of the beam was calculated.

Using the same procedure, the calculated displacement response was wavelet transformed. The results are presented in a three-dimensional plot for scaled 1 to 25 as shown in Figure 5. It is clear that, at x = 60 mm and x = 120 mm, the wavelet coefficients exhibit variations analogous to this obtained for a beam with one crack, thus indicating the locations of the two cracks.

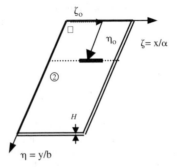

Figure 6. Cracked plate in non-dimensional coordinates.

3.2 Cracked plates

In this section the wavelet based approach is investigated in case of a cracked plate.

An elastic rectangular plate of thickness H and in-plane dimensions $a \times b$ contains a surface crack lying parallel to one side of the plate as shown in Figure 6. The crack is $2C$ in length with variable depth. The centre of the crack is located at (x_0, y_0) and has a depth of h_0. The following non-dimensional parameters are introduced: $\zeta_0 = x_0/a$, $\eta_0 = y_0/b$, position of the crack centre; $\xi_0 = h_0/H$, relative crack depth at the centre of the crack; $2c = 2C/a$, a measure of the length of the crack.

The cracked section is represented as a continuous line spring with a varying stiffness along the crack. The flexibility of the crack is modelled by calculating the additional rotation Θ along the crack to bending stress σ_b as (Rice & Levy 1972)

$$\Theta = \frac{12(1 - \nu^2)}{E} \sigma_b a_{bb} , \qquad (4)$$

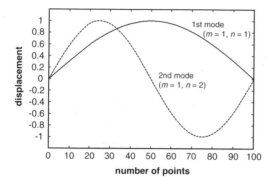

Figure 7. Calculated displacement response of the over-all cracked plate along a vertical line $\zeta = 0.5$ for the first and second mode. Crack location $\eta_0 = 0.2$, relative crack depth 20%.

where E is the Young's modulus and ν is the Poisson's ratio. The compliance coefficient a_{bb} is given by

$$a_{bb} = \frac{1}{H}\int\limits_0^h g_b{}^2 dh \qquad (5)$$

where g_b is the dimensionless function depending on the relative crack depth.

Using the usual boundary conditions and the slope discontinuity along the crack, the equation of motion of the plate can be solved and the mode shapes can be calculated (Khadem & Rezaee 2000a,b).

As a first example, a plate simple supported at all four edges with an all-over part through crack at $\eta_0 = 0.2$ running parallel to its horizontal edge is considered. The depth of the crack is assumed constant along its length. The following dimensions and mechanical characteristics are considered: $a = 0.2$ m, $b = 0.3$ m, $H = 0.004$ m, $E = 200$ GPa, $\nu = 0.3$, $\rho = 7860$ Kg/m^3. Numerical calculations were performed for different values of the relative crack depth ξ_0. Here we refer to the case where $\xi_0 = 0.4$

Response data of the plate displacement along vertical lines ($\zeta = $ const.) at different locations are generated. This way, the problem becomes practically one-dimensional. The displacement response of the cracked plate against the coordinate η along a vertical line $\zeta = 0.5$ is shown in Figure 7. It can be seen that the displacement data reveal no local features that directly indicate the existence of the crack.

The structural response data are analyzed using the continuous wavelet transform. The wavelet transform is implemented for scales 1 to 25 using a "symm 4" as the analyzing wavelet. The results of the wavelet analysis are presented in Figure 8 for scales 2, 5, 10 and 15.

It can be seen that the wavelet transform coefficients exhibit a maximum at point $\eta = 0.2$. This implies the presence of a singularity. To be certain about the

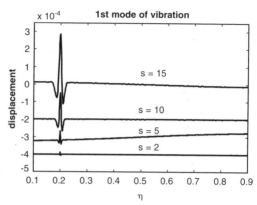

Figure 8. Wavelet analysis of different scales based on the displacement response of the cracked plate in Figure 6 (first mode of vibration).

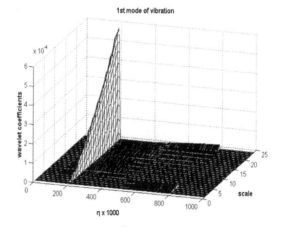

Figure 9. Three-dimensional plot of the wavelet transform showing the trend of the wavelet modulus maxima at crack location.

existence of the crack the trend of the wavelet maxima at this point has been investigated with decreasing scale (Fig. 9). It can be seen that the absolute value of the wavelet maxima decreases in a regular manner as the scale decreases.

The second example refers to a simply supported plate containing a finite crack of relative length $c = 0.2$. The center of the crack is located at $\zeta_0 = 0.3$, $\eta_0 = 0.4$ and has relative depth $\xi_0 = 0.4$. The dimensions and mechanical characteristics of the plate are considered the same as in the previous example.

Based on the two-dimensional vibration mode shapes calculated in the (ζ, η) coordinates, response data of the plate displacement is generated. The data is represented by a two-dimensional array of 41×41

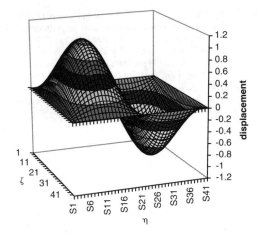

Figure 10. Calculated two-dimensional displacement response of a plate with a finite crack of length $2c = 0.2$. Crack's centre location $\zeta_0 = 0.3$, $\eta_0 = 0.4$, relative crack depth 40%.

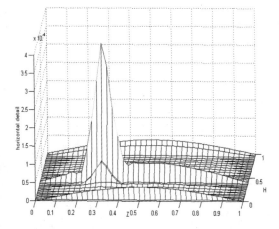

Figure 11. The horizontal detail of the wavelet transforms at decomposition level 2. Crack centre located at (0.3, 0.4), relative crack length $2c = 0.2$.

points using the same spatial sampling distance of 0.025 for both directions. The displacement response of the cracked plate is shown in Figure 10.

In this case the two-dimensional wavelet transform is employed. Using the "symm 4" wavelet the wavelet transform is performed at a level (or scale) 2 of decomposition. The results are shown in Figure 11. It can be seen that the coefficient exhibits a maximum value at the point $\zeta_0 = 0.3$, $\eta_0 = 0.4$. This identifies the centre of the crack. In moving away form the crack centre, along the line containing the crack, the value of the coefficient decreases as the distance

increases, reaching a relatively small value at the tow ends of the crack. This gives a good estimate about the relative length of the crack.

4 CONCLUSIONS

A method for crack detection in structures has been presented. The viability of the method has been demonstrated by analyzing the vibration modes of beam and plate structures using the "symm 4" wavelet. The location of the crack was determined by the variation in the spatial response of the transformed signal at the site of the crack. Such local variations usually do not appear from the measured data they are, however, discernible as singularities when using wavelet analysis.

In conclusion, the presented results provide a foundation for using wavelet analysis as efficient crack detection tool. Further work is needed to validate and advance this damage detection tool. For example, the effect of crack size and the sensitivity of the method to random noise. Work is already under progress on the above-mentioned issues.

REFERENCES

Cawley, P. & Adams, A.D. 1979. The location of defects in structures from measurements of natural frequencies. *Journal of Strain Analysis* 14: 49–57.

Chondros, T.G. & Dimarogonas, A.D. 1980. Identification of cracks in welded joints of complex structures. *Journal of Sound and Vibration* 69: 531–538.

Deng, X. & Wang, Q. 1998. Crack detection using Spatial measurements and wavelet. *International Journal of Fracture* 91: L23–28.

Dimarogonas, A. 1996. Vibration of cracked structures. *Engineering Fracture Mechanics* 55: 831–857.

Hong, J.C., Kim, Lee, H.C. & Lee, Y.W. 2002. Damage detection using Lipschitz exponent estimated by the wavelet transform: applications to vibration modes of a beam. *International Journal of Solids and Structures* 39: 1803–1816.

Narkis, N. 1994. Identification of crack location in vibrating simple supported beam. *Journal of Sound and Vibration* 172: 549–558.

Mallat, S. & Hwang, W. 1992. Singularity detection and processing with wavelets. *IEEE Transactions of Information Theory 38*: 617–643.

Kim, T.J., Ruy, Y.S., Cho, H.M. & Stubbs, N. 2002. Damage identification in beam-type structures: frequency-based method vs mode-shape-based method. *Engineering Structures* 25: 57–67.

Khadem, S.E. & Rezaee, M. 2000. An analytical approach for obtaining the location and depth of an all-over part-through crack on externally in-plane loaded rectangular plate using vibration analysis. *Journal of Sound and Vibration* 230: 291–308.

Khadem, S.E. & Rezaee, M. 2000. Introduction of modified comparison functions for vibration analysis of a rectangular cracked plate. *Journal of Sound and Vibration* 236: 245–258.

Paipetis, S.A. & Dimarogonas, A.D. 1986. Analytical methods in rotor dynamics. *Elsevier Applied Science.* London.

Quek, S., Wang, Q., Zhang, L. & Ang, K. 2001. Sensitivity analysis of crack detection in beams by wavelet analysis. *International Journal of Mechanical Sciences* 43: 2899–2910.

Rice, J.R. & Levy, N. The part-through surface crack in an elastic plate. *Journal of Applied Mechanics* 3: 183–194.

Rizos, P., Aspragathos, N. & Dimarogonas, A.D. 1990. Identification of crack location and magnitude in a cantilever beam from the vibration modes. *Journal of Sound and Vibration* 138: 381–388.

Stubbs, N. & Kim, T.J. 1996. Damage localization in structures without base line modal parameters. *American Institute of Aeronautics and Astronautics* 34: 1644–1649.

Wang, Q. & Deng, X. 1999. Damage detection with spatial wavelets. *International Journal of Solids and Structures* 36: 3443–3468.

Vibration damping as an NDE tool for woven carbon fabric laminates

C. Kyriazoglou, R.D. Adams & F.J. Guild
Department of Mechanical Engineering, University of Bristol, UK

B.H. Le Page
School of Engineering, University of Surrey, UK

ABSTRACT: The results reported in this paper explore the use of vibration damping as a potential non-destructive evaluation method to detect early cracks in woven fabric laminates. The experimental work consists of measurements of energy loss during the resonant vibration of woven fabric laminates as well as measurements of Dynamic Flexural Modulus and specifically demonstrates the sensitivity of Specific Damping Capacity (SDC) to the presence of early damage, whilst the values of Dynamic Flexural Modulus are hardly affected. The value of SDC is defined as the ratio of the energy loss in one cycle of resonant vibration to the total energy stored within that cycle. Closed form analysis using the solution of the classical wave equation can be applied to undamaged beams and to beams containing homogeneous damage. Some key features of the classical solution are denoted, and presented.

1 INTRODUCTION

Woven carbon fabric laminates are receiving more attention as potential materials for engineering applications due to their excellent drape-ability, allowing for complex shapes to be made, and resistance to impact damage (Ishikawa & Chou 1982). Despite these benefits, their wider use is restricted by a lack of understanding of issues relating to their structural integrity. Quantification and modeling of damage accumulation possesses inherent difficulties due mainly to the complexity of the fiber architecture. Effective detection of early damage in woven carbon fabric laminates still remains an elusive and challenging task. However, such detection constitutes a vital step in allowing wider use of these materials in critical applications. This work is a part of a wider program, which aims to develop sound physical models to describe the relationship between damage and residual properties in woven fabric composites, and to develop proposals for a non-destructive evaluation tool to monitor damage in service.

2 EXPERIMENTAL PROCEDURE

In the steady state dynamic method, beams are vibrated in free-free flexure in their first mode of resonant vibration, as shown in Figure 1. Two thin wires are used to support the vibrating beam, at the nodal positions.

Two coils, using magnets attached at each end of the beam, drive the vibration.

A sinusoidal input signal is produced from a frequency generator. The oscillating signal is amplified through a power amplifier and current flows to the two coils, which are connected in series with each other, and the power amplifier. Current flowing into the magnetic fields of the two end magnets produces a sinusoidal force that drives the beam symmetrically. At resonant vibration the maximum displacement, at the center of the beam, is measured using a laser device. At resonance, values of resonant frequency,

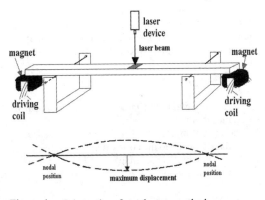

Figure 1. Schematics of steady state method.

maximum displacement and current flowing into the coils are recorded.

3 CLASSICAL SOLUTION

The classical solution of the wave equation has been well documented by Guild & Adams (1981). However a number of key features should be emphasized.

Making the usual assumptions for a Bernoulli-Euler beam, the equation of motion for free-free lateral vibration is

$$EI\frac{\partial^4 u}{\partial x^4} + \rho A\frac{\partial^2 u}{\partial t^2} = 0 \tag{1}$$

where, E = Dynamic flexural modulus, I = beam second moment of area, x = distance along beam length, u = transverse displacement, ρ = density of beam material, A = cross sectional area of beam, t = time.

Having obtained the experimental values of resonant frequency, f_{res}, of the vibrating beam in its first mode of resonant vibration, the value of the wavelength number λ including the rotary inertia J of the end magnets (Guild & Adams 1981) and measured the period T of the resonant vibration, flexural modulus, E_{flex}, is given by:

$$E_{flex} = \frac{m}{Il}(\frac{2\pi}{T})^2(\frac{l}{\lambda})^4 \tag{2}$$

where λ = wavelength number, T = period of resonant vibration. The energy stored by the beam in bending, U is given by:

$$U = \frac{E_{flex}I}{2}\int_0^l(\frac{\partial^2 \upsilon}{\partial x^2})^2 \partial x \tag{3}$$

It is assumed that the energy stored by the beam in shear is negligible.

From equation (3), using integration by parts, it may be shown that U is an implicit function of the maximum lateral vibration of the beam in its first mode of resonant vibration, and its resonant frequency:

$$U = f(u_{max}^2, f_{res}^2, \lambda) \tag{4}$$

where u_{max} = maximum lateral displacement during resonant vibration.

The energy input, ΔU, is calculated by measuring the voltage induced to the coils when they were excited externally to vibrate in the magnetic field at known frequency and amplitude.

$$\Delta U = f(u_{max}, \Gamma_0, \Gamma_J, i) \tag{5}$$

where Γ_0 and Γ_J are the calibration factors of the two coils, i = amperes that flows through the coils, to induce a certain voltage.

Having calculated the total strain energy stored in one cycle, U, from eq. (4) and the energy dissipated per that cycle, ΔU, from eq. (5), the *specific damping capacity (SDC)*, ψ, is then defined by:

$$\psi = \frac{\Delta U}{U} \tag{6}$$

4 MATERIALS

4.1 *Material preparation*

For all GFRP coupons the resin matrix was an Astor Stag epoxy resin system with a MNA Hardener curing agent and a K61B Accelerator. The reinforcement was E-glass continuous fibers with a density of 2.56 gm/cm^3. For the Woven GFRP systems, the resin matrix was a Shell Epikote 828 epoxy resin system with a MNA curing agent and K61B accelerator. The fiber reinforcement was a Fothergill Engineered Fabrics Ltd Y0227 E-glass continuous eight-harness satin weave fabric. The cloth had an approximate weight of 297 gm/m^2 and the density of the E-glass fibers was taken as 2.56 gm/cm^3. The Woven CFRP systems had a Ventico M7750 epoxy resin system and the fiber reinforcement was a T300 carbon fiber continuous five-harness satin weave cloth.

4.2 *Laminate construction*

All laminates investigated in this study were in the form of simple beam coupons. The types of lay ups of those laminated coupons varied. The lay-ups and materials tested were: (0°/90°)$_S$ continuous GFRP; (90°/0°)$_S$ continuous GFRP; woven (0°, 90°) GFRP reinforced with 4 layers of 8 harness satin weave; woven (0°,90°) CFRP reinforced with 4 layers of 5 harness satin weave; and woven quasi-isotropic reinforced with 4 layers of 5 harness satin weave (0°, 90°/±45°)$_S$. Furthermore a notch was introduced to the Woven quasi-isotropic CFRP systems, by drilling a circular hole of a 5 mm diameter at the center of the coupons.

5 MECHANICAL AND DYNAMIC TESTING

The overall assessment of the mechanical and dynamic behavior of a series of undamaged and damaged coupons was considered.

For the case of the cross ply continuous GFRP coupons, damage was introduced by quasi static loading, using a strain rate of 0.1 mm/sec. The effect of damage to a series of coupons was investigated; each coupon was loaded in tension up to a different maximum strain in order to obtain a series of increasing crack densities. The same methodology for introducing damage was used in the case of woven cross ply GFRP coupons. A series of woven cross ply CFRP coupons were subjected to an increasing number of cycles of fatigue loading in tension, at 60% of their static failure load. The incrementation of the increasing number of cycles differed for each of the Woven CFRP coupons.

The effect of the localized damage in a area adjacent to a notch for a series of Woven quasi-isotropic CFRP coupons, was considered. A circular notch at the center of the test coupons was introduced. The notched woven quasi-isotropic CFRP coupons were subjected to different number of cycles of fatigue loading in tension, in order to understand the mechanics and formation of localized damage around the notch.

All coupons were tested dynamically using the steady state technique, in both their undamaged and damaged status. Experimental observations of resonant frequencies and maximum displacement were made. By comparing the dynamic behavior of the tested coupons, prior to and after the initiation of damage, trends of dynamic behavior sensitive to the presence of cracks, were sought.

6   RESULTS AND DISCUSSION

Observations of damage patterns in the GFRP cross-ply laminates indicated that damage was initiated in the form of matrix cracking in the transverse direction while the longitudinal layers remained intact. The values of dynamic flexural modulus for the damaged beams were reduced, as expected, but were not dependent on the amplitude of the vibration for both undamaged and damaged beams. Values of specific damping capacity increased with increasing damage; for the damaged beams, the measured values were found to be dependent on vibration amplitude. Examples of this behaviour are shown in Figure 2. Changes in normalised values, with respect to undamaged values, for the dynamic flexural modulus and the specific damping capacity are presented in Figures 3 and 4 in relation to increasing crack densities. The normalised ratios of specific damping capacity were all calculated for the same energy input during the test. The results show the expected behaviour for the different laminates. When the cracks are in the outer plies, $(90/0)_s$ laminates, the values of both dynamic flexural modulus and specific damping capacity are far more sensitive to crack density. When the cracks are in the inner plies, $(0/90)_s$ laminates, the values of flexural modulus show a small decrease for high crack density. However, the values of specific damping capacity show a small but significant increase for low crack density but the values, when compared for equal energy inputs, appear unaffected by higher crack densities. The damage in the woven cross ply GFRP coupons was observed using both optical microscopy and scanning electron microscopy

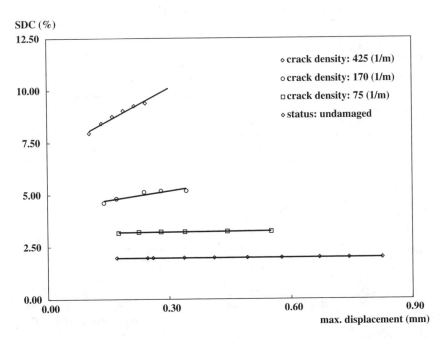

Figure 2.   Comparison of values of Specific Damping Capacity for undamaged and damaged $(90^\circ/0^\circ)_s$ continuous GFRP laminates.

(SEM). It was found that damage was initiated as matrix cracking in the transverse tows and was mainly concentrated in the crimp area.

The normalised values in relation to increasing applied maximum strain in the mechanical test, for flexural modulus and specific damping capacity, are presented in Figure 5. The results are comparable to those from the cross-ply laminates. The presence of damage, which was observed above around 0.5% strain, leads to increase in specific damping capacity and decrease in dynamic flexural modulus. However, when compared for the same energy input, the values are not

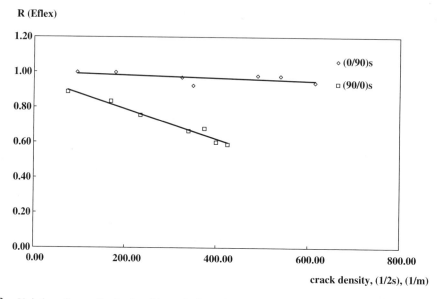

Figure 3. Variation of normalized ratio of dynamic flexural modulus, R (E_{flex}), with transverse crack density for GFRP cross-ply laminates.

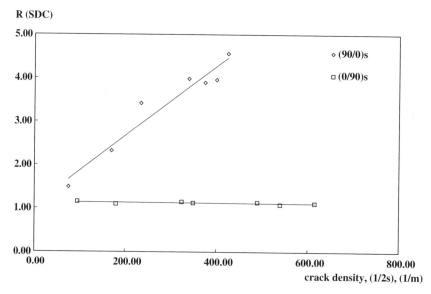

Figure 4. Variation of normalized ratio of Specific Damping Capacity, R (SDC), with transverse crack density for GFRP cross-ply laminates.

sensitive to the increasing level of damage at higher strains.

The damage in the woven cross ply CFRP coupons was observed on the polished edges of the specimens using optical microscopy. Damage was in the form of matrix cracking within the transverse tows, similar to the damage observed for the woven GFRP specimens. The values of dynamic flexural modulus were not changed even for the highest number of cycles. Normalised ratios for specific damping capacity showing the change in relation to the increasing number of

cycles of loading are presented in Figure 6. Results were obtained for 5 different specimens for different fatigue levels. There is an overall trend of increasing values of specific damping capacity with number of cycles for all the specimens.

Damage in the notched woven cross ply CFRP coupons was assumed to initiate as matrix cracking in the transverse tows in the area surrounding the notch and it was restricted in the outer (0°/90°) layers whereas the inner (±45°) layers remained intact. This assumption follows the associated damage observations made

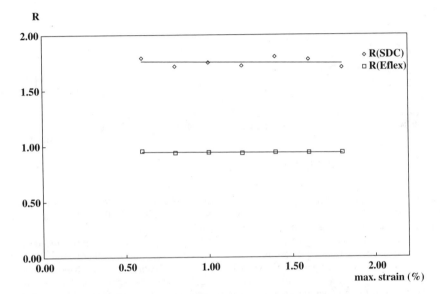

Figure 5.   Variation of normalized ratios with maximum applied strain for GFRP woven laminates.

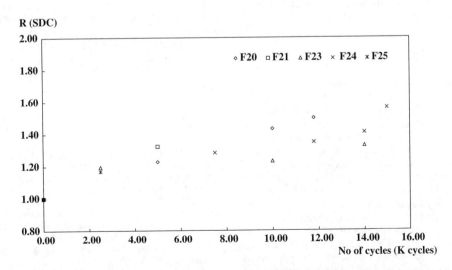

Figure 6.   Variation of normalised ratio of Specific Damping Capacity R (SDC), with fatigue cycles for CFRP woven laminates.

max. displacement (mm)

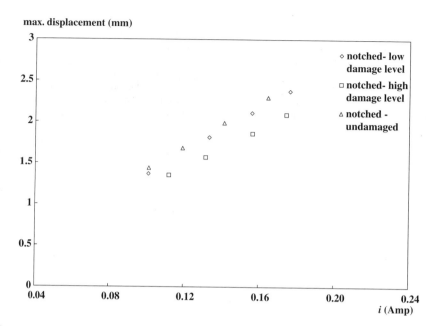

Figure 7.   Comparison of experimental results for notched woven CFRP laminates at different damage levels.

in the case of an analogous notched Woven quasi-isotropic GFRP coupon (Belmonte et al 2001).Comparisons for the force maximum displacement relationships for notched specimens with varying levels of damage are presented in Figure 7. These preliminary results indicate an effect of damage on the global measurements. However, since the damage is localized, further analysis of the notched structures could not be provided by the classical solution using continuum mechanics. These results can be analysed using the finite element analysis method.

7   CONCLUSIONS

A series of test coupons were considered and tested in both their undamaged and damaged states. Increased values of Specific Damping Capacity were found for all damaged coupons, whilst values for flexural modulus demonstrate less sensitivity to increasing damage. Values of SDC for the $(90°/0°)_S$ GFRP coupons were found to increase for increasing crack density; coupons with a $(0°/90°)_S$ lay-up demonstrate less sensitivity. That is expected, since only the outer layers would degrade as crack density increased and the dynamic behavior of vibrating beams is largely defined by the characteristics of the outer layer. The same patterns of increasing normalised damping values in relation to damage levels, were observed for the woven cross ply GFRP 8HSW test coupons and

the Woven cross ply CFRP 5HSW coupons whilst the normalised values for flexural modulus of the coupons are hardly affected.

An association between the localized damage around the notch and the overall measurements was found. The quantification of the localized damage level, which requires analysis using the finite element analysis method, may lead to more reliable use of notched CFRP structures. Overall, alteration in dynamic behaviour could be conveniently utilized to signify the intensity of the accumulated damage. These experimental observations strongly support the concept of monitoring changes in SDC values in order to detect successfully preliminary damage in woven fabric laminates before the material experiences other (more potentially damaging) failure modes.

REFERENCES

Belmonte, H. M.S., Manger, C. I. C., Ogin, S. L., Smith, P. A. & Lewin, R., 2001, Characterization and modeling of the notched tensile fracture of woven quasi-isotropic GFRP laminates, *Composites Science and Technology*, 61, 585–597.
Guild, F. J. & Adams, R. D., 1981, A new technique for the measurement of the specific damping capacity of beams in flexure, *J. Physics: Sci. Instrum,* 14, 355– 363.
Ishikawa, T. & Chou, T. W., 1982, Stiffness and Strength behaviour of woven fabric composites, *Journal of Materials Science*, 17, 3211–3220.

Radiography

Emerging Technologies in Non Destructive Testing, Van Hemelrijck, Anastasopoulos & Melanitis (eds)
© 2004 Swets & Zeitlinger, Lisse, ISBN 90 5809 645 9

Quantitative determination of micro crack density in composites and polymers by X-ray refraction topography

K.-W. Harbich

Federal Institute for Materials Research and Testing, BAM Berlin, Germany

ABSTRACT: X-ray refraction topography is used to determine the internal surface density of impacted CFRP laminates and PE/PP polymer samples. The method is based on refraction of X-rays caused by interfaces of microstructures in heterogeneous materials. This study presents a series of micro crack investigations which are visualized by two-dimensional spatially resolved X-ray refraction topographs. As the experimental set up allows of continuous sample rotation, it is sensitive to separate orthogonal arranged fibre directions and, in addition, the detection of micro crack orientations. The inspection of both, fibre debonding and micro crack distribution give a more detailed information about the mechanical performance of CFRP laminates. The investigation of impacted PE/PP polymer samples illustrates the characterization of micro crack surface density and their spatial as well as orientation distribution. For analytical requirements the determination of the ratio of orthogonal arranged inner surfaces allows to estimate the total inner surface properties of micro structural defects.

1 INTRODUCTION

Failure analysis of impact damaged materials as well as analytical investigation of mechanical loading of CFRP components request a reliable knowledge of micro structural properties to characterize the damage state. In particular the investigation of the influence of micro cracking in correlation with mechanical properties is important for residual strength prediction and the description of fatigue behaviour (Trappe et al. 2002). The non-destructive characterization of micro structural defects down to nanometer dimensions can be performed by X-ray refraction topography (Harbich et al. 1997, 2001). Quantitative X-ray refraction techniques are most practical for improved evaluation of high performance composites and other heterogeneous materials. In contrast to X-ray diffraction methods the used technique is based on the scattering detection of the refraction of X-rays due to differences in X-ray refractive index at interfaces of microstructures. By direct measurement the method determines the amount and the location of internal surfaces within dimensions ranging from micrometers to nanometers. The intensity of the refracted radiation is proportional to the inner surface density Σ in units m^2/cm^3, i.e. the surface to unit-volume of a sample. In the case of micro cracks Σ can be determined by means of a calibration standard consisting of a stack of parallel aligned thin foils of known internal surface (Harbich, in prep.). In scanning a sample each individual scattering volume represents the local integral interface properties. The combination of simultaneously measured X-ray refraction and X-ray absorption data allows the analysis of the correlation between the local internal surface and local density fluctuations of the material.

The micro crack propagation in a damaged sample frequently is irregular. For this reason a precise detection sometimes is more complicated than the detection of the debonding of well oriented fibres in composites. Due to different micro crack orientations the refraction intensity varies significantly on the samples position with respect to the cross section of the X-ray beam. A refraction signal only occurs when the micro crack orientation is parallel to the incident X-ray beam. In order to get quantitative results a sample rotation therefore is necessary to record the relevant refraction data.

2 PHYSICS AND EXPERIMENTAL SETUP

The physics of X-ray refraction is analogical to the well known refraction of visible light by optical lenses and prisms which generally is governed by Snell's law. The major difference from optics is given by the refractive index n of X-rays in matter. As n takes the value of nearly 1, deflection is caused at very small angles up to a few minutes of arc. Deflection of X-rays at interfaces is ruled by the real part of the complex index of refraction

$$n = 1 - \varepsilon \qquad (1)$$

where the increment ε depends on the radiation wavelength λ and the electron density ρ of the material

$$\varepsilon \sim \rho \, \lambda^2 \qquad (2)$$

The angular scattering distribution $I^*_R(\theta)$ of refracted X-ray radiation by cylinder-like structures can be expressed by (Hentschel et al. 1994)

$$I^*_R(\theta) \sim \varepsilon^2 / \theta^3 \qquad (3)$$

At a fixed angle θ and a constant wavelength λ the dependence on the refractive index leads to a scattering intensity I^*_R which increases with the square of the electron density difference at interfaces, for example at the interface of a void or micro crack and the surrounding solid material. This is in sharp contrast to density measurements which are based on the physics of X-ray absorption. Depending on the electron density the attenuated signal I_A decreases exponentially due to the well-known law of absorption $I_A = I_0 \, e^{-\mu \rho d}$. I_0 represents the intensity of the primary X-ray beam and μ denotes the mass absorption coefficient, respectively.

With the abovementioned restrictions the scattering intensity I^*_R only depends on the internal surface density Σ (inner surface/unit volume), the X-ray path d penetrating the sample and the transmitted intensity I_A:

$$I^*_R = I_R - I_A = C \, d \, I_A \text{ with } C = k \, \Sigma \qquad (4)$$

where I_R is the detected refraction intensity. The essential refraction value C contains both, the inner surface density Σ and the instrumental constant k. The latter quantity is simply obtained by a calibration standard of known inner surface density Σ_C. Its relation to the corresponding refraction value C_C results in $k = C_C/\Sigma_C$. Finally this leads to a practical expression for quantitative determination of the inner surface density Σ. It is calculated by means of the measured intensities I_R and I_A:

$$\Sigma = 1/d \, [(I_R/I_{R0}) \, (I_{A0}/I_A) \, -1] \, \Sigma_C/C_C \qquad (5)$$

I_{R0} and I_{A0} indicate the relevant background intensities. Both signals are estimated when the sample position is out of the X-ray path.

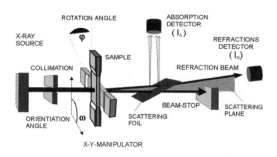

Figure 1. Instrumental set up for simultaneous 2-dim registration of refraction and absorption intensities.

The experimental instrumentation consists of a standard fine structure X-ray generator providing monochromatic Mo k_α-radiation. The refraction effect is measured using a commercial small angle X-ray camera of the Kratky type in combination with two scintillation detectors for simultaneous detection of X-ray refraction intensity I_R and the sample absorption I_A. Except for evaluation of equation (5) the absorption data set is transformed into a two-dimensional scanning radiograph. This topographic image is used to analyze local density properties. A personal computer stores and processes the scattering intensity data and controls the sample scanning system. The position of the X-ray beam is fixed and the sample is moved stepwise. A schematic representation of the experimental instrumentation is shown by Figure 1.

The primary beam penetrates the sample, where it is partly absorbed. Using a scattering foil a proportional scattering intensity corresponding to I_A is registered by an absorption detector. Correlatively the refraction detector registers the signal I_R. The refraction beam passes a detection slit which is fixed to a small scattering angle close to 2 minutes of arc. The cross section of the primary beam can be selected and is usually limited to 0.2 mm $\times$ 0.05 mm. In order to separate 0°/90°-arranged micro-structural defects, the sample orientation is modified with respect to the orientation angle ω. In addition, the rotation by φ permits to detect different interface alignments, particularly with regards to micro cracks.

3 RESULTS

As an instance of characterizing impact damaged unidirectional CFRP laminates some exemplary results of X-ray refraction measurements are shown in Figure 2. The scanned area is 4 cm $\times$ 4 cm. Both the left and the right topograph reveal the inner surface density Σ of the same sample but at orthogonal orientation. In consideration of identical scaling the topographs significantly differ from each other regarding the mean

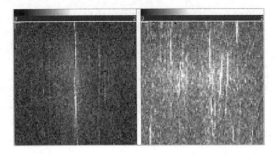

Figure 2. X-ray refraction topographs at ω = 0° (left, fibre debonding)) and at ω = 90° (right, micro crack detection) of an impact damaged unidirectional CFRP laminate. The 2-dim scanned area constitutes 4 cm × 4 cm.

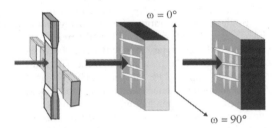

Figure 3. Sample positioning for separate detection of orthogonally arranged micro structural defects by X-ray refraction.

value of Σ. The left image depicts two-dimensional scanning at the orientation angle of ω = 0°. In this case the fiber direction is aligned parallel to the beam stop, thus the detected refraction signals primarily are caused by fibre/matrix debonding. The light line in the center corresponds to a visible crack which intersects the impacted sample. It coincides with the fibre orientation. The X-ray refraction topograph in the right of Figure 2 reproduces the measurement at ω = 90°. As the sample orientation has changed by a right angle all refraction signals detected at the previous positions are suppressed. The light lines now are arranged perpendicular to the fibre direction. They clearly indicaate refraction signals exclusively as a result of intense matrix damage by micro crack propagation (Bullinger et al. 2002).

With respect to the image in the left the detected micro crack propagation obviously dominates the effect of fibre debonding. Except for a distinct heterogeneousness of local Σ-fluctuation the major difference between both topographs arises from the mean refraction values: C(left) = 0.23 mm^{-1} and C(right) = 0.46 mm^{-1}. Micro cracks cause a total inner surface density Σ(crack) which in this case seems to be about two times larger than the corresponding value Σ(fibre) given by fibre/matrix debonding. This has the

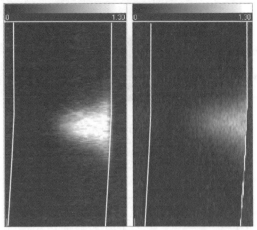

Figure 4. X-ray refraction topographs at ω = 90° (left) and at ω = 0° (right) of an impacted PE/PP sample. The scanned area at φ = 0 is 1 cm × 2 cm. Separation of 0°/90° micro crack alignment: horizontal orientation (left), vertical orientation (right).

remarkable implication to characterize different types of micro structural defects quantitatively. Accordingly, the ratio Σ(crack)/Σ(fibre) may also be useful to control various material parameters, especially the influence of the chemical composition of CFRP laminates.

The X-ray refraction topographs pictured in Figure 4 shows an example of an impacted PE/PP specimen. The white outlines indicate the sample dimension, and the scanning area inside is 1 cm × 2 cm. Just as explained before, the topographs in Figure 4 as well represent the inner surface density S of the same sample area but measured at orthogonal orientation. Because of the absence of any fibres in both cases the total damage region solely consists of coherently disposed micro cracks. The clear difference between the micro crack densities possibly may result from geometric restrictions of the sample. However, we have to consider that the main micro crack orientation at ω = 90° is strictly parallel to the impact direction. In order to obtain a more precise characterization of the micro crack arrangement the instrumental application allows to modify the sample position with regards to the rotation angle φ. The principle of the corresponding manipulation is exemplified in Figure 5. It is important to realize that a refraction signal exclusively is detectable if the micro crack orientation is almost parallel to the X-ray beam. Consequently, perfectly parallel aligned micro cracks should cause a sharp refraction signal occurring only at a single angle increment of rotation. On the other hand, a real isotropic crack arrangement results to a refraction value that should be constant within the complete range of φ = 0° to φ = 360°. Actually the latter is correct

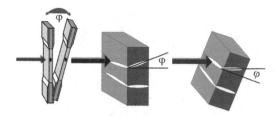

Figure 5. The rotation by φ permits to detect diverse interface orientation with reference to the surface of the sample. A refraction signal only occurs if the interface is in horizontal adjustment (right position).

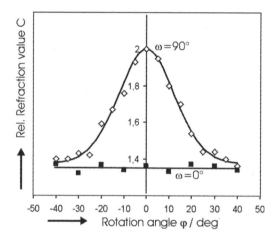

Figure 6. Verification of anisotropic ($\omega = 90°$) and isotropic ($\omega = 0°$) micro crack arrangement. See topographs in Figure 4.

for the sample position at the orientation angle of $\omega = 0°$. In contrast with this result the φ-inspection at $\omega = 90°$ suggests the characteristics of an anisotropic micro crack behaviour. A summarized representation of the rotation measurements is shown in Figure 6.

Since the refraction value C is proportional to the inner surface density a comparison of the different areas enclosed by the particular curves with the abscissae gives relief to interprete the data from

Figure 6. Generally it is sufficient to determine the respective areas in the range from 0° to 180°. In consideration of the anisotropic offset the ratio of the corresponding inner surface densities is given by $\Sigma(\omega = 90°)/\Sigma(\omega = 0°) = 1.105$. This small variance of about 10 percent may surprise concerning the significant differences between the topographs in Figure 4. It follows that the inspection by means of sample rotation is indispensable for a reliable quantitative determination of micro crack surfaces.

4 CONCLUSIONS

Two-dimensional X-ray refraction topography combines quantitative non-destructive detection of micro structures with the advantage of imaging their spatial distribution. The separation of orthogonally arranged interfaces within unidirectional CFRP laminate characterizes different types of micro structural defects quantitatively due to material treatment. The investigation of an impacted PE/PP specimen has shown that sample rotation is needed to get a more exact determination of the corresponding micro crack surface density. Furthermore, the established φ-inspection generally is qualified for evaluating the orientation distribution of inner surface micro structures.

REFERENCES

Bullinger et al. 2002. X-ray refraction topography of impact damage of CFRP laminates; *Proc. 11th intern. symp. CNDE, Berlin, 26–28 June 2002.*
Harbich et al. 1997. X-ray refraction for nondestructive investigations of micro cracks and impacts; *Proc. intern. conf. Micro Mat '97, Berlin, 16–18 April 1997.*
Harbich et al. 2001. X-ray refraction characterization of non metallic materials; *NDT & E International 34: 297–302.*
Hentschel et al. 1994. Nondestructive evaluation of single fibre debonding in composites by X-ray re fraction; *NDT & E International 27 (5): 275–280*
Trappe et al. 2002. Micro cracking and stress under fatigue loading of CFRP; *Proc. 11th intern. symp. CNDE, Berlin, 26–28 June 2002.*

Emerging Technologies in Non Destructive Testing, Van Hemelrijck, Anastasopoulos & Melanitis (eds)
© *2004 Swets & Zeitlinger, Lisse, ISBN 90 5809 645 9*

High energy X-ray imagery on large or dense objects: definition and performance studies of measurement devices by simulation

J.L. Pettier, D. Eck & R. Thierry
CEA DEN DED/SCCD Cadarache France

ABSTRACT: High energy X-ray applications for Radiography Testing (RT) and Computerized Tomography (CT) on large or dense radioactive objects require X or γ rays of at least 1-MeV to reach sufficient S/N Ratio in attenuation measurements. These high-energy photons are produced either by a ^{60}Co isotopic source (as a monochromatic continuous beam) or by a Linear Accelerator of electrons (Linac) coupled with a Bremsstrahlung target (polychromatic pulsed beam). Numerical simulation is used to great extend through parametrical studies in order to design the measurement device, the mechanical set-up and to define acquisition paths. Simulation is also a very helpful tool to optimize the performances according to object characteristics and data acquisition time. The simulation code named MODHERATO ("Modélisation Haute Energie pour la Radiographie et la Tomographie") takes into account statistical noise, geometrical blurring effect and detection dynamic range to simulate physical phenomena and electronics behavior which damage the attenuation measurements. This article presents the benefits of the simulation through a few examples after a short description of the different modules.

1 INTRODUCTION

X-ray imagery in nuclear field has many potential applications such as: control and characterization of activated components, fuel assembly and waste drums. But high cost of the equipments and the fact that those measurements are quite time consuming in an hostile environment are playing against the use of these non destructive techniques.

A simulation code becomes essential for a better use of these equipments and to estimate performances (Hugonnard & Glières 1999, Freud & al. 2000). In the high energy field, beam characteristics (angular anisotropy) justify the development of the MODHERATO code dedicated to X or gamma-ray collimated imagery between 1 to 15 MeV energy range.

2 DESCRIPTION

2.1 *Beam modeling*

MODHERATO uses the results of MCNP code (Briesmeister 2000) in Electron–Photon mode to describe the X-ray beam generated by accelerated electrons within a Bremsstrahlung target as a function of the energy and the angle ($\Phi(E, \theta)$). This statistical code has permitted to optimize the thickness of tungsten target in order to increase photonic production.

The evaluation of beam anisotropy is based on the discretization of each spectrum as function of angular classes, themselves defined from the central axis. Attenuation calculations depend on the number and width of photon histories. For a Linac, available flux (photons/steradian/pulse) is computed according to peak intensity and pulse time ($<5\,\mu s$). The maximal flux is directed forward the target. The useful emission cone is about ten degrees around the central axis.

For an isotropic distribution beam (^{60}Co source), the flux depends on its activity, its age and its decreasing period. The photon histories fit with the emission peaks. The angular isotropy of the beam allows the use of a larger emission cone (several tens degrees).

2.2 *Object description*

Among radioactive objects, waste drums and containers are the largest and heaviest. In broad outline they are made of mix of materials compacted or not then holded up by an hydraulic cement. The container is composed of 8 to 15-cm thick concrete shell or a 0.1 to 0.5-cm thick metallic shell.

The interfaces between waste, binder and container have to be checked in order to make sure that there is no empty space and the waste should be devoid of liquid.

A Constructive Solid Geometry using basic forms (parallelepiped, ellipsoïd and cylinder) allows the

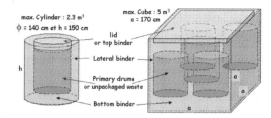

max. Cylinder : 2.3 m³
φ = 140 cm et h = 150 cm

max. Cube : 5 m³
a = 170 cm

lid
or top binder

Lateral binder

Primary drums
or unpackaged waste

Bottom binder

h

a

a

a

Figure 1. Cylindrical and cubic geometries of conrete objects.

definition of three dimensional objects. The center of this geometry is used as the center of the 3D space origin for all calculations.

The object βγ-activity generates its own gamma-ray radiation. The MCNP code in Photon mode allows to quantify this passive gamma-emission. Results show that the passive flux is about $10^5 \gamma/cm^2/s$ in front of a collimated detector set near an irradiating waste drum with a contact dose about 50 Gy/h. With pulse time around 5 μs, the passive flux is lower than 1 γ/cm^2 during measuring time. That demonstrates the interest to use a pulsed source and a collimated detector for testing highly irradiating objects.

2.3 Detector model

Collimator apertures and focal distances lead to suppose that all photons hit perpendicularly the surface area of detector pixel. The photon flux at the entry of the detector is limited by the surface of the post-collimator aperture.

In the case of a semiconductor crystal detector, its modelization consists in computing the energy deposition generated by each photon. This calculation is made at the mean energy of each spectrum class. The current conversion is carried out according to the pair effect production coefficient and the charge collection ratio. The output signal is digitized on the dynamic range of the Analog-Digital Converter after electric charge integration and current voltage conversion. This dynamic range can reach 10^6 for a CdTe pixel (Glasser & al. 1992).

For a scintillator pixel (NaI, BGO) coupled with an integrator-amplificator stage (photodiode, photomultiplicator), integration or threshold-counting modes can be defined. The combined dynamic range with this technology is lower (from 10^3 to 10^4).

2.4 Attenuation computing

Photonic attenuation measurements through the object are the basis for these imagery techniques. The source and the detector are set on both sides of the object (photonic transmission). It is the ratio between the attenuation-free counting (I_0) and the attenuated

photon counting (I) on the source-detector path that defines an elementary measurement:

$$\ln\left(\frac{I_0}{I}\right) = \ln\left(\frac{\int_E \Phi_0(E,\theta)}{\int_E \Phi_0(E,\theta) e^{-\int_L \mu(x,y,z,E) dl}}\right) \quad (1)$$

where E = mean energy in each photon class, θ = angle between ray-path and the central axis, L = ray-path between source center and detector center, dl = step on ray-path in cm, μ = lineic attenuation coefficient in cm^{-1}, I_0/I = inverse relative attenuation.

The calculation integrates the product of crossing lenghts in each object by the lineic attenuation coefficient of the matching material at the mean energy of each photon class. The attenuation coefficients are tabulated and available in a reduced data base completed as one goes along from XCOM code (Berger & al. 1999) between 0.5 to 15 MeV.

A statistical noise is applied on each computed value to simulate the independent random processes (photon production, photon-mater interaction and photon detection). Lower the photon number reaching the detector (I) greater the noise is (SNR = $\sqrt{I}$).

In CT, it is particulary important to adapt the direct signal and the most attenuated signal within the detector dynamic range to preserve an optimum quantitative imagery (Hammersberg & Mangard 1999).

2.5 Sampling and paths

Attenuation calculations are achieved in the object fixed reference system while the source-detector system linked is in rotation. If the beam is limited in space (aperture of a few degrees) or if the detector pixels are not contiguous it is necessary to carry out a transaxial scanning motion of the source-detector reference system to complete all the object full width or the space between each detector edge. The transaxial sampling step (Δu) has to be smaller or equal to the post-collimator width. This sampling step defines the number of pixels by projection ($M_{/proj}$) and the spatial frequency domain [ν_{min}, ν_{max}] such as:

$$\nu_{min,max} = \pm \frac{1}{2\Delta u}$$

For a radiography path, projections are collected on the vertical axis The axial sampling step (Δz) is lower or equal to the post-collimator height.

In tomography, projections are collected in the transaxial slice which contains the source and the detector. The angular sampling step has to be such that the projection number (N_{proj}) is included in π/4 to π/2 times M_{proj}. This condition is necessary in compliance with an analytical reconstruction algorithm.

212

Helical path combines source rotation with object lifting so two successive projections are collected on different heights. This path is defined by the axial pitch (Δh) matching with the excursion on the vertical axis after a complete revolution of the object-detector set. The distance between two successive projections is equal to Δh/N$_{proj}$.

The accuracy on the mechanical positioning of the set-up is assumed ideal in MODHERATO.

The data set related to a CT scanning is named a sinogram. The coordinates of each measurement are not saved but they are computed thanks to parameters stored in a descriptor file linked with the sinogram file.

2.6 Sinogram processing

The set of numerical attenuation values on a dedicated path and with ad hoc sampling spaces leads to a direct (RT) or a postponed imagery (CT). CT needs an analytical or iterative image reconstruction adapted to acquisition geometry. The reconstructed parameter is the opacity in radiography and the lineic attenuation coefficient in tomography. Pixel size (Δp) has to be lower or equal to the expected spatial resolution.

The analytical reconstruction matches with the inversion of Radon Transform. This numerical processing computes the 1D Fourier Transform of each projection then multiply this transform data set by the ramp filter. This process has to balance with an apodization window (Hanning, Hamming, Butterworth) on all or a part of the frequency space. Statistical noise and sudden variations generated by different materials indeed match with these high spatial frequencies. At the end of filtering, Inverse Fourier Transformation and back-projection processing are applied by interpolation on the image grid. These numerical processes are adapted to fan beam and parallel grometries for circular detectors (Turbell 1999).

In 3D reconstruction, the same processing is used and the sinogram is computed by interpolation between the nearest projections of each reconstruction plane. In this case radial resolution is deliberately decreased in order to approach axial resolution and to reconstruct the object on a regular 3D grid.

2.7 Device performances

The performances of an X-ray imagery apparatus are linked to the spatial resolution and the accuracy. These indicators are integrated in the 1D Modulation Function Transfer (MTF):

$$MTF(\nu)=\frac{\sin\left(\frac{\pi\nu d}{M}\right)}{\frac{\pi\nu d}{M}}\cdot\frac{\sin\left(\frac{\pi\nu a(M-1)}{M}\right)}{\frac{\pi\nu a(M-1)}{M}}\cdot\frac{\sin(\pi\nu\Delta u)}{\pi\nu\Delta u} \qquad (2)$$

where d = detector width, a = source width, M = enlargement, Δu = transaxial sampling interval and ν = spatial frequency.

The first two terms describe the geometrical blurring and the third the convolution generated by the transaxial sampling.

Improving the performances would consist of optimizing the parameters to enlarge the MTF on its frequencial space and to increase the SNR. The SNR stays often low in our applications and therefore the filtering has to be accurately adjusted if we wish to preserve the precision without compromising the intrinsic spatial resolution.

3 MONOCHROMATIC APPLICATIONS

3.1 Sampling optimization in fan beam geometry

The device TRANSEC (Pettier 1996) is available for Non Destructive Evaluation (NDE) of cylindrical waste drum (or another object) with volume lower or equal to 220-l. A 300-GBq cobalt-60 source ($\bar{E}$ = 1.25 MeV) generates a fan beam with a 2.1-mm focal spot. Ten detectors are set at 120 cm of the spot on an arc of circle of $\pm$30 degrees around the central axis. Each detector is set behind a 0.42-cm width by 0.21-cm height aperture.

The first object matches a 1.42-density bitumen matrix in which metallic parts have been set (from 0.5 to 0.1-cm diameter). The metal container is 0.1-cm thick. The attenuation dynamic range is lower than 10^2 in all directions which means a minimum SNR of 26 in the sinogram.

Figure 2 shows the CT-images obtained from real acquisition (left) and from simulation (right) in the

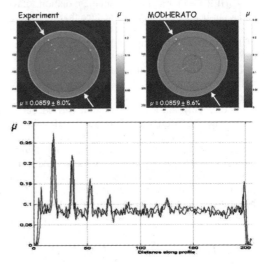

Figure 2. Experiment (left) and simulation (right) CT-images and profiles on the 46-cm diameter cylindrical bitumen phantom.

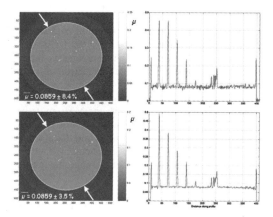

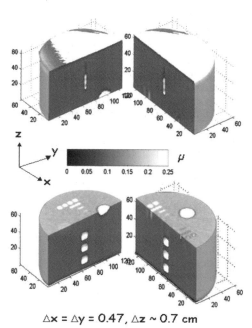

Figure 3. CT-images and profiles with an optimal sampling and a light filtering (top) and a strong filtering (bottom).

same conditions. The sinograms are 180 times 250 attenuation rays and the reconstruction is carried out on 256 times 256 pixels with a strong filtering.

The lineic attenuation coefficients of bitumen and iron at 1.25-MeV are respectively equal to 0.0852 and 0.4205-cm^{-1}. We note on the profiles that these conditions allow only the detection of metal parts but not their identification (60% contrast on 0.5-cm diameter tube). The spatial resolution is lower than 1 lp/cm for these sampling conditions. The cylindrical artefacts match with efficiency differences between detection channels. The external artefact is reproduced by simulation using an efficiency gap of 15% on the corresponding data.

This application demonstrates the validity of MODHERATO for a monochromatic radiation source and a fan-beam geometry. We deduce that the current sampling is insufficient to get the intrinsic resolution of 1.5-lp/cm for a 0.3-cm width beam near the object center.

Figure 3 shows the simulated CT-images obtained with a sinogram of 450 times 600 attenuation rays (more than 6 times the previous data volume with $\Delta u = 0.21$-cm) and a reconstruction carried out on 512 times 512 pixel grid. This transaxial sampling and a light filtering allow to detect the 0.4-cm diameter metal tube with 100%-contrast. This study demonstrates the interest to adjust sampling and ramp filtering when we wish to favour the spatial resolution. Of course this gain, with the same detector number, induces 6 times more data and collecting time.

3.2 Helical path optimization in fan beam geometry

In the case of a monochromatic beam, an helical path and a 3D reconstruction can produce a density matrix.

$\Delta x = \Delta y = 0.47,\ \Delta z \sim 0.7$ cm

Figure 4. CT-image on a 57-cm diameter cylindrical concrete phantom.

This matrix is useful for attenuation corrections in gamma assay on total or partial volume of radioactive objects to limit quantification errors. This 3D tomography is an alternative (in term of ratio spatial resolution/measurement time) between DR and 2D-CT.

A 220-l concrete phantom of 57-cm diameter and 88-cm height allows to estimate the performances of an helical path in the fan-beam geometry. This phantom is made up by a 0.1-cm thick metallic container filling with a 2.35-density concrete. Sets of three equidistant cylinders filled with air (5, 2.5 and 1.25-cm height, diameter and distance) are placed on the central vertical axis. Sets of four square calibrated iron and air flaws (from 2 to 0.125-cm edges) allow to characterize the FTM and to deduce the spatial resolution in the median slice. Two 10-cm diameter spheres are also defined in the same plane, one filled with aluminum and the other with water.

Figure 4 shows the simulated 3D CT-image divided on horizontal and vertical median planes. Isovalue near zero is plotted in white color. The voxel size is 0.175 cm^3. This helical CT needs 22 revolutions of 180 projections thus 9 times as many data as a radiography carried out with a 0.2 cm axial step. Each projection is distant of $\Delta z = 0.022$ cm.

The axial and radial resolutions are almost homogeneous about 0.2 lp/cm overall the object.

The concrete μ-accuracy in the median slice is 2% (Figure 5) in spite of a low SNR in the sinogram.

214

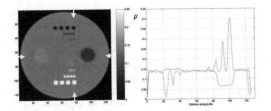

Figure 5. Median plane extract from helical-CT on the 57-cm diameter cylindrical concrete phantom.

The 3D CT allows a better detection efficient than a radiography which has a poor contrast on objects placed nearby the center of a concrete drum.

The $\mu_{1.25\,MeV}(x,y,z)$ matrix can be interpolated on others energies (Compton field) to produce an attenuation matrix with a good accuracy (Alvarez & Macovski 1976) for gamma emission tomography.

4 POLYCHROMATIC APPLICATIONS

4.1 *Polychromatic radiation optimization*

The following applications concern our EXCAL-IBURHE project on large and dense objects. First of all the source sizing required the making of a 1:5 scale phantom representing a waste container of 170-cm edges filling with five 220-l primary drums (Figure 1). The phantom allows the introduction of 12-cm diameter cylindrical inserts (aluminum, iron and lead) representing an attenuation dynamic range up to 10^7 on diagonals as in a real container.

The source used for the experiment is a Minilinatron Varian. The radiation dose is equal to 23 Gy/min-m at the 9-MeV electron energy. The focal spot is equal to 0.2-cm large. The pulsed flux in the beam axis is about 10^8 X/cm^2/pulse at 1 meter of the focal spot. The detector is made of CdTe crystals associated with a 10^6 digitizing electronics (Rizo & al. 2000).

Figure 6 shows CT-images obtained from real acquisition (left image) and from simulation (right image) in the same conditions.

The phantom set-up with three Al inserts and one Fe insert has an attenuation range between 10^2 and $3\ 10^2$ on the diagonals. The sinogram is composed by 1125 projections of 1400 pixels ($\Delta u = 0.05$ cm $< d = 0.08$ cm). The image is carried out on 1024 × 1024 pixel grid ($\Delta p = 0.035$ cm).

The SNR is greater than 25 and thus allows the use of a light filtering to preserve an optimal spatial resolution. The 1, 0.5 and 0.25-cm diameter holes in each insert are accurately reproduced which induces a spatial resolution better than 2 lp/cm. The μ-accuracy is better than 7% on the Al matrix and Al inserts and greater than 4% on the Fe insert.

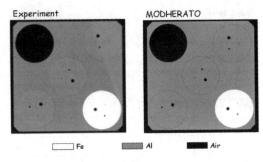

Figure 6. Experiment-simulation comparison on 1:5 scale phantom "Al-Fe" (46-cm diagonals).

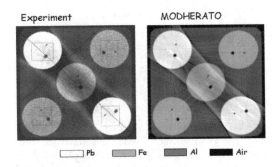

Figure 7. Experiment-simulation comparison on 1:5 scale phantom "Al-Fe-Pb" (46-cm diagonals).

In huge attenuation conditions (three iron and two lead inserts: "Al-Fe-Pb") some artifacts appear directly below the most attenuated paths (Figure 7) in proportion to missing data in the sinogram. The CT-images stay useful to count and locate the primary drums.

These 1:5 scale comparisons allow us to define the working energy field of the source to set on EXCAL-IBURHE process. The loss of flux due to the real focal length has to be compensated by a greater radiation dose. We need a Linac with an adjustable energy between 9 to 15-MeV-electrons with a radiation dose equal to 100 Gy/min-1 m with a maximal 0.2-cm focal spot.

4.2 *Beam hardening correction*

Optimizing a polychromatic beam induces also the reduction of the beam hardening phenomenon. To do that the most part of lower energy photons are removed with a physical filter set at the source output. This part decreases of course the measurement range. If this filtering is incompatible with the dynamic range needed it is still possible to calculate a numerical correction. This correction applied on the projections according to thickness and crossing materials improves the accuracy on CT-images.

215

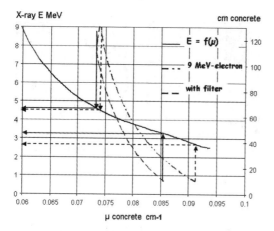

X-ray E MeV   cm concrete

Figure 8. X-ray energy (continue line and left scale) and crossing thickness of concrete (dotted lines and right scale) according to $\mu(E)$ and energy range projected on left scale from intersections with μ-curve.

A concrete step wedge allows an estimation of the mean energy variation of a 9-MeV-electron beam without or with a 3-cm thick iron beam filter (Figure 8).

The mean energy varies from 2.75 to 4.5 MeV and from 3.25 to 4.6 MeV respectively without and with the filter as a function of the crossed concrete thickness. The beam filter reduces by 30% the beam hardening range. When the object is made up of several materials it is necessary to estimate the beam hardening with a step-wedge phantom which represents the mixture.

These simulated results allow to calculate a correcting polynomial to apply on the sinogram, which is strictly suitable for the electron beam energy and the material composition.

5 CONCLUSION

We have shown here the interest to have at our disposal a simulation tool devoted to high energy X-ray imagery on volumic and dense objects. The collimated attenuation measurements are often carried out in detection limit and then the low SNR decreases the performances. So it is essential to select the optimal parameters for the data acquisition chain and for the numerical processing according to the wanted objective: spatial resolution or quantitative analysis. This approach is quite justified by the fact that data collection time lasts from one to several hours and to optimize beam hardening correction when the object is made up with material mixtures.

The inner structure of MODHERATO, the simple Constructive Solid Geometry and the limited energetic cutting (slowly μ-evolution in Compton field) induce moderate computing time going from several minutes to several hours on a standard UNIX workstation according to the object complexity, the sampling and the reconstruction matrix dimensions.

This software tool will benefit of a coupling with AVS Express environment in order to display and to interactively or automaticaly process 2D and 3D images.

Current studies are under work to enhance the code with iterative algorithms likely to take into account *a priori* knowledges to limit the artifacts due to incomplete sinogram (short scan or detection limit rays).

ACKNOWLEDGEMENT

The authors wish to thank C. Mennessier (ESCPE Lyon) for her very useful contribution to the development of MODHERATO and the DRT/LETI/DSIS of CEA Grenoble for providing the irradiation facilities and their large experiment in the field of X-ray NDT applications.

REFERENCES

Alvarez, R. E. & Macovski, A. 1976. Energy-selective Reconstructions in X-ray Computerized Tomography. Phys. Med. Biol. Vol. 21 n° 5: 733–744

Berger, M. J. & al. 1999. XCOM: Photon Cross Section Database (version 1.2), [Online]. Available: http://physics.nist.gov/xcom [2003, January 20]. National Institute of Standards and Technology, Gaithersburg, MD.

Breismeister, J. F. 2000. MCNP a general Monte Carlo Code Los Alamos, New Mexico

Freud, N. & al. 2000. Simulation of X-Ray NDT Imaging Techniques. Roma: Proceedings of the 15th WCNDT Physics Laboratory, National Institute of Standards and Technology

Glasser, F. & al. 1992. Application of Cadmium telluride detectors to high energy computed tomography. Nuclear Inst. & Methods A322: 619–622

Hammersberg, P. & Mangard, M. 1999. Optimal computerized tomography performance. Berlin: DGZfP-Proc. BB 67-CD: paper 6

Hugonnard, P. & Glière, A. 1999. X-ray Simulation and Applications, Berlin: DGZfP-Proc. BB 67-CD: paper 17.

Pettier, J. L. 1996. Determination of physical homogeneity by X-ray Imaging. Nuclear Technology Vol. 115

Rizo, P. & al. 2000. Application of transmission tomography to nuclear waste management. Roma: Proc. of the 15th WCNDT LETI CEA France

Turbell, H. 1999. Three-dimensional image reconstruction in circular and helical computed tomography. Linköping University: Thesis n° 760

Magnetic techniques

Emerging Technologies in Non Destructive Testing, Van Hemelrijck, Anastasopoulos & Melanitis (eds)
© 2004 Swets & Zeitlinger, Lisse, ISBN 90 5809 645 9

Non-destructive method for the measurement of stress in bent pipes

S. Iimura
Tokyo Gas Co., Ltd., Tokyo, Japan

Y. Sakai
JFE Engineering Corporation, Tsu, Mie-prefecture, Japan

ABSTRACT: It is essential to know the stresses generated in pipelines in order to maintain their safety from external forces. Pipelines are in service, so it must be possible to do this by non-destructive diagnosis. The highest stresses often occur in bent rather than straight sections. We have developed a method of non-destructively measuring stresses occurring in bent pipes using the principle of magnetostriction. The values measured by this magnetic method contain the stress generated during manufacture and the stress generated by external forces. This study examined the measurement of bent pipes subject to external force to distinguish the stress generated during manufacture from that generated by external forces. The values were regressed to Von Karman's theoretical equation for a bent pipe. From the regression curve obtained, it was possible to obtain the external force moment that purely applies bending to the bent pipe.

1 INTRODUCTION

In order to safely maintain and protect buried pipelines against ground subsidence and other external forces, it is essential to know the stresses to which such pipelines are subjected. Bent pipes are generally used in pipelines extending across rivers and at the entrances to multi-purpose utility tunnels, and maximum stresses often occur in bent rather than straight sections. We have developed a method of non-destructively measuring stresses occurring in bent pipes, improving on the method of Sakai et al. (1991) using the principle of magnetostriction. In this paper, stress measurement applying the principle of magnetostriction is called magnetic stress measuring method or simply, "magnetic method".

Magnetic stress measuring methods have only been used on straight pipes, not bent pipes, for two reasons. First, values measured by a magnetic method are often widely scattered because of the lack of uniformity in the manufacture of bent pipes. However, the stress generated in the surface of a pipe when bending moment is applied by an external force to a straight pipe can be represented by a simple equation called the primary equation of a cosine function. Therefore, the value of bending stress corresponding to the external force can be accurately determined by regressing the measured value to its cosine function. In contrast, the stresses generated in a bent pipe are represented by a complicated

equation, and the value of stress generated by external forces cannot be accurately determined by regressing highly scattered data to a complex equation. A second reason is the high probability that the large stress generated in the bent pipe when it is manufactured remains as residual stress. This study investigated these two problems.

The results of applying a magnetic method to measure the initial stress in a bent pipe confirmed the predicted presence of residual stress of the large stress generated by the manufacturing process.

When a magnetic method is used to measure the stress in a bent pipe that has been subjected to external forces such as ground subsidence over the years following installation, it detects the total of the stress generated during manufacture and the stress generated purely by external force. If this total stress exceeds a stipulated level, it is necessary to release the stress by excavating and exposing the pipe, then correcting the bending caused by the subsidence etc. If the stress measured by the magnetic method is assumed to be completely generated by subsidence, which is an external force, the stress is likely to be released in the incorrect direction. To correctly release stress, only the stress generated purely by external force must be identified.

In this study, we have examined the measurement of bent pipes subject to external forces to distinguish the stress generated during manufacture and that generated

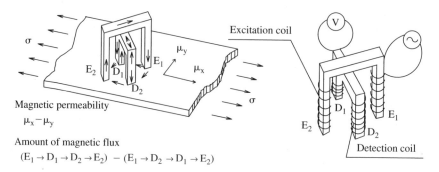

σ

E₂ D₁ E₁ μ_y μ_x

D₂

Excitation coil

V

~

D₁ E₁

E₂

D₂

Detection coil

σ

Magnetic permeability

$\mu_x - \mu_y$

Amount of magnetic flux

$(E_1 \rightarrow D_1 \rightarrow D_2 \rightarrow E_2) - (E_1 \rightarrow D_2 \rightarrow D_1 \rightarrow E_2)$

Figure 1. Principle of a magnetic stress measurement using magnetic anisotropy sensor.

by external forces. The values measured by a magnetic method were regressed to Von Karman's (1911) theoretical stress calculation equation for a bent pipe. From the regression curve obtained, it was possible to obtain only the external force moment that purely applies bending to the bent pipe.

2 PRINCIPLE OF MAGNETIC STRESS MEASUREMENT

When stress exists in a ferromagnetic material, the magnetic permeability changes. This phenomenon is called the magnetostrictive effect. Figure 1 shows the principle of stress measurement by the magnetic method. In this case, the magnetic permeability in the direction of tensile stress, μ_x, is slightly higher compared to μ_y in the orthogonal direction, causing an anisotropic magnetic field.

In the magnetic method, the stress level and stress direction are measured using a magnetic anisotropy sensor (Sakai & Tamura 2000). The sensor is assembled with two cores: an excitation core and a detector core. Each core consists of a wound coil. The excitation coil is connected to a power supply that supplies alternating current, and the detection coil is connected to a specialized voltmeter.

When the sensor is placed over the object to be measured, most of the magnetic flux emitted from the sensor E_1 travels directly to E_2. However, some travels along the path indicated by the arrows because the magnetic permeability between E_1 and D_1 is larger than that between E_1 and D_2. Thus, magnetic flux is induced in the detection core, resulting in a current and voltage.

The voltage V is expressed by the following equation

$$V = K_0 \cdot (\mu_x - \mu_y)$$

where K_0 is a constant determined by the excitation conditions, magnetic characteristics, etc; μ_x is the

magnetic permeability in the direction of the X-axis; and μ_y is the magnetic permeability in the direction of the Y-axis.

Anisotropy $(\mu_x - \mu_y)$ of the magnetic permeability is proportional to the stress difference $(\sigma_x - \sigma_y)$, so the above equation can be written as the equation

$$V = K_1 \cdot (\sigma_x - \sigma_y)$$

where K_1 is a constant determined by the excitation conditions, magnetic characteristics, etc; σ_x is the stress in the direction of the X-axis; and σ_y is the stress in the direction of the Y-axis.

Accordingly, once the proportionality constant "K_1" (magnetic sensitivity) is calibrated, the stress difference can be estimated by the output voltage of the sensor. Moreover, the direction of the principal stresses can be determined from the output voltage by rotating the sensor on the surface of the object because the direction of the maximum output voltage is the direction of the maximum principal stress σ_1, and the direction of the minimum output voltage is the direction of the minimum principal stress σ_2.

3 VON KARMAN'S BENT PIPE STRESS ANALYSIS THEORY

Von Karman's theory (below called "Karman's equation") explains the stress generated in a bent pipe when it has been subjected to a bending moment by an external force. Karman's equation is defined as shown in Figure 2 in terms of ϕ, the circumferential angle of the pipe.

Assuming that v is Poisson's ratio, R is the radius of curvature of the bent pipe, r is pipe radius, t is pipe thickness, I is the geometrical moment of inertia, and M is the moment acting on the bent pipe, then the longitudinal stress σ_L and circumferential stress σ_c of the

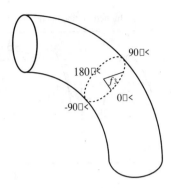

Figure 2. Definition of the angular position.

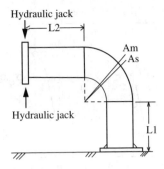

Figure 3. Illustration of experiment.

pipe can be represented as

$$\sigma_L = \frac{kMr}{I}\left\{\sin\phi - \frac{6}{5+6\lambda^2}\sin^3\phi + \frac{9v\lambda}{5+6\lambda^2}\cos 2\phi\right\}$$

$$\sigma_c = \frac{kMr}{I}\left\{v\sin\phi - \frac{6v}{5+6\lambda^2}\sin^3\lambda + \frac{9\lambda}{5+6\lambda^2}\cos 2\phi\right\}$$

$$\lambda = \frac{tR}{r^2} \quad, \qquad k = \frac{10+12\lambda^2}{1+12\lambda^2}$$

where λ is a pipe coefficient and k is a flexibility factor.

4 EXPERIMENTAL PROCEDURE

Figure 3 shows the loading method. Straight pipes were welded to both ends of the bent pipe and flanges with a thickness of 50 mm were welded to the ends of the straight pipes. One flange was attached to the base and the other flange was loaded in the inner bend and outer bend directions with a hydraulic jack.

The test was conducted using two kinds of bent pipes. Table 1 shows their specifications. Figure 4 shows the wall thickness distribution on the middle section of the bent pipes. In the case of the welded elbow, its thickness is higher at 90° or −90° because of the welding seams. In the induction pipe bend, the pipe wall is slightly thicker at an angle of −23°, but this is also a welding seam. In the case of the induction pipe bend, it is shaped using high frequency induction heating, so the wall thickness is higher at −90° and thinner at +90°.

The magnetic stress measurements at a section in the middle of the bent pipe were performed along the entire circumference at intervals of 5° in the circumferential direction. Near the location of the magnetic stress measurement, strain gauges were installed on the half section at intervals of 15° to verify the results. These strain gauges were two-axis strain gauges (longitudinal and circumferential directions).

In Figure 3, magnetic stress measurements were conducted at section Am and strain gauges were attached at section As.

Table 1. Specimen specifications.

Type	Welded elbow	Induction pipe bend
Material	API-X60	API-X65
External diameter	610 mm	610 mm
Radius of curvature	900 mm	1800 mm
L1	1590 mm	2090 mm
L2	1810 mm	2070 mm
λ	0.189	0.414
Yield stress	410 MPa	450 MPa

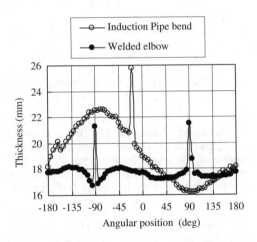

Figure 4. Comparison of wall thickness distributions.

5 RESULTS AND DISCUSSION

In Figure 5, white circles represent the distribution of the principal stress differential (obtained by subtracting the circumferential stress from the longitudinal stress of the pipe) measured by the strain gauges when an inner bend load of 80 kN (welded elbow) and 100 kN (induction pipe bend) was applied. In the following figures, these values are used unless otherwise specified. Next, the solid line is obtained by regressing the

221

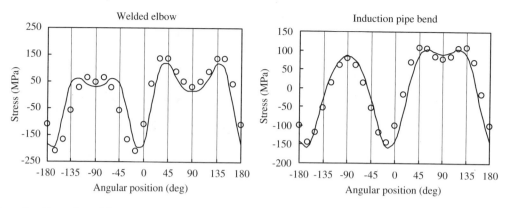

Figure 5.   Comparison of the stress measured by the strain gauges.

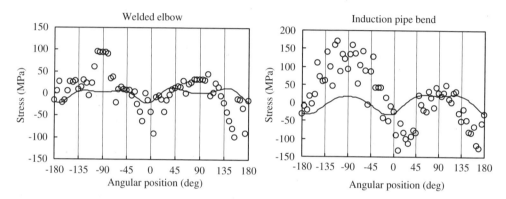

Figure 6.   Results of principal stress measurement by the magnetic method with no load.

principal stress differential obtained by this strain gauge to Karman's equation. The figure shows that Von Karman's theory accurately represents the stress distribution. The results for the induction pipe bend conform particularly closely with the theory. In both cases the thickness of the wall of the bent pipe is the average thickness excluding the seams.

Figure 6 shows the stress distribution measured by the magnetic method when the load is 0 N (no-load state). This figure shows that in the bent pipe, the residual stress generated during manufacture is ±100 MPa for the welded elbow and is ±150 MPa for the induction pipe bend. If external force is applied to a bent pipe in this state, the stress corresponding to this external force is added to the residual stress, making it difficult to precisely determine only the external force. Therefore, the value measured by the magnetic method was regressed to Karman's equation to obtain the stress distribution, as shown by the solid line. By regression to Karman's equation, due to the filter effect of the theory, the external force component is reduced to about 1/5 of the maximum measured

stress: at the maximum/minimum, ±20 MPa for the welded elbow, and ±30 MPa for the induction pipe bend. This is possible because the phase of the stress distribution generated by the external force differs from the phase of the stress distribution generated during manufacture of the bent pipe.

White circles in Figure 7 represent the stress differential measured by the magnetic method. The solid line represents the stress differential distribution obtained by regressing this value to Karman's equation. There are large variations between the data and the regression curve.

Figure 8 shows stress divided into the longitudinal and circumferential directions based on Von Karman's regression curve obtained by the method in Figures 5 and 7. The broken lines represent the stress distribution obtained by regressing the stress measured by the strain gauges to Von Karman's equation, and the solid lines that obtained by regressing the values measured by the magnetic method to Karman's equation. The value measured by the magnetic method is the principal stress differential, however it is possible to separate

222

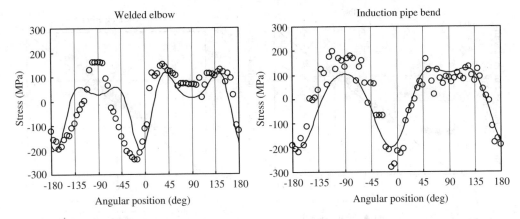

Figure 7. Comparison of the principal stress differential distribution measured by the magnetic method with the regression curve to Karman's equation.

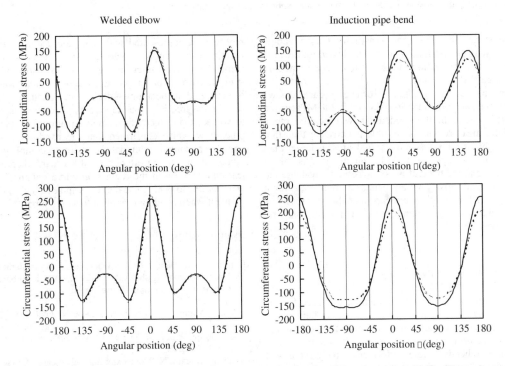

Figure 8. Comparison of Von Karman's regression curve of the stress based on the strain gauges with Von Karman's regression curve of the stress distribution based on the magnetic method.

the longitudinal and circumferential directions by regression to Karman's equation. The values agree very closely for the welded elbow, but there is a discrepancy of about 30 MPa for the induction pipe bend.

Figure 9 shows stresses divided into the longitudinal stress and circumferential stress at each loading step measured by strain gauges and the magnetic method.

In the magnetic method, absolute maximum stress at each step of loading is calculated by regressing to Karman's equation.

Through experiments on applying the magnetic method to various kinds of straight pipe, the magnetic sensitivity is reduced when the stress differential exceeded 70% of the yield stress, and the greater

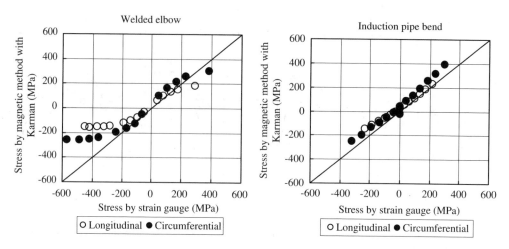

Figure 9. Comparison of the absolute maximum stress based on strain gauges with the absolute maximum stress obtained by regressing the value measured by the magnetic method to Karman's equation.

the stress, the lower the sensitivity of the gauge. Applying this finding to the material used for the bent pipes, the application range of the magnetic method is ±290 MPa for the welded elbow and ±310 MPa for the induction pipe bend. However, Figure 9 shows that the range of linearity is about ±200 MPa in the former case. In the welded elbow case, the range where measurement is possible is narrower than in a straight pipe, because the residual stress is larger than that of a straight pipe. Although in the case of an induction pipe bend, it is not possible to evaluate the drop in sensitivity without measurement data for a higher strain level, it appears that the sensitivity is maintained up to a higher stress than in the case of a welded elbow.

6 CONCLUSIONS

In this research, a magnetic method was used for the non-destructive measurement of stress in bent pipes. Bending tests of bent pipes made using two different manufacturing methods – welded elbows and induction pipe bends – were performed. Although the wall thickness distribution of the two bent pipes differed and both had thick welding seams, the research confirmed that the stress distribution generated in a bent pipe can be represented with adequate accuracy for practical use by applying Karman's stress equation and using the average thickness without the seams. The study confirmed that, as predicted, large residual stress is generated in bent pipes during manufacture. The stress measured by a magnetic method was regressed to Von Karman's theoretical stress equation for a bent pipe. It was also confirmed that the residual stress component could be eliminated and almost completely ignored thanks to the filter effect of Karman's theoretical equation. From the regression curve obtained, it was possible to distinguish the stress and external force moment when bending is applied to a bent pipe.

REFERENCES

Karman, Th. Von, 1911. Uber die formanderung dunwandgerrohre, insbesondere federnder Ausgleichrohre, *Zeitschrift des Vereines deutscher Ingenieure*, Vol. 55: 1889–1895

Sakai, Y., Matoba, Y., Suzuki, M., Ogawa, Y., Oka, S. 1991. Development of stress level meter using magnetic anisotropy, *Proc. 23rd symp. Stress-strain measurement*: 132–137: JSNDI

Sakai, Y., Tamura, N. 2000. Experimental research for estimating erection stress of steel bridge using magnetic anisotropy sensor, *Proc. non-destructive testing in civil engineering 2000*: 61–70. Oxford: Elsevier

Emerging Technologies in Non Destructive Testing, Van Hemelrijck, Anastasopoulos & Melanitis (eds)
© 2004 Swets & Zeitlinger, Lisse, ISBN 90 5809 645 9

Relaxation times in magneto-thermal NDT

N.J. Siakavellas
Department of Mechanical Engineering and Aeronautics, University of Patras, Greece

ABSTRACT: The time available to obtain thermal images after an electromagnetic excitation in a conducting material under inspection, i.e. the time period over which the temperature differences around a crack are still detectable, is a key parameter in a magneto-thermal NDT. The dependence of the relaxation times on the shape, the dimensions and the material properties of the inspected specimen are investigated, as well as the dependence on other characteristic times of the problem, such as the times related to the heat transfer by conduction and convection. By proper selection of some parameters, it is possible to have relaxation times as long as needed.

1 INTRODUCTION

Eddy current testing is the most usual NDT method for the evaluation of conductive metal structures (Takagi et al. 1996, Albanese et al. 1998). The process consists of measuring the change in the impedance of a coil, as it is moved over a defect. Thus, the material excitation and the material response are of the same nature (electromagnetic) and almost simultaneous. However, the procedure may be time consuming, especially if there is no previous information about the possible crack orientation and the entire surface must be scanned point by point. An ideal diagnostic method should combine two-dimensional information from the area under inspection during a time period sufficiently long, so that the information obtained is interpreted correctly.

Recent developments in infrared cameras and computer technology have made infrared thermography attractive for various engineering applications (Chrysohoos 2002, Ay et al. 2002), including non-destructive evaluation (Shepard 1997, Bates et al. 2000, Maierhofer et al. 2002). An alternative method for non-destructive testing in conducting materials has been proposed (Siakavellas 2000), that combines electromagnetic excitation and transient infrared thermography: A time-varying magnetic field induces eddy currents in the conducting material under inspection. The heat generated by the eddy currents, flows mainly from the border to the center of the specimen, crossing the current flow. Thus, the specimen is "swept" by current flow and heat flow in two different directions. Because of this property, a crack with arbitrary orientation, will modify the heat flow either directly or

indirectly and, consequently, the temperature distribution. By employing infrared thermography, it is then possible to visualize in 2-D the temperature distribution over the whole surface of the tested model. Contrary to the classical eddy current testing, the material excitation and the material response are not simultaneous, since the time constants related to the thermal process are some orders of magnitude greater than the electromagnetic ones. So, it is possible to have detectable temperature differences not only during the heating period, but also for a sufficiently long time after the shutdown of the exciting field.

A critical parameter arising from the proposed method is the time available to obtain thermal images, i.e. the time period over which the temperature differences around the crack are still detectable. The relaxation time is defined as the time taken after the electromagnetic excitation for the sample to reach the final steady state temperature. It is then desirable to have relaxation times as long as possible, by properly selecting the experimental parameters.

In the present work the relation of the relaxation times to other characteristic parameters of the problem is investigated. Obviously the relaxation times depend not only on the heating period and the intensity of the heating source (i.e. the exciting field), but also on the shape and the dimensions of the specimen, the material properties etc. Since the number of parameters involved in the problem is very large, dimensionless parameters will be used. In the next section, the electromagnetic and thermal equations are derived. Then, they are transformed in dimensionless form; so, the number of free parameters is drastically reduced.

2 ELECTROMAGNETIC-THERMAL MODEL

Here a two-dimensional analysis is presented. It is considered that the specimen is thin in the z direction and lies in the x-y plane. Furthermore it is assumed that the exciting field is uniform in space and perpendicular to the faces of the specimen, i.e. it is parallel to the z-axis. Under these assumptions, the density of the eddy current induced in the specimen by the exciting field has two components, j_x and j_y, that may be expressed in terms of a function u (Krawczyk & Tegopoulos 1993) as:

$$j_x = \frac{\partial u}{\partial y}; \quad j_y = -\frac{\partial u}{\partial x} \tag{1}$$

· The function u (stream function), a scalar quantity used in two-dimensional problems (instead of the electric vector potential used in 3-D), is a solution of the following equation:

$$\frac{\partial^2 u(x,y,t)}{\partial x^2} + \frac{\partial^2 u(x,y,t)}{\partial y^2} = \sigma \frac{\partial B}{\partial t} \tag{2}$$

where B = the exciting field. It is assumed then that the magnetic field induced by the eddy currents is small compared to the exciting one and may be neglected. This is true if the mean value of the specimen thickness and the crack length is much smaller than the penetration depth of the exciting field (Yuan 1981). Equation (2) is subjected to the boundary condition $u = 0$ on the edges of the plate.

The generated power (per unit volume) by the eddy currents, that is used as the thermal source of the process, may be expressed according to equation (1) as:

$$p(x,y;t) = \frac{j_x^2 + j_y^2}{\sigma} = \frac{1}{\sigma}\left[\left(\frac{\partial u}{\partial x}\right)^2 + \left(\frac{\partial u}{\partial y}\right)^2\right] \tag{3}$$

For the thermal field analysis, the assumption that the plate is thin in the z direction implies that the temperature $T(x,y,t)$ is uniformly distributed over the thickness w of the plate. Then, the energy balance equation for a small control volume dV, is:

[Rate of heat gain by conduction (in x,y)] + [Rate of heat gain by convection (in z)] + [Rate of energy generation in dV] = [Rate of storage of energy in dV].

When the appropriate mathematical expressions are introduced for each of these terms, we obtain the following equation for the temperature field, $T(x,y,t)$:

$$\frac{\partial^2 T}{\partial x^2} + \frac{\partial^2 T}{\partial y^2} - \frac{h_1 + h_2}{kw}(T - T_\alpha) + p(x,y,t) = \frac{1}{\alpha}\frac{\partial T}{\partial t} \tag{4}$$

where h_1, h_2 = mean heat transfer coefficients for the two faces of the specimen exposed to the magnetic

flux, k = thermal conductivity, α = thermal diffusivity and $p(x,y,t)$ is given by equation (3). Equation (4) is subjected to homogeneous boundary conditions the 2nd kind on the circumferential surface, since the implicit assumption for its derivation is that the heat losses through the circumferential surface are negligible. The initial condition is taken as:

$$T(x,y,t) = T_0 = T_\alpha \quad \text{at } t = 0 \tag{5}$$

i.e. it is assumed for simplicity that initially the temperature of the specimen is equal to the ambient one.

The type of the excitation and the heating time are key parameters of the problem. Here an exponential variation of the exciting field is assumed:

$$B(t) = B_0 e^{-t/\tau_m} \tag{6}$$

where B_0 = the magnetic flux density at $t = 0$ and τ_m = a decay time constant. Since τ_m is very small, compared to the thermal time constants, the heating time for this type of excitation is very short ($t \approx 5\tau_m$). By introducing equation (6) to equation (2) and separating the variables, it is easy to demonstrate that

$$u(x,y,t) = u_0(x,y)e^{-t/\tau_m} \tag{7}$$

where $u_0(x,y)$ is the solution of equation (2) at $t = 0$:

$$\frac{\partial^2 u_0(x,y)}{\partial x^2} + \frac{\partial^2 u_0(x,y)}{\partial y^2} = -\sigma \frac{B_0}{\tau_m} \tag{8}$$

3 DIMENSIONLESS FORM OF EQUATIONS

The electromagnetic and thermal equations are now made dimensionless, by employing the non dimensional parameters of distance (X, Y), stream function (U) and temperature (θ), where:

$$X = \frac{x}{l}; \quad Y = \frac{y}{l}; \quad U(X,Y,t) = \frac{u(x,y,t)}{\sigma \dfrac{B_0}{\tau_m} l^2} \tag{9}$$

$$\theta(X,Y,t) = \frac{T(x,y,t) - T_0}{T_f - T_0} \tag{10}$$

Here l is a characteristic dimension of the specimen and $T_f - T_0$ is the final rise in temperature of an isolated plate under the action of the exciting field.

According to equation (9), equations (7) and (8) are now written as:

$$U(X,Y,t) = U_0(X,Y)e^{-t/\tau_m} \tag{11}$$

$$\frac{\partial^2 U_0}{\partial X^2} + \frac{\partial^2 U_0}{\partial Y^2} = -1 \tag{12}$$

Equation (12) is subjected to the boundary condition $U_0 = 0$ on the edges of the plate.

In equation (4) the term multiplying $(T - T_\alpha)$ has dimensions of $[L^{-2}]$. If we put

$$\lambda^2 = \frac{kw}{h_1 + h_2} \qquad (13)$$

and take into consideration equations (9), (10) and (5), the equation for the thermal field is written as:

$$\frac{1}{l^2}\left(\frac{\partial^2 \theta}{\partial X^2} + \frac{\partial^2 \theta}{\partial Y^2}\right) - \frac{\theta(X,Y,t)}{\lambda^2} + \frac{p(X,Y,t)}{k(T_f - T_0)} = \frac{1}{\alpha}\frac{\partial \theta}{\partial t} \qquad (14)$$

The generated power, given by equation (3), may be expressed in terms of X, Y and U_0 as:

$$p(X,Y,t) = \sigma \frac{B_0^2}{\tau_m^2} l^2 \left[\left(\frac{\partial U_0}{\partial X}\right)^2 + \left(\frac{\partial U_0}{\partial Y}\right)^2\right] e^{-2t/\tau_m} \qquad (15)$$

On the other hand, the energy balance equation for a plate of mass density ρ and specific heat c is:

$$\rho c V (T_f - T_0) = \int_{t=0}^{\infty} P(t)dt \qquad (16)$$

The total power, $P(t)$, dissipated to the plate material by the exciting field, is given analytically, for a plate of surface S, as (Siakavellas 1997):

$$P(t) = \frac{1}{4}\sigma w \frac{S^2}{f}\left(\frac{dB}{dt}\right)^2 \qquad (17)$$

where f is a shape factor, that may be expressed analytically for plates of various shapes (Siakavellas 1997). Otherwise, f is computed numerically.

By using equation (6) for $B(t)$ and introducing equation (17) to equation (16), we finally obtain:

$$T_f - T_0 = \frac{1}{8}\frac{\sigma}{\rho c}\frac{S}{f}\frac{B_0^2}{\tau_m} \qquad (18)$$

If equations (15) and (18) are used in equation (14), the latter is written as:

$$\frac{1}{\tau_\kappa}\left(\frac{\partial^2 \theta}{\partial X^2} + \frac{\partial^2 \theta}{\partial Y^2}\right) - \frac{1}{\tau_\lambda}\theta(X,Y,t) +$$

$$\frac{8\gamma f}{\tau_m}\left[\left(\frac{\partial U_0}{\partial X}\right)^2 + \left(\frac{\partial U_0}{\partial Y}\right)^2\right]e^{-2t/\tau_m} = \frac{\partial \theta}{\partial t} \qquad (19)$$

where U_0 = solution of equation (12) and $\gamma = l^2/S$. Obviously, if the specimen is a rectangular plate, γ is

the aspect ratio. The τ_κ, τ_λ are characteristic times of the problem, relative to the heat transfer by conduction and convection respectively. They are related to the characteristic lengths l and λ by:

$$\tau_\kappa = \frac{l^2}{\alpha}; \qquad \tau_\lambda = \frac{\lambda^2}{\alpha} \qquad (20)$$

The boundary condition for equation (19) is $\partial\theta/\partial n = 0$ on the circumferential surface and the initial condition $\theta(X,Y,t) = 0$ at $t = 0$, according to equations (5) and (10). As far as the boundary conditions at the plate defects are concerned, it is assumed that a flaw acts as a barrier for the transport of heat and the current. Thus, the components of both the current density and the heat flow normal to the flaw surface are set equal to zero.

4 RELAXATION TIMES

The relaxation time is the time taken after the electromagnetic excitation to reach the final steady state temperature. Strictly speaking, this is accomplished only in the limit $t \to \infty$ (i.e. a time interval at least tenfold the time constant τ_κ or τ_λ), since the specimen temperature tends asymptotically to its final steady state value. However, for practical purposes, we may consider that thermal equilibrium is established when the temperature variation is smaller or at least equal to a critical value ε. This quantity is determined from the experimental restrictions.

In fact, we may define relaxation times relative to the time and the spatial derivatives of the temperature. For example, let us consider the conditions:

$$\left|\frac{\partial T}{\partial t}\right| \leq \left(\frac{\Delta T}{\Delta t}\right)_{min} \equiv \varepsilon_t$$

$$\left|\frac{\partial T}{\partial x}\right| \leq \left(\frac{\Delta T}{\Delta x}\right)_{min} \equiv \varepsilon_x; \quad \left|\frac{\partial T}{\partial y}\right| \leq \left(\frac{\Delta T}{\Delta y}\right)_{min} \equiv \varepsilon_y \qquad (21)$$

For an infrared camera, the critical values ε_t, ε_x and ε_y depend on the sensitivity, the image update rate (frames per second), the displayed image (number of pixels in horizontal and vertical direction) etc. The relaxation times, R_t, R_x and R_y, at a given position (x,y) in the specimen, are then defined as the elapsed time after which the corresponding condition (21) is continuously satisfied.

If the dimensionless parameters are used, the conditions (21) are written as:

$$\left|\frac{\partial \theta}{\partial t}\right| \leq \left(\frac{\Delta \theta}{\Delta t}\right)_{min} \equiv E_t$$

$$\left|\frac{\partial \theta}{\partial X}\right| \leq \left(\frac{\Delta \theta}{\Delta X}\right)_{min} \equiv E_X; \quad \left|\frac{\partial \theta}{\partial Y}\right| \leq \left(\frac{\Delta \theta}{\Delta Y}\right)_{min} \equiv E_Y \qquad (22)$$

The critical values ε_i, E_i, are related by:

$$\varepsilon_t = (T_f - T_0)E_t$$

$$\varepsilon_x = \frac{T_f - T_0}{l}E_X; \quad \varepsilon_y = \frac{T_f - T_0}{l}E_Y \qquad (23)$$

5 APPLICATIONS AND RESULTS

The numerical solution of the dimensionless equations (12) and (19), with the appropriate boundary and initial conditions, yields the dimensionless temperature $\theta(X,Y,t)$. The relaxation times R_t, R_X and R_Y are obtained from the conditions (22). This model has been applied to numerous test problems, by considering horizontal plates of various shapes. For the estimation of the mean heat transfer coefficients, h_1 for the top and h_2 for the bottom surface of the plate, and consequently the characteristic time τ_λ, simplified equations are used for free convection from horizontal plate to air at atmospheric pressure (Holman 1992), resulting to: $h_1 + h_2 = 1.91(\Delta T/L)^{1/4}$, where $\Delta T = T - T_\alpha$ and L a characteristic length of the plate. Here we consider $\Delta T = (T_f - T_0)/2$ and $L = S/P$, where S is the area and P the perimeter of the surface, as suggested by Goldstein et al. (1973) and Lloyd & Moran (1974). Note that the above formula for L is also applicable to unsymmetrical planforms.

The results presented here concern a test problem with aspect ratio $\gamma = 1$ (square plate), $t_\kappa = t_\lambda = 1000\,s$ and $t_m = 5\,ms$. A crack is situated (in dimensionless coordinates) between (0.38, 0.15) and (0.46, 0.15), i.e. it is parallel to x-axis. The relaxation times R_t, R_X and R_Y (for $E_t = 0.02$, $E_X = E_Y = 1.0$), are presented in the form of contour maps in Figures 1, 2 and 3 respectively. From Figures 1b and 2b it is clear that the relaxation times R_t and R_X have their maximum (local or solute) values in the region around the crack. Note that, in a plate without defects (Figs. 1a and 2a) the corresponding values are relatively small. On the contrary, R_Y although it is relatively large in the corresponding region for a plate without defects (Fig. 3a), it is small around the crack (Fig. 3b). This is due to the fact that the crack is parallel to x-axis, so it forces the heat flux component parallel to y-axis, to be equal to zero.

On the other hand, in a major part of the plate surface the relaxation times are smaller for the plate with the crack than a plate without defects. This is clearer from Figures 1c, 2c and 3c, where the difference between the relaxation times (plate with crack minus plate without crack) is illustrated. This may be explained by the fact that the heating of the plate with the crack by the time-varying magnetic field is less effective than the heating of a plate without defects by the same exciting field, since the resistance to the current flow is greater in the

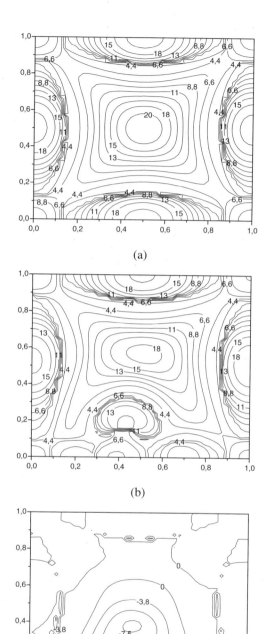

(a)

(b)

(c)

Figure 1. Relaxation times (in sec), relative to the time derivative of temperature. (a) Plate without defects. (b) Plate with a crack. (c) Difference between the relaxation times (b) − (a).

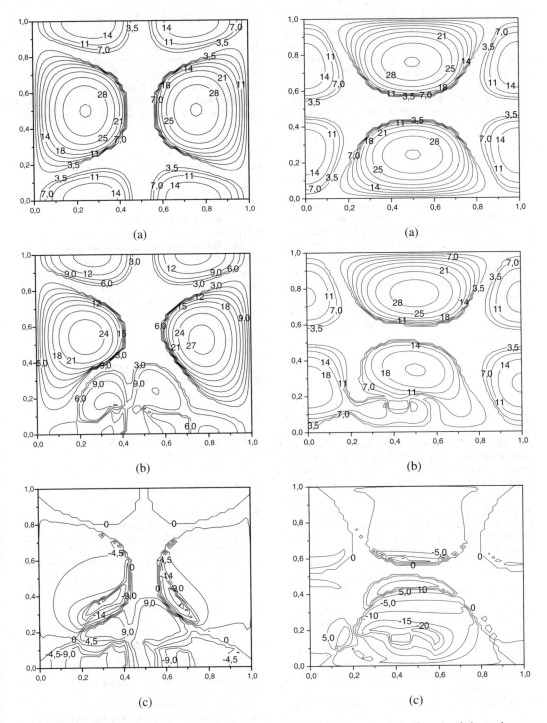

(a)

(b)

(c)

Figure 2.   Relaxation times (in sec), relative to the space (X) derivative of temperature. (a) Plate without defects. (b) Plate with a crack. (c) Difference between the relaxation times (b) – (a).

Figure 3.   Relaxation times (in sec), relative to the space (Y) derivative of temperature. (a) Plate without defects. (b) Plate with a crack. (c) Difference between the relaxation times (b) – (a).

former case. However, within a relatively large region around the crack this difference is positive (Figs. 1c, 2c). Consequently, the time available to take thermal images in the crack region is greater. In the case of R_Y the region with positive difference is displaced towards the plate center (Fig. 3c).

For the dependence of the relaxation times on the shape of the specimen there is not a general rule. In fact, the relaxation times around a crack depend not only on the specimen shape, but are also a function of the crack position and orientation with respect to the specimen geometry.

As far as the dependence of the relaxation times on the characteristic times τ_κ and τ_λ is concerned, the calculations in various cases have shown that the relaxation times around the crack increase linearly with τ_κ (when τ_λ is constant). This is so because greater τ_κ means either greater dimensions of the specimen, or smaller thermal diffusivity of the inspected material, so thermal equilibrium is established later. On the other hand, the relaxation times increase slowly with τ_λ (when τ_κ is constant), since, when τ_λ increases, the rate of heat transfer by convection decreases (for $\tau_\lambda \to \infty$, the plate is an isolated one, i.e. the 2nd term in equation 19 vanishes).

6 CONCLUSIONS AND DISCUSSION

In magneto-thermal NDT the heating of the inspected material by a time-varying magnetic field results to measurable temperature gradients for a sufficiently long time. Obviously, the temperature gradients are more important during the heating period than later. However, in experimental conditions it may not be possible to take thermal images during this period, especially when we have access to only one of the faces of the specimen. So, the relaxation time, i.e. the time period for which the temperature differences around the crack are still detectable, is a critical parameter for the proposed method.

The results from the numerical investigation for plates heated by a fast-decaying magnetic field uniform in space, may be summarized as follows:

- Within a relatively large region around the crack the relaxation times are greater for the plate with the crack than for a plate without defects.
- Around the crack, the relaxation times increase with both the characteristic times τ_κ and τ_λ that are relative to heat transfer by conduction and by convection respectively.
- The relaxation times increase when $T_f - T_0$ is increasing.

Of course, the relaxation times depend mainly on the sensitivity and other characteristics of the infrared camera in use (Eqs. 21–23). However, when the performances are relatively low, the relaxation times may be improved by increasing for example $T_f - T_0$ (Eqs. 18 and 23). This may be achieved by either increasing the intensity of the heating source, or by elongating the heating time. Thus, it is possible to have relaxation times as long as needed.

The present work was limited to 2-D and to an exciting field uniform in space. In practice, the shape and the dimensions of the inspected objects permit the use of exciting fields uniform in space only in a few cases. In general, a coil moving over the inspected object produces an exciting field non-uniform in space. So, only a part of the whole surface is exposed to the magnetic flux at a given instant and the material is not heated simultaneously. Thus, we switch to a dynamic situation, where the infrared camera must follow the exciting field that acts as a moving heat source.

Note that moving heat source problems are also encountered in other engineering applications, such as manufacturing and tribology (Hou & Komanduri 2000). Of course, new variables appear now: the shape of the moving heat source, its velocity etc., doing the problem more complicated.

The numerical investigation in a first phase, and, the experimental one, later, is our priority on the above item, together with an analysis in 3-D, in order to verify whether is it possible to determine by this method, the position and the shape of defects deep in the specimen.

ACKNOWLEDGMENT

Work supported by C. Caratheodory Program of University of Patras.

REFERENCES

Albanese, R., Rubinacci, G., Takagi, T. & Udpa, S.S. (eds). 1998. Electromagnetic Non-Destructive Evaluation, in: *Studies in Applied Electromagnetics and Mechanics, Vol. 14*. Amsterdam: IOS Press.

Ay, H., Jang, J.Y. & Yeh, J. 2002. Local heat transfer measurements of plate finned-tube heat exchangers by infrared thermography. *Int. J. of Heat & Mass Transfer* (45): 4069–4078.

Bates, D., Smith, G., Lu, D. & Hewitt, J. 2000. Rapid thermal non-destructive testing of aircraft components. *Composites: Part B* 31:175–185.

Chrysochoos, A. 2002. La Thermographie Infrarouge, un outil en puissance pour etudier le comportement des materiaux. *Mecanique & Industries* 3: 3–14.

Goldstein, R.J., Sparrow, E.M. & Jones, D.C. 1973. Natural convection mass transfer adjacent to horizontal plates. *Int. J. Heat Mass Transfer* 16:1025.

Holman, J.P. 1992. *Heat Transfer*. London: McGraw Hill.

Hou, Z.B. & Komanduri, R. 2000. General solutions for stationary/moving plane heat source problems in manufacturing and tribology. *Int. J. Heat Mass Transfer* 43:1679–1698.

230

Krawczyk, A. & Tegopoulos, J. 1993. *Numerical Modelling of Eddy Currents*. Oxford: Clarendon.

Lloyd, J.R. & Moran, W.R. 1974. Natural convection adjacent to horizontal surface of various planforms. ASME Pap. 74-WA/HT-66.

Maierhofer, Ch., Brink, A., Rolling, M. & Wiggenhauser, H. 2002. Transient thermography for structural investigation of concrete and composites in the near surface region. *Infrared Physics & Technology* 43: 271–278.

Shepard, S.M. 1997. Introduction to active thermography for non-destructive evaluation. *Anti-Corrosion Methods and Materials* 44(4): 236–239.

Siakavellas, N.J. 1997. Two simple models for analytical calculation of eddy currents in thin conducting plates. *IEEE Trans. on Magnetics* 33: 2245–2257.

Siakavellas, N.J. 2000. A proposal for magneto-thermal NDT in conducting materials. In D.V. Hemelrijck, A. Anastasopoulos, T. Philippidis (eds), *Emerging Technologies in NDT*: 179–186. Rotterdam: Balkema.

Takagi, T., Bowler, J.R. & Yosida, Y. (eds). 1996. Electromagnetic Non-Destructive Evaluation, in: *Studies in Applied Electromagnetics and Mechanics*, *Vol. 12*. Amsterdam: IOS Press.

Yuan, K.Y. 1981. Finite element analysis of magnetoelastic plate problems. Dept. of Structural Engineering, Report No. 81-4, Cornell University.

Application of the magnetic Barkhausen noise to determination of hardness and residual-stress variations in hardened surface layers

J. Grum & P. Žerovnik

Faculty of Mechanical Engineering, University of Ljubljana, Slovenia

ABSTRACT: The paper treats the application of the magnetic Barkhausen noise for the assessment of microstructure, hardness, and residual stresses. Measurements were performed on induction surface-hardened specimens, the thickness of hardened layers being different in each case. A change in the microstructure of the hardened surface layer produces changes in specific electric conductivity σ and relative permeability μ_r. The two parameters and chosen analysing frequencies influence the depth sensing of micromagnetic changes. The most commonly chosen characteristic of the Barkhausen noise is the V_{RMS}^2 value of a captured signal. The choice of analysing frequency determines the depth in which the Barkhausen noise will be analysed to determine microhardness or residual stresses.

1 · INTRODUCTION

In the assessment of machine parts, the quality of a surface-hardened layer, i.e., microhardness variation and residual-stress gradient in the surface layer, is very important. Recently to this aim various non-destructive testing methods have been increasingly applied, particularly because of the direct applicability of the methods to material testing. The automated production of machine parts requires on-line monitoring of the state of material; therefore, the methods applied should be sufficiently reliable, fast and reproducible. One of such methods is the micromagnetic method based on the Barkhausen noise. From the viewpoint of physics, the micromagnetic method is based on the fact that a ferromagnetic material, when magnetised by the alternating current, will contain small magnetic domains. When an external magnetic field affects a ferromagnetic material, a movement of magnetic-domain walls will occur, which produces changes in the size and shape of the latter. A variation in the magnetic flux will induce voltage in the measuring coil, which can be registered, and then processed (Jiles et al., 1994; Theiner et al., 1987; Mitra et al., 1996). Numerous studies have shown that relatively small differences in mechanical properties of ferromagnetic materials can be efficiently detected by the micromagnetic method based on the Barkhausen noise.

2 EXPERIMENTAL PROCEDURE

2.1 *Experimental setup and testing procedure*

For the investigations, an experimental setup was arranged to capture voltage signals of the magnetic Barkhausen noise (BN). It consisted of a magnetisation unit, a sensor for capturing voltage signals, a signal amplifier with a relevant band-pass filter, and a computer-aided unit for determination of microstructure or microhardness and residual stresses. Figure 1 shows a block scheme of the experimental setup for micromagnetic testing based on the Barkhausen noise.

Before starting experiments, optimum magnetising parameters producing the movement of magnetic domains characteristic of the Barkhausen noise were to be determined. Very important parameters in magnetisation are the magnetising frequency (f_e), the magnetising current (I), and the magnetic field strength (H) in the specimen. The magnetic field strength depends on the magnetising current (I), the number of windings (n) of the yoke, and the mean path length of the magnetic flux in the yoke and the specimen (L).

The captured magnetic BN signal is composed of a series of abrupt changes of voltage produced by movements of the magnetic domains. In most cases the captured voltage signals cannot be directly related to individual parameters to assess the state, i.e., properties, of the surface layer. The parameters most frequently

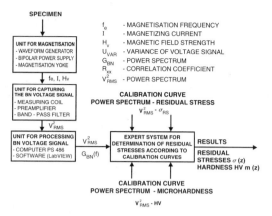

Figure 1. Experimental setup for capturing voltage signal of the magnetic Barkhausen noise.

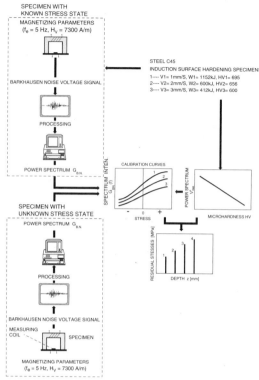

Figure 2. Block scheme of calibration of experimental setup.

applied to assessment of the surface layer are the microstructure, microhardness, and residual stresses. For further efficient analysis of the voltage signals, an appropriate method for signal processing should be chosen in order to use the characteristic value of the voltage signal for elaboration of calibration curves. Finally the relationship between the voltage-signal characteristic (V_{RMS}^2, G_{BN}, Δt) and the chosen surface characteristic (HV_m, σ_{RS}) was assessed by means of selected statistical methods. Different methods may be applied to processing of the magnetic BN voltage signal such as:

a. power frequency spectrum V_{RMS}^2,
b. spectrum intensity with given frequency $G_{BN}(f)$,
c. measurement of time delays of the signals Δt.

The method being comparative, first, calibration measurements were made at etalons establishing the dependence of the microstructure, microhardness, and residual stresses on the characteristic value of the BN voltage signals. Calibration was carried out with specimens having known properties and called etalons. When a calibration curve was known, a measurement could be performed at an unknown specimen, and then the microhardness or residual stresses from the calibration curve determined. The etalons had different thicknesses of the hardened layers. In the first phase of calibration, the specimens were in unstressed condition and it was assumed that they contained no residual stresses. Then selected tensile forces were gradually applied to the specimens at a testing machine. At the same time BN voltage signals were being captured. The captured voltage signals were then processed by an adequate method. Finally calibration curves were elaborated. The same procedure was applied to compression-stressed specimens.

The calibration of the experimental system and elaboration of the calibration curves were followed by measurements, and later a classification of the specimens with the unknown properties. A block scheme of the individual operations is shown in Figure 2. The upper part of Figure 2 shows the procedure of elaboration of calibration curves using the etalons with the known microhardness and residual stresses, whereas the lower part shows the procedure of testing of an unknown specimen, i.e., determination of the characteristics of the surface-hardened layer.

2.2 Specimen preparation

For flat specimens C45 carbon structural heat-treatment steel having $150 \times 30 \times 6$ mm in size was used. The specimens were annealed at a temperature of 680°C to relieve internal stresses. After annealing all the specimens showed minimum and approximately the same residual stresses, which indicated that the original properties at the specimen surfaces were comparable.

Figure 3 shows the specimen through-thickness microhardness variation. Different heat inputs were obtained with a double-winding induction loop with the same inductor and the same gap size but with different specimen shifts. The V_{RMS}^2 characteristic value of the voltage signal and its through-thickness variation were determined as well. The microhardness determination

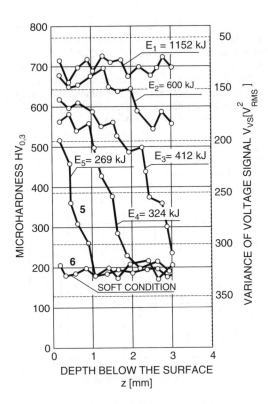

Figure 3. Specimen through-thickness microhardness variation under different induction-hardening conditions.

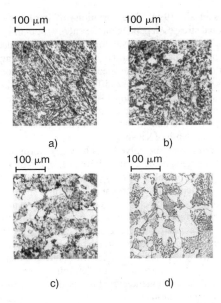

Figure 4. Microstructures of individual zones of induction surface-hardened specimen.

in relation to the V_{RMS}^2 voltage-signal value was accomplished by means of a calibration curve.

The different energy inputs in induction heating produced changes in the microstructure of C45 steel at the transition from the hardened to unhardened part of the material. At the surface martensite microstructure was obtained. Depending on the energy inputs, it transformed, in different depths, to bainite and pearlite-ferrite microstructures respectively. Figure 4 shows micrographs obtained in the chosen depths and with an energy input of E = 324 kJ.

The first micrograph (4a) shows the martensite microstructure in a depth of 150 μm, the second (4b) a transitional martensite-bainite microstructure in a depth of 900 μm, the third (4c) a pearlite-ferrite microstructure with a small portion of bainite in a depth of 1500 μm, and the fourth (4d) coagulated lamellar pearlite-cementite in a depth of 2000 μm where no transformation occurred.

2.3 Variations of the BN characteristic value and through-thickness microhardness

The microhardness variation, i.e., the depth of the hardened layer, depends on the conditions used in surface hardening. The different energy inputs in induction heating of the surface ensured the desired temperature variation in heating so that after quenching the desired depth of the hardened layer, and the desired profiles of micorhardness and residual stresses in the thin surface layer were obtained. The changed material properties in the thin hardened surface layer affected the movement of magnetic domains, i.e., the voltage induced in the detection coil.

In the analysis of the Barkhausen noise in the chosen depth, it was important to filter the BN voltage signals. A frequency filter provided a signal of the voltage induced in the detection coil sensing the changes during the process of specimen magnetisation. Thus a Butterworth filter of the fourth order in a series connection with a low-pass and high-pass filters of the second order was applied. In the experimental setup thus four different band-pass Butterworth filters of the fourth order, passing signals in different frequency ranges, i.e., from 0.7 to 20, 0.7 to 40, 0.7 to 100, and 0.7 to 200 kHz, were built in. With the same analysing frequency, the sensing depth was affected only by substance properties of the material in the thin surface layer of the specimen due to heat treatment.

For the individual specimens, sensing depths, for which HV and σ_{RS} were determined from the captured voltage signal, were calculated. Eq. (1) indicates that with the same analysing frequency f_a and a higher specific electric conductivity σ and relative permeability of the material μ_r, a smaller depth of the micromagnetic changes was obtained and vice versa

235

(Jiles et al., 1994). The variations of hardness and residual stresses depended on the energy input in specimen heating and the quenching conditions to ensure the microstructure required. A lower energy input resulted in heating of the specimen to lower temperatures so that between T_{A3} and T_{A1} lower microhardness was obtained at the specimen surface. The lower microhardness at the surface, in turn, resulted in a better conductivity σ and permeability μ_r and, consequently, a smaller depth of sensing.

A question to be solved was how to determine the depth related to the captured BN voltage signal. In case the depth of sensing was the same or smaller than the depth of the hardened layer, the exponential law of Barkhausen emission damping was presumed. The damping of the Barkhausen emission to the given depth depended on the electric conductivity and material permeability in the surface specimen layer. How strong the damping of the Barkhausen emission in certain depths was could be determined by changing analysing frequencies, which is indicated in the following equation:

$$\delta = \sqrt{\frac{1}{\pi . f_a . \sigma . \mu_0 . \mu_r}} \qquad (1)$$

Figure 5 shows the influence of energy input on the depth of sensing micromagnetic changes in the surface-hardened layer. It can be seen that a lower energy input resulted in specimen heating to lower temperatures so that lower microhardness was obtained at the specimen surface. The lower microhardness in the surface-hardened layer produced a better conductivity σ and permeability μ_r and thus a smaller depth of sensing micromagnetic changes, and vice versa.

The data on the specific electric conductivity of steel σ in the soft state (ferrite-pearlite) and the quenched state (martensite) were obtained in Ref. 4. If linear varying of electric conductivity is presumed, the variation of electric conductivity can be determined on the basis of the known through-thickness microhardness of the hardened layer. The relative permeability μ_r of the analysed material was calculated from the ratio of the magnetic flux density B to the magnetic field strength H under the presumption that all the specimens showed the same microhardness in the entire depth of sensing. Both values were determined from the magnetisation curves captured at the specimens with the determined microhardness; therefore, several specimens were induction surface-hardened. Thus different variations of through-thickness microhardness in the hardened layer were obtained.

Table 1 shows the values calculated for individual depths of sensing micromagnetic changes with reference to the changes of specific electric conductivity σ and relative permeability μ_r of the surface-hardened layer.

From the chosen characteristic value V^2_{RMS} of the Barkhausen noise, Vickers microhardness was determined using the calibration curves. Table 2 shows the results of evaluation of microhardness using the calibration curves and the results obtained in measurement of Vickers microhardness. An average microhardness value to the calculated depth of sensing was read from the calibration curves. Figure 5 shows the Vickers-microhardness variation across the entire specimen cross-section. The diagram indicates that the through-depth microhardness strongly depended on the energy input.

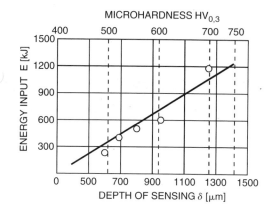

MICROHARDNESS $HV_{0,3}$

Figure 5. Influence of energy input in the depth of sensing.

Table 1. Calculated depths of sensing under different induction-hardening conditions.

Spec.	Energy input E [kJ]	Elec. conduc. σ [$\Omega^{-1} m^{-1}$]	Perme. μ_r	Depth of sensing δ [μm]
1	1152	$1{,}05 \cdot 10^6$	209	1265
2	600	$1{,}35 \cdot 10^6$	225	1049
3	412	$1{,}76 \cdot 10^6$	263	836
4	324	$2{,}10 \cdot 10^6$	295	707
5	269	$2{,}39 \cdot 10^6$	327	632

Table 2. Variation of measured Vickers-microhardness at specimen surface and varition of microhardness obtained from calibration curves.

Spec.	Measured vickers microhardness $HV_{0.3}$	Microhardness from calibration curves HV/V^2_{RMS}
1	695	673
2	656	634
3	600	567
4	554	511
5	510	441

In the assessment of quality of the surface-hardened layer, the magnitude of residual stresses achieved and the gradient of residual stress were important too. The variation of residual stresses was determined by means of a variance of voltage signal which was, in turn, determined by calibration at the etalons. Different tensile forces were gradually applied to the individual etalons at the testing machine. At the same time BN voltage signals were being captured. Then the characteristic value V_{RMS}^2 was determined. The compressive and tensile stresses in the etalons were calculated from the known load forces. After signal processing, V_{RMS}^2 and G_{BN} characteristic values were obtained, and a calibration curve was elaborated. The calibration curve is a curve indicating the relationship between the mechanical stress of the given etalon and the V_{RMS}^2 ali G_{BN} values of the voltage signal. Thus for every etalon a calibration curve was elaborated. In the analysis of an unknown specimen, the size and variation of residual stresses in the surface layer of the material could be determined from the captured Barkhausen noise and the characteristic values calculated from the calibration curves.

Figure 6 shows the average residual stresses determined in the surface-hardened layer to the depth of sensing. In the hardened layers of the individual specimens compressive residual stresses were obtained regardless of the conditions of heating and quenching.

3 CONCLUSIONS

The efficiency of induction surface hardening was assessed by means of the non-destructive micromagnetic method based on the captured BN voltage signals. The analysis was focused on three characteristics of the hardened layer, i.e., microstructure, microhardness, and residual stresses. It was found that changes of energy input in the specimen surface layer affected the microstructure variation and, consequently, microhardness and the residual stress. In order to determine HV and σ_{RS}, the etalons of the calibration curve related to the data on the Barkhausen noise and the V_{RMS}^2 characteristic value were produced. The depth of measurement depended on the chosen analysing frequency f_a, electric conductivity σ and relative permeability μ_r. The smaller the depth of sensing micromagnetic changes, the greater the accuracy of determination of the relationship between HV–G_{BN} and σ_{RS}–G_{BN} respectively. In the determination of microhardness and residual stresses from the magnetic Barkhausen noise, account should be taken that the results obtained with a certain analysing frequency represent average values to the depth of sensing.

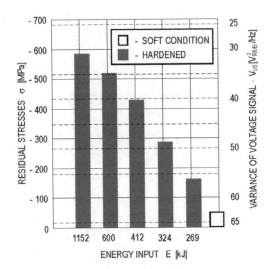

Figure 6. Variation of residual stresses after specimen surface hardening.

REFERENCES

Jiles, D.C. & Suominen, L. 1994. Effects of Surface Stress on Barkhausen Emissions – Model Predictions and Comparison with X – Ray Difraction Studies, Division of Material Sciences.

Theiner, W.A. & Deimel, P. 1987. Non-destructive Testing of Welds with 3MA-Analyzer, Nuclear Engineering and Design, Vol. 102, 257–264.

Mitra, A., Ravikumar, B., Mukhopadhyay, N.K., Murthy, G.V.S., Das, S.K. & Bhattacharya, D.K. 1996. Residual Stress Evaluation in Service Exposed Shot Peened Spring by Micromagnetic and XRD Techniques, 14th World Conference on Non Destructive Testing, New Delhi, 879–882.

Rohlfing, H., Schmid, H., Templ, A. 1976. Tabellenbuck fur Elektrotecnik, Ummler – Bonn, 289–290.

Grum, J., Žerovnik, P. 2002. Assessment of a Shot-peened Surface with the Micro-magnetic Method, 32nd International Conference and NDT Technique Exposition-Defectoscopy Liberec, 63–72.

Grum, J., Žerovnik, P. 2000. Use of the Barkhausen effect in the measurement of residual stresses in steel, Insight Vol.42, No.12, 796–800.

Material characterisation

Emerging Technologies in Non Destructive Testing, Van Hemelrijck, Anastasopoulos & Melanitis (eds)
© 2004 Swets & Zeitlinger, Lisse, ISBN 90 5809 645 9

Stress wave propagation in fresh mortar

D.G. Aggelis & T.P. Philippidis
Department of Mechanical Engineering and Aeronautics, University of Patras, and
Institute of Chemical Engineering and High Temperature Chemical Processes, Patras, Greece

ABSTRACT: In the present paper wave propagation characteristics of mortar i.e., phase velocity and attenuation are examined by means of a series of through-transmission ultrasonic measurements. Specimens of various water to cement ratios (w/c) and aggregate (sand) to cement ratios (a/c) were produced and tested in order to study the influence of these mix design parameters on the propagating wave. The introduction of wave packets of different frequencies revealed interesting velocity dispersion features. Discrepancies concerning the amplitude of the signals can also be observed, implying that discrimination between materials as to different constituent proportions is possible. This could be a step towards quality estimation of fresh concrete, since its ultimate mechanical properties depend strongly on the mix proportions and especially on the w/c of the fresh material.

1 INTRODUCTION

Stress wave propagation has been used since long to assess quality of cementitious materials (Kaplan 1959). Recently a number of efforts has seen publicity concerning ultrasound monitoring of fresh concrete aiming mainly at the set point estimation (Grosse 2001, Garnier 1995) or the influence of admixtures (Rapoport 2000), showing only the qualitative influence of water dosage on the propagating wave. Parameters as the wave velocity and energy in through-transmission measurements or reflection coefficient when a reflection method is used, provide valuable information as to the hardening state of the material. Nevertheless, the inhomogeneous nature of fresh concrete consisting of cement grains, fine and coarse aggregates and air bubbles suspended in water allows only for qualitative conclusions about the state of the material. By the use of narrow band tone bursts at several frequencies the dispersive and attenuative nature of freshly mixed cementitious material is highlighted in this work. Different mix parameters i.e. water to cement ratio, w/c and aggregate (sand) to cement ratio, a/c by mass were employed in order to estimate their influence on the propagating wave. The results show that the a/c and sand fineness control mainly the attenuative behavior, while frequencies of about 1 MHz propagate with velocity 3 times that of 20 kHz.

2 EXPERIMENTAL DETAILS

The experimental setup is similar to the one described in (Philippidis 2002) used for w/c determination of hardened concrete. It consists of a waveform generator board (Physical Acoustics, PAC WaveGen 1410), two broadband transducers of center frequency 500 kHz (Panametrics V413), a PAC preamplifier 1220A and a PAC Mistras 2001 data acquisition system. A 10 cycle sinusoidal wave in sinusoidal envelope is fed to the driving transducer. The center frequencies of the pulses vary from 20 kHz to 1 MHz in increments of 100 kHz for high frequencies or smaller in order to capture in more detail the rapidly increasing trend of velocity and attenuation at lower frequencies. In Figure 1, an example of a narrow band pulse is depicted with center frequency 250 kHz. The distance between the sensors is 10 mm since in greater distances the signal is severely attenuated and not suitable for analysis. The sensors were in direct contact with the material while rubber plates were used to constrain the fresh mortar.

The constituent materials (cement II 32.5, limestone, sand and water) were weighted using a 0.1 mg accuracy balance, mixed and stirred for approximately 1 min. Then the mortar was poured between the transducers and compacted for another min by means of a stick, which resulted in the release of air bubbles from the surface of the specimen. Ultrasonic measurements

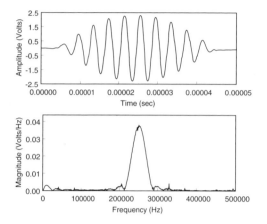

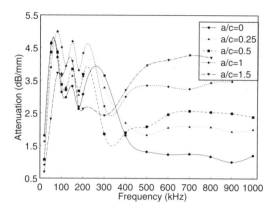

Figure 1. Input signal of 250 kHz center frequency, (a) time domain, (b) frequency.

Figure 2. Effect of a/c on the attenuation vs frequency curve.

were recorded 3 min after mixing of the ingredients with water.

3 ATTENUATION MEASUREMENTS

The four most important sources of attenuation in suspensions are: absorption losses in each of the individual components, visco-inertial losses due to density discrepancies of the constituent phases, thermal dissipation losses and scattering (McClements 2000). These mechanisms contribute to the overall attenuation and cannot be directly separated. To obtain a measure of overall material attenuation, the absolute maximum voltage of the received waveform was compared to that of pure water. This way, for each frequency used the attenuation coefficient was calculated as:

$$a = -\frac{20}{x} \cdot \log\left(\frac{A_X}{A_0}\right) \qquad (1)$$

where A_0 is the amplitude of the wave after it has traveled distance x through water and A_X is the amplitude after the same distance through mortar.

3.1 Effect of sand content

The effect of aggregate to cement ratio on the attenuation vs frequency curves of specimens with w/c = 0.55 is shown in Figure 2. It is observed that for high frequencies the increase of sand amount adds to the attenuation while the situation is approximately inversed for low frequencies.

The increasing attenuation of high frequencies with inclusion content can be attributed to the combined effect of scattering mechanisms (Farrow 1995)

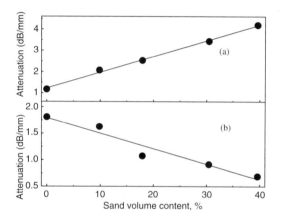

Figure 3. Attenuation vs sand content curve for (a) 800 kHz and (b) 20 kHz pulse for mortar with w/c = 0.55.

along with visco-inertial and thermal effects (Hipp 1999) and is common in suspensions of solids in liquids. The effect of individual grains, depending also on the wave frequency, is additive for a range of volume contents, attenuation being proportional to it. This seems to be the case for the examined material as the attenuation of high frequency pulses, i.e. in the range of 500 kHz to 1 MHz is proportional to the sand content as can be seen in Figure 3a for the case of a 800 kHz pulse. This behavior is typical of isolated particle mechanisms (Urick 1948), while multiple particle interactions are manifested through non-linearity between attenuation and concentration being expected generally at higher concentrations.

Therefore, the influence of inclusion amount in overall attenuation is clear at high frequencies. At the low frequency band though, the effect of sand (or inclusion) content is not clear (Dukhin 2001) and in

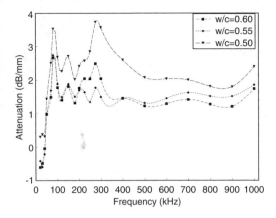

Figure 4. Effect of w/c on attenuation of mortar with a/c = 1.

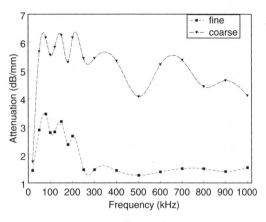

Figure 5. Effect of sand fineness on attenuation.

fact only the attenuation of 20 kHz pulses seems inversely proportional to sand content, see Figure 3b.

From Figure 2 it is noted that the attenuation coefficient obtains a maximum value at frequencies around 100 kHz. Since the sand grain characteristic dimension averages at about 2 mm, this is approximately the frequency that the wavelength becomes comparable to the grain size (McClements 2000).

3.2 w/c influence

The w/c of fresh mortar has an impact on the density of the suspending medium (cement paste) while affecting also the sand content since water, cement and sand are the only ingredients of mortar. Indeed for a constant a/c of 1, w/c ratios of 0.50, 0.55 and 0.60 correspond to sand volume fractions, s, of 31.7%, 30.4% and 29.3% respectively. The variation of attenuation versus frequency curves for these specific mixtures is shown in Figure 4. The trend is similar to that of Figure 2 showing that the increase of w/c lowers the attenuation of high frequencies probably due to the decrease of sand content.

3.3 Sand fineness influence

The aggregate size is in direct relation with attenuation. The variation of attenuation vs frequency curves with different grain size is shown in Figure 5 for two cases of w/c = 0.50 and a/c = 1 mortar. The "coarse" sand specimen is assumed to have mean grain size of more than 2 mm while the fine less than 1 mm. Except for the frequency of 20 kHz there is a great discrepancy, with coarse sand mortar being much more attenuative than the fine one. This is expected since coarse sand grain is larger in size and closer to the ultrasonic wavelengths used.

4 DISPERSION

A suspension of rigid particles in liquid exhibits certain dispersive behavior (Harker 1991). At high frequencies the velocity asymptotically approaches a maximum defined mainly by the velocity of the hosting medium. The volume content and size of the inclusions though dominate the increasing trend from the low frequency value of velocity to the maximum, without however exhibiting the same strong influence as in attenuation.

In the present case velocity was measured from the time difference between the first threshold crossing of the input electrical signal and that of the received waveform. This threshold was adequately selected higher than the standard system noise but low enough to be sensitive to the first signal oscillations. Values of longitudinal wave velocity start lower than 500 m/s for the frequency of 20 kHz climbing over 1000 m/s at about the frequency of 300 kHz staying approximately constant afterwards, see Figure 6. These low values of velocity are typical for fresh concrete examined with low frequencies (Grosse 2001) and have been attributed mainly to the air void content of the fresh material (Boutin 1995).

In Figure 6 the effect of different sand content can be observed for specimens of w/c = 0.55. It seems that the asymptotic maximum value is higher for paste and reduced with an increase of inclusion content, fact usually met in suspensions (Harker 1991).

The variation of aggregate size, although not a key factor for velocity, results in somewhat increased maximum values for the coarse mortar of paragraph 3.3 relatively to the fine sand specimen (approximately 80 m/s).

The water content of mortar does not seem to have a strong influence on the dispersion curves as seen in Figure 7, although the w/c = 0.50 (corresponding to

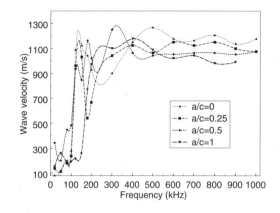

Figure 6. Effect of a/c on dispersion curves of mortar with w/c = 0.55.

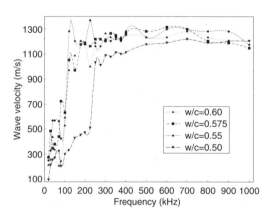

Figure 7. Effect of w/c on dispersion curves of mortar with a/c = 1.

the aggregate richer mix) exhibits constantly lower values than the other mixes.

5 BROADBAND EXAMINATION

The response of the material at different frequencies was also examined using sine sweep pulses of 1 ms duration, starting at 20 kHz and ending at 1 MHz.

Therefore using a single pulse, excitation of this wide band is accomplished yielding information about the propagation of different frequencies through mortar. Figures 2, 3a suggested that high frequencies are attenuated increasingly with aggregate content while the propagation of lower ones is facilitated. This can also be observed in Figure 8 where the FFT of sine sweep signals through mortar with different a/c ratios is depicted. One can mention the lower magnitude of high frequency components for increasing a/c while

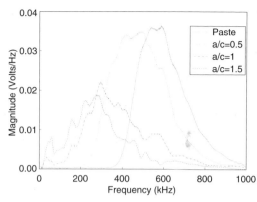

Figure 8. Effect of a/c on FFT of broadband signals.

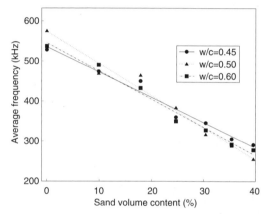

Figure 9. Effect of sand content on average frequency of broadband signals.

the low frequency (<100 kHz) magnitude rises. Nevertheless, more striking is the frequency shift as the peak frequency for paste is approximately 600 kHz reducing to 280 kHz for mortar with a/c = 1.5 (s = 39.6%).

This clearly shows the effect of sand content. Indeed the average frequency defined as:

$$A_f = \frac{\int f \cdot V(f)\,df}{\int V(f)\,df} \qquad (2)$$

where f = frequency; and V(f) = Fourier transform of the signal after it has propagated through the specimen, reduces linearly with aggregate content as can be seen in Figure 9. It seems that this parameter, among others of frequency and time domain, is quite insensitive to the water content variations being defined exclusively by the amount of sand used. It has also been proven sensitive to variations in aggregate content and size of hardened concrete (Otsuki 2000).

It is observed that for each 3 specimens sharing the same s%, the average frequency is approximately the same regardless the w/c while all three lines follow the same trend as to the s% and no relation with the w/c has been observed.

6 CONCLUSIONS

The ultrasonic examination of fresh mortar specimens at different frequencies, has revealed in the present work the strongly dispersive and attenuative nature of this material. Alterations in mix proportions allowed for estimation of their effect on the propagating wave.

It is verified herein, that cementitious slurries exhibit an increase of high frequency attenuation with the increase of both the sand content and the grain size. The water content although considered the most important factor concerning the quality of concrete seems to have only secondary influence on both the attenuation and velocity mainly through the control of aggregate content. From the above it is clear that a single wave parameter is hardly enough to characterize fresh mortar as to constituent proportion determination and especially w/c. Therefore combined information from different wave parameters is required along with higher reproducibility and accuracy of experimental measurements for reliable characterization of the material.

REFERENCES

Boutin, C. & Arnaud, L. 1995. Mechanical characterization of heterogeneous materials during setting. *European Journal of Mechanics, A/Solids* 14(4): 633–656.

Dukhin, A. S. & Goetz, P. J. 2001. New developments in acoustic and electroacoustic spectroscopy for characterizing concentrated dispersions. *Colloids and Surfaces A: Physicochemical and Engineering Aspects* 192: 267–306.

Farrow, C. A., Anson, L. W. & Chivers R. C. 1995. Multiple scattering of ultrasound in suspensions. *Acustica* 81: 402–411.

Garnier, V., Corneloup, G., Sprauel, J. M. & Perfumo, J. C. 1995. Setting time study of roller compacted concrete by spectral analysis of transmitted ultrasonic signals. *NDT&E International* 28(1): 15–22.

Grosse, C. U. & Reinhardt, H. W. 2001. Fresh concrete monitored by ultrasound methods. *Otto Graf Journal* 12: 157–168.

Harker, A. H., Schofield, P., Stimpson, B. P., Taylor R. G. & Temple J. A. G. 1991. Ultrasonic propagation in slurries. *Ultrasonics* 29: 427–438.

Hipp, A. K., Storti, G., Morbidelli, M. 1999. On multiple-particle effects in the acoustic characterization of colloidal dispersions. *J. Phys. D: Appl. Phys.* 32: 568–576.

Kaplan, M. F. 1959. The effects of age and water/cement ratio upon the relation between ultrasonic pulse velocity and compressive strength, *Magazine of Concrete Research* 11(32): 85–92.

McClements, D. J. 2000. Ultrasonic measurements in particle size analysis. In R. A. Meyers (ed.), *Encyclopedia of Analytical Chemistry*. Chichester: John Wiley & Sons Ltd.

Otsuki, N., Iwanami, M., Miyazato, S., Hara, N. 2000. Influence of aggregates on ultrasonic elastic wave propagation in concrete. In T. Uomoto (ed.), *Non-Destructive Testing in Civil Engineering*, Amsterdam: Elsevier.

Philippidis, T. P. & Aggelis, D. G. 2002. An acousto-ultrasonic approach for the determination of water to cement ratio in concrete. *Cement and Concrete Research*: in press.

Rapoport, J. R., Popovics, J. S., Kolluru, S. V. & Shah S. P. 2000. Using ultrasound to monitor stiffening process of concrete with admixtures, *ACI Materials Journal* 97(6): 675–683.

Urick, R. J. 1948. The absorption of sound in suspensions of irregular particles, *Journal of the Acoustical Society of America*, 20(3): 283–289.

Emerging Technologies in Non Destructive Testing, Van Hemelrijck, Anastasopoulos & Melanitis (eds)
© 2004 Swets & Zeitlinger, Lisse, ISBN 90 5809 645 9

Low temperature transitions used as a non-destructive indication of durabilty in fibre-reinforced polymers

R.D. Adams & M.M. Singh

Department of Mechanical Engineering, University of Bristol

ABSTRACT: Absorbed moisture affects the low temperature transitions of polymer composite systems. By measuring the longitudinal shear modulus and loss factor over a range of temperature, from $-150°C$ to $+20°C$, both before and after conditioning in a hot, wet environment it was noted that the presence of absorbed water reduced the temperature at which the transition took place, and increased the peak loss factor. Even allowing for differences in measurement frequencies, the transition temperatures of the composites were found to be greater than those of the corresponding matrix resins.

1 INTRODUCTION

The damping of polymeric materials is determined by their molecular structure. As the temperature of a polymer is raised, it passes from a glassy state, in which it behaves essentially as an elastic solid, and is relatively stiff and non-dissipative, to a rubbery state, in which it behaves more like a highly viscous fluid, with a very low stiffness, and also a low loss factor. The transition from the glassy to the rubbery state is accompanied by a rapid fall in modulus, and a peak in the damping, or loss factor, of the material. Other peaks in the loss factor – temperature curve occur at lower temperatures and are associated with relaxations, or re-organisations of the molecular structure[Ward 1983, Roberts & White 1973]. The main glass transition temperature is usually referred to as the α-transition, and is associated with large-scale motions of the polymer chains [Hall 1989]. The transitions occurring at lower temperatures are thought to be due to small-scale motions of side chains.

The glass transition temperature is a material parameter of prime importance for polymer composites. From an engineering point of view, operating at temperatures in the main (α) glass transition region is generally to be avoided. Because the focus is usually on the α-transition, the low temperature transitions are often neglected. The β-transition may, however, be of practical interest, since the loss factor peaks without a dramatic loss of stiffness. The enhanced damping properties of the β-peak could be put to use, particularly in, say, composite structures employed in low temperature aerospace applications. Moreover, mechanical properties of polymers, such as their ductility and the

fracture toughness, G_{Ic}, have been shown to be related to both the magnitude of the β-transition, and the temperature at which it occurs, at a given frequency [David & Etienne 1992, Heijboer 1968, Boyer 1976].

A knowledge of the variation of the dynamic properties with moisture is of importance, not only for design purposes, but also to gain a better understanding of how water affects the microstructure of polymer composite materials. In an earlier study[Adams & Singh, 1996] the changes in the dynamic shear properties of several polymer composite systems caused by the absorption of moisture were investigated. It was of interest to produce some moisture-induced damage, so steam was chosen as a suitably aggressive environment, although composites in service are unlikely ever to be exposed to such harsh conditions. Steam represents the worst hot, wet conditions at atmospheric pressure, and therefore any composite materials that were not affected by steam should also maintain their properties under less severe conditions. In that study, all measurements were made under ambient conditions. In order to put the results into a wider context, it was considered that measurements should be made over a range of temperatures on samples exposed to the same aggressive environment. Measuring the dynamic properties near the α-transition temperature, T_g, is not easy. At this temperature, the loss factor becomes very high so that accurate determination of both the resonant frequency and the damping become technically difficult. At temperatures above T_g, the modulus is so low that the materials may well deform under their own weight, and it is difficult to acquire reliable values of modulus. In addition, any examination of the effect of water will require some means of preventing drying of the specimens. Low temperature dynamic tests also offer the possibility of investigating the state of cure of part-cured polymers, without altering the cure state in the very act of testing the material, as happens when such studies are carried out near the α-transition. A study was carried out, therefore, to investigate the effect of the presence of absorbed moisture on the β-transition of several polymer composite systems, and the results are presented in this paper.

2 MATERIALS

Three different carbon fibre reinforced epoxy resins manufactured by Ciba (now Hexcel Composites) and denoted by 913C, 914C and 924C, respectively, were studied together with a glass fibre reinforced epoxy, 913G, also produced by Ciba, and carbon fibre reinforced polyether ether ketone (PEEK), APC2, manufactured by ICI. The fibre-volume fractions and matrix mass fractions were determined by methods recommended by CRAG[Curtis 1988]: the matrix resins of the carbon fibre epoxy composites were digested in concentrated sulphuric acid, and that of the glass fibre composite, 913G, was burned off at 590°C. The fibre volume fraction of the carbon-fibre reinforced PEEK was found by the density method, assuming there was no voids present. All the carbon fibres were low modulus PAN-based fibres. According to the manufacturers, those used to reinforce the epoxy resins were coated with an epoxy-based size, while the E-glass fibres were treated with a silane coupling agent. The surface treatment of the AS4 fibres in the APC2 was not described in the manufacturer's data.

The laminates were all unidirectional, reinforced with continuous fibres, and were nominally 2 mm thick. The thickness of the unreinforced epoxy samples was 1.2 mm, and that of the unreinforced PEEK 3.21 mm. Specimens were cut approximately 150 mm by 8 mm, with the fibres parallel to the long axis. The edges were not sealed in any way.

3 EXPERIMENTAL PROCEDURE

3.1 Conditioning

The materials were first dried in a vacuum oven at 50°C for a minimum of 7 h, to constant weight. They were then exposed to steam, removed at intervals for weighing, and maintained in steam until they became saturated with moisture. The conditioning times were, therefore, determined by the time taken for the moisture content to become almost constant.

An exception was made for the fibre reinforced 913 composites, however, because it had been found

that these did not reach a true equilibrium level [Adams & Singh 1996]. They were, therefore, removed for testing after approximately 140 h, during which period, the moisture uptake had been shown to be predominantly by diffusion. When these intermediate tests were completed, the samples were returned to the steam chamber, until the total exposure time was around 650 h. The samples were then re-tested, for β-peak determination. Similarly, the 913 resin was tested at the time the moisture content first began to level out, and then again, later after approximately 650 h.

Some samples of unreinforced 913 and 914 resins were "recovered". These were exposed to steam for 100 and 400 h, respectively, tested, and then dried at 100°C until there was no further change in their mass. They were then tested again.

The moisture content, M, of the samples was defined by

$$M = \frac{m_t - m_0}{m_0} \times 100\% \tag{1}$$

where m_0, m_t are the initial mass and that after an exposure time t, respectively. The moisture content was plotted against the square root of the conditioning time in steam, to determine the moisture absorption characteristics of the materials.

3.2 Dynamic torsion tests at cryogenic temperatures

A small torsion pendulum [Adams & Singh 1990, Adams & Maheri 1994] was used to measure the dynamic longitudinal shear modulus, G, and loss factor, η.

The shear modulus, G, was determined from the resonant frequency, f_{nt}, of the test-piece in torsion by:

$$G = \frac{4\pi^2 f_{nt}^2 I_m L}{k_2 bh^3} \tag{2}$$

where I_m is the moment of inertia of the inertia bar about the longitudinal axis of the specimen, L is the length, b the width and h the thickness of the specimen. The shape factor, k_2, is dependent on the aspect ratio b/h and is given by:

$$k_2 = \frac{1}{3}\left(1 - \frac{192h}{\pi^5 b}\sum_{n=1,3,5.....}^{\infty}\frac{1}{n^5}\tanh\left(\frac{n\pi b}{2h}\right)\right) \tag{3}$$

The loss factor, η, was determined by the half-power bandwidth method:

$$\eta = \frac{f_2 - f_1}{f_{nt}} \tag{4}$$

where f_2, f_1 are the frequencies at which the amplitude of the vibration falls to $1/\sqrt{2}$ of the maximum amplitude, which occurs at the resonant frequency, f_{nt}.

When the temperature had reached the desired minimum, the resonant frequency was determined, and the test program was then started. For each test, the frequency-amplitude data were logged, and the temperatures just before and just after each test were also recorded, together with the resonant frequency and the loss factor. The modulus was later calculated from the above equations.

4 RESULTS AND DISCUSSION

4.1 Moisture absorption characteristics

Two stages were observed in the uptake of moisture by the 913 composites. In the first stage, the absorption was primarily by Fickian diffusion and an apparent saturation level was reached. After about 250 h exposure, however, the moisture content began to increase further, and continued to increase throughout the duration of the conditioning. This second stage was found to be associated with delamination, collection of water in the cracks thus created, and the formation of surface blisters. The blisters were clearly visible, and the cracks were observed on polished sections of 913G samples exposed to steam for about 700 h.

The absorption of water by polymers depends both on the availability of holes (or free volume) within the polymer structure into which water molecules can diffuse, and on the presence of polar sites at which water molecules can form hydrogen bonds.

It can be seen that under the severe conditions provided by the steam, both of the epoxy resins exhibited non-Fickian behaviour. The 913 resin showed a loss of mass after prolonged exposure, which suggests that some component of the resin was leaching out. No extractions were performed, however, so the exact nature of the leachate was not determined.

The 914 resin showed an apparent two-stage absorption process, although its carbon-fibre composite reached a single plateau (Fig. 1). Scanning electron microscopy of samples of 914 resin produced no evidence of the existence of a second distinct toughening material phase, which might have accounted for the appearance of a second plateau.

4.2 Dynamic shear properties at cryogenic temperatures

4.2.1 Unreinforced resins

The shear loss factor of the unreinforced 913, 914 and PEEK resins are shown in Figures 2 and 3. The two epoxy resins, 913 and 914, behaved in a similar manner, but the depression of the transition temperature of the

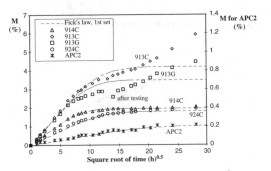

Figure 1. Moisture absorption by fibre reinforced composites (0° orientation) exposed to steam.

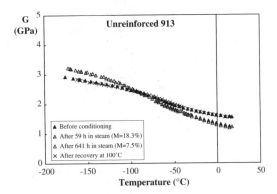

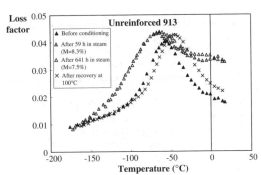

Figure 2. Dynamic shear properties of unreinforced 913 resin as a function of temperature.

913 resin was less than that of the 914 at comparable moisture contents.

A low temperature transition was detected in the thermoplastic PEEK (Fig. 3) at −69°C, a lower temperature than in the epoxies. This compares with the value of −89°C at 1 Hz, reported by David and Etienne [1992] and corresponds to an Arrhenius activation energy of 50 kJ/mol, which is in close agreement with the values they determined.

249

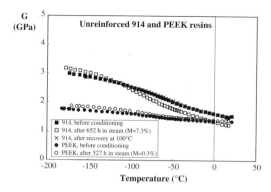

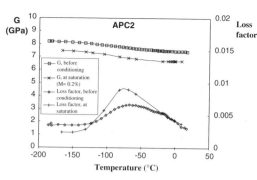

Figure 4. The longitudinal shear modulus and loss factor of APC2 before and after conditioning in steam.

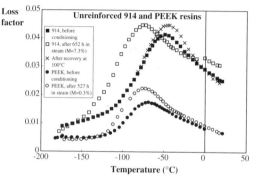

Figure 3. Dynamic shear properties of unreinforced 914 and PEEK resins as a function of temperature.

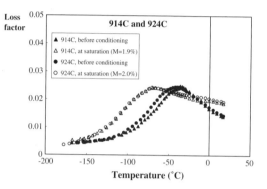

Figure 5. The longitudinal shear loss factors of 914C and 924C as a function of temperature, before and after conditioning in steam.

Furthermore, they found that, for a semi-crystalline sample of PEEK, the β-peak had a broad, asymmetric shape, as is also evident in Figure 6. Whereas they found that the presence of water had an insignificant effect on the β-peak, our results showed that water plasticises this resin, too.

There was an increase in the height and width of the loss factor peak, and a depression of T_β, although the moisture content was only 0.3% of the dry mass. Sasuga and Hagiwara [1985] report a similar effect on this transition peak in amorphous PEEK in the presence of moisture.

4.2.2 Carbon fibre reinforced PEEK

Figure 4 shows the variation of dynamic shear modulus and loss factor with temperature of carbon fibre reinforced PEEK (APC2), before and after conditioning in steam. The results for the dry material compare well with the peak specific damping capacity of 4.5% (loss factor, 0.0072) at a temperature of −63°C obtained by Adams and Gaitonde [1993].

Their values were derived from flexural tests on ±45° laminates. They estimated that the maximum shear damping loss factor of unreinforced PEEK would be between 0.011 and 0.024 at −63°C.

The results of this work show the magnitude of the loss factor to be of this order (0.0175), but that the temperature was six degrees lower, at −69°C. The difference is probably due to differences in the test frequencies.

4.2.3 Fibre reinforced epoxies

The variation of loss factor with temperature of the carbon fibre reinforced composites, 914C and 924C is shown in Figure 5, and the corresponding curves for the 913 composites is given in Figure 6.

The results show that all the dry epoxy matrix composites were very similar in their behaviour, with T_β at around −40°C, and η_p close to 0.025. The 913 composites were, however, markedly more affected by exposure to steam than either 914C or 924C. Whereas the equilibrium moisture contents of the latter two materials were both around 2%, the 913 composites absorbed more than twice as much after the same period in steam. Correspondingly, the changes in the dynamic properties were more significant. Tests on another material with the same apparatus indicated that

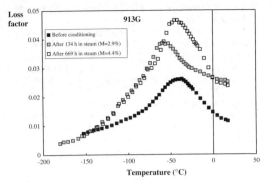

Figure 6. The longitudinal shear loss factor of 913G as a function of temperature, before and after conditioning in steam.

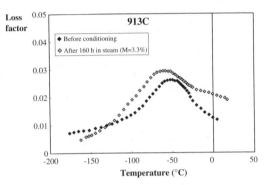

Figure 7. The longitudinal shear loss factor of 913C as a function of temperature, before and after conditioning in steam.

modulus measurements were repeatable to within less than 3%, but damping measurements (over the same range of loss factor) varied by an average of 6%, due predominantly to materials variation.

It is possible to compare the responses of 913C and 914C, since their unreinforced matrix materials were exposed to the same conditions, and both were reinforced with carbon fibre. Assuming that all the water absorbed by the composites was taken up by the matrix alone then, when the first wet tests were performed, i.e. at 3.25% and 2% moisture in 913C and 914C, respectively, the water content of their resins, M_r, were 8.5% and 6.9% respectively, approximately equal to the first "equilibrium" levels in the unreinforced resins. The 913 resin with the lower T_g would have a relatively lower free volume fraction. However, the uptake of moisture is higher than in the 914, which implies that processes other than simple diffusion are occurring, such as osmosis and hydrolysis. The fact that the residual mass of the 913 resin, on drying at 100°C was higher than that of the 914 resin, suggests that more water may be bound to the polymer chains. The presence of bound water molecules will hinder reorganisation of the

network, requiring not only a higher temperature for the transition to take place, but also more energy dissipation. Thus the depression of T_β in 913 would be smaller than in 914, both for the unreinforced and carbon fibre reinforced materials, as was observed.

5 CONCLUSIONS

The experimental technique has been shown to provide a means of observing the nature of transitions occurring in fibre-reinforced composites as well as in unreinforced resins, when these have absorbed moisture. It could be extended to measurements at temperatures above ambient temperature, and in humid conditions. The method has the advantage that relatively large specimens may be tested, so that the effects of flaws and of end-conditions are reduced.

Some aspects of the microstructure of the materials have been deduced from observations of the β-transitions. The similarities of the three epoxy resins, and their difference from the thermoplastic PEEK resin have been demonstrated. The thermoplastic polymer was shown to undergo a similar low temperature transition, but it was less pronounced than in the epoxies. The relatively small amplitude of the β-peak, together with the fact that it occurs some 90K below ambient temperatures, means that changes in the dynamic properties of the material due to the presence of water are not significant at room temperature.

The increase in loss factor and reduction of shear modulus observed in the first series of experiments[Adams & Singh 1996] in which measurements were made only at ambient conditions can now be seen in the context of the changes in the nature of the β-transition caused by the presence of water. The increases in height and width of the transition peak due to the presence of water affect the room temperature properties. Thus at a given temperature the proximity to the transition temperature determines the observed changes in the dynamic properties of polymer composite materials due to water absorption.

REFERENCES

Adams, R.D. & Singh, M.M. 1990. The effect of exposure to hot, wet conditions on the dynamic properties of fibre-reinforced plastics. Proceedings of Durability '90 Conference. Brussels.

Adams, R.D. & Gaitonde, J.M. 1993. Low temperature flexural dynamic measurements on PEEK, HTA and some of their carbon fibre composites. Comp Sci Tech. 47. 271–287.

Adams, R.D. & Maheri, M.R. 1994. Dynamic flexural properties of anisotropic fibrous composite beams. Comp Sci Tech. 50. 497–514.

Adams, R.D. & Singh, M.M. 1996. The dynamic properties of fibre-reinforced polymers exposed to hot, wet conditions. Comp Sci Tech. 56. 977–997.

Boyer, R.F. 1976. Mechanical motions in amorphous and semi-crystalline polymers. *Polymer*. 17. 996–1008.

Carter, H.G. & Kibler, K.G. 1978. Langmuir-type model for anomalous moisture diffusion in composite resins. *J Comp Mat* 12. 118–131.

Curtis, P. (ed.). 1988. CRAG test methods for the measurement of the engineering properties of fibre-reinforced plastics. Royal Aerospace Establishment Technical Report No. TR88012.

David, L. & Etienne, S. 1992. Molecular mobility in para-substituted polyaryls. 1. Sub-T_g relaxation phenomena in poly(aryl ether ether ketone). *Macromolecules* 25. 4302–4308.

Hall, C. 1989. *Polymer Materials* 2nd Edition. UK: Macmillan Education Ltd.

Heijboer, J. 1968. Dynamic mechanical properties and impact strength. *J Poly Sci*: Part C. 16. 3755–3763.

Roberts, G.E. & White, E.F.T. 1973. Relaxation processes in amorphous polymers. Haward, R.N. (ed.). *The Physics of Glassy Polymers*. London: Applied Science Publishers. 153–222.

Sasuga, T. & Hagiwara, M. 1985. Molecular motions of non-crystalline poly(aryl ether-ether-ketone) PEEK and influence of electron beam irradiation. *Polymer*. 26. 501–505.

Ward, I.M. 1983. *Mechanical Properties of Solid Polymers* 2nd Edition. Chichester: John Wiley and Sons.

Emerging Technologies in Non Destructive Testing, Van Hemelrijck, Anastasopoulos & Melanitis (eds)
© 2004 Swets & Zeitlinger, Lisse, ISBN 90 5809 645 9

Determining the in-situ tensile properties of pipelines using the automated ball indentation (ABI) technique

A.C. Russell, B.L. Jones & L. Manning
GE Power Systems PII Limited, Cramlington, United Kingdom

ABSTRACT: Measuring the tensile properties of linepipe accurately where there is no available information (usually old pipelines), without taking the pipeline out of service, is valuable when assessing the criticality of features detected in pipelines. This allows the level of conservatism used in a Fitness-For-Purpose assessment (FFP) to be minimised, the alternative being expensive cut-outs or the assumption of very low properties. PII operate an instrument, which can measure the tensile properties of pipelines in-situ using the Automated Ball Indentation (ABI) technique, leaving only a small insignificant surface indentation. Validation work has therefore been completed to relate the results of ABI tests in the field to those from API 5L tensile specimens, used originally to qualify the line pipe. Instrument constants used to calculate tensile properties have been derived, and a methodology for field use determined, which minimise the difference between results of the two methods, improving accuracy.

1 INTRODUCTION

1.1 *Need for in-situ measurements*

The ability to determine the tensile properties of linepipe accurately where there is no information (usually old pipelines), without taking the pipeline out of service, is valuable when assessing the criticality of features detected by in line inspection vehicles (Intelligent Pigs). Accurate determination of these properties allows the level of conservatism used in a Fitness-For-Purpose (FFP) assessment to be minimised. This allows the life of the pipeline to be maximised and the level of repair to be minimised, whilst keeping the risk "As Low As Reasonably Practicable" (ALARP). As an example of the alternative, the US 49 CFR Part 192 regulations suggest that in the absence of tensile records for a pipeline, a yield strength of 24 ksi (165 MPa) should be assumed unless an expensive tensile test programme is completed. In many cases, this will lead to a very conservative assessment resulting in needless costly maintenance activity.

Specified Minimum Yield Strength (SMYS), or some function of it, is the criterion for flow stress in the majority of FFP codes. If we consider the levels of conservatism involved in the use of SMYS, it is important to understand how it is measured. In addition, we need to understand how it varies locally due to the metallurgical variations, which occur during the production of linepipe, affecting the local yield strength of the material.

In the western world, the vast majority of linepipe is qualified for strength in accordance with the requirements of American Petroleum Institute (API) 5L, Specification for Line Pipe. The design codes are based on these results. Many projects have imposed more stringent test values over and above those required by API 5L, but in few, and probably no cases is the manner or the frequency of testing altered. The API test results have therefore been used as the "standard" against which the ABI results have been compared, as design and FFP codes have been validated against them.

API 5L defines the yield strength as "the tensile stress required to produce a total elongation of 0.5% of the gauge length, as determined by an extensometer" (42nd edition, 2000, section 6.2.1). There are other definitions of yield strength, but this is a useful and conveniently measured parameter. It is important however to consider the frequency with which this is carried out and the methods which are employed.

In the pipeline industry, as in many other industries, the prime criterion for selecting a mechanical test method is not that it gives an accurate and reproducible measure of the property we need to know, but that it is easy to carry out. This is perhaps best exemplified by the Charpy test which tells us very little quantitatively of the material's ability to resist the propagation of longitudinal cracking, particularly in the high strength super-tough steels we produce today, but it is invariably required and measured as a criterion in that regard.

Similar considerations apply in respect of strength testing. The pipeline profession talks constantly of "the API strap test" and uses it as the means by which the yield strength and the ultimate tensile strength of linepipe materials are assessed. To comprehend the implications of the values measured, and the relationships which they may have to actual local yield or flow stress values in the region of a corrosion defect in a pipeline, it is necessary to understand how pipe is made, and particularly how it is qualified in terms of its mechanical strength.

Linepipe is produced in the west to a series of "grades", set down by API. Qualification of a "heat" of steel, subsequently processed and manufactured into linepipe, means that a sample ring has been cut from one joint of the linepipe so produced, and that a strap sample has been flattened, and when tensile tested, has given a result for the yield strength in excess of that required for the given grade. The grade gives little indication of the precise local mechanical properties of the pipeline steel in service, other than a vague reassurance that it is likely to exceed, for reasons made clear elsewhere, (Streisselberger et al (1991)) that indicated by the grade level to which it is qualified.

1.2 Description of the tool

PII operate an (Advanced Technology Corporation) ATC instrument, which can measure the tensile properties of materials, leaving only a small insignificant surface indentation. This Automated Ball Indentation (ABI) technique is therefore non-destructive.

A tungsten carbide ball tipped indenter is mounted in a chuck. The chuck is mounted on a shaft, which is driven into the specimen by a linear actuator via a load cell. The displacement is measured by a Linear Variable Differential Transformer (LVDT). The tungsten carbide ball (typically 0.762 mm) is pressed about 0.11 mm into the surface at a constant rate to establish the load/deflection relationship. The ball is partially unloaded and then reloaded several times during the cycle to enable the plastic deformation to be quantified. The information enables calculation of the yield strength of the material, as well as estimation of the engineering Ultimate Tensile Strength (UTS). It is important to stress that the tensile properties are derived directly from the test results and material specific constants: it does not simply correlate hardness values with UTS.

A similar machine is available from Frontics of Korea, though there are some differences in the method of calculating tensile properties.

1.3 Purpose of test programme

This work was completed by PII to relate the results of ABI tests completed on pipelines in the field to those from API 5L tensile specimens. ABI instrument constants have been derived which minimise the differences between the two measurement processes. This enables more accurate non-destructive determination of the local tensile properties of in-service linepipe, where these are unknown. A methodology has also been determined for use of the instrument in the field to further minimise variability and hence errors.

Line pipe from a mill is produced to meet minimum yield and UTS requirements as set out in API 5L. In order to achieve this, the properties are assumed to be distributed normally. Distribution data have been analysed from several mills and various grades of pipe in order to establish typical relationships. The ABI results have been analysed further, in order to determine from the local results the API grade to which the sample linepipe is most likely to have been qualified. This work is in progress and is not detailed in this paper.

2 YIELD STRENGTH MEASUREMENT BY ABI

The principles behind the ABI technique are discussed in detail elsewhere (Murty, 1998). In order to illustrate the factors influencing the accuracy of the results however, a brief description of the methodology is given in this section and section 3.

It is assumed that the material behaves similarly under tensile and compressive loading. As the ball is pressed into the test material, both elastic and plastic compressive deformation takes place with an increasing volume of strain-hardened material. The load/depth relationship at each partial cycle is used along with the Meyer's coefficient and ball diameter to obtain a constant (A) from a regression analysis.

2.1 Formula for calculating yield strength

2.1.1 Recommended general coefficients for steels
The yield strength is calculated by multiplying "A" by a further material class specific constant, β_m (sometimes referred to as the "yield strength slope"). The following equation is used to calculate the yield strength:

$$\sigma_y = \beta_m A \qquad (1)$$

where σ_y = yield strength; β_m = yield strength slope; and A = constant derived from the ABI test

It is claimed that a single value of β_m has been found to be applicable to all carbon steels in cold or hot rolled conditions, namely 0.2285. With no information other than that the steel is ferritic, the manufacturer suggests using this form of the equation and these values of β_m.

2.1.2 "Generic" form of the equation

In some references (Haggag, 1993) a further parameter b_m is added as shown below to give a "generic" form of the equation.

$$\sigma_y = \beta_m A + b_m \qquad (2)$$

where b_m is termed the "material yield constant" (the offset in the equation of a straight line)

In tests on transmission pipelines (Haggag et al., 1999), β_m was calculated as 0.3585 with a b_m of −239 MPa for the 0.508 mm diameter indenter, while the 0.762 mm indenter gave values of 0.4273 and −285 MPa respectively.

2.2 Coefficients for pipelines

Line pipe is produced using one of a number of processes, which may include hot rolling, cold bending, cold expansion and welding. The resulting pipe is therefore anisotropic, with tensile properties varying through wall, circumferentially and axially. The ABI instrument samples a small amount of the material at the surface, however the API 5L standard measures bulk properties. The through wall effects may therefore have a significant effect on the apparent differences between the two measurements. This test programme was therefore aimed at establishing the optimum coefficients for use with pipelines.

3 UTS MEASUREMENT BY ABI

True strain is calculated from the measured plastic depth of the indenter (the depth of indentation following elastic recovery). The equation for true stress contains a parameter δ relating to the constraint effect for plastic deformation. The value of this parameter changes as the plastic zone develops and is itself a function of both α_m (the "constraint value") and true stress. Solution is therefore by iteration. α_m is a material dependent parameter which varies for different structural steels, depending on the strain rate sensitivity and triaxial hardening. In the tests on pipe materials (Haggag, 1999) it was taken as 1.3 and 1.2 for the 0.508 mm and 0.762 mm diameter indenters respectively. It is also therefore a function of indenter diameter.

A regression analysis is used to determine the strain hardening coefficient n and the strength coefficient K, expressing true stress as a function of true strain. The true stress may be calculated from equation 3.

$$\sigma_{ts} = K(\varepsilon_p)^n \qquad (3)$$

where σ_{ts} = true stress; ε_p = plastic strain; n = strain hardening exponent; and K = strength coefficient

Under compressive loading the material does not experience necking, therefore the UTS is determined indirectly. It is stated (Murty et al. 1998) that n equals the true uniform strain at the UTS of the material under tensile loading. The true UTS can therefore be found using equation 3 substituting n for ε_p. The nominal engineering UTS is found by substituting ε_p in equation 3 with (n/e). (where e ≈ 2.718)

4 TEST PROGRAMME

4.1 Factors investigated

The accuracy of the instrument is dependent on a number of factors. In order to determine which are significant, the following were investigated:

- Instrument attachment method
- Indenter ball diameter
- Surface preparation
- Constants used to calculate yield strength
- Constants used to calculate UTS
- Material through wall properties
- Existing stress on the specimen.

The results are dependent on indentation rate; therefore, this was held constant for the duration of the test programme.

4.2 Instrument attachment method

The instrument is fitted with two magnetic mounts, having "V" mounts at their base. These are only suitable for certain pipe diameters and the smaller indenters (0.508 mm and below). Other magnetic mounts are available for other pipe diameters, however it is inconvenient to carry several of these for a field operation.

The magnetic mounts were found to be unreliable in practice and were therefore used only to position the instrument before restraining it using other methods. The mounts need:

- to provide a rigid base for the instrument
- to prevent the instrument lifting off the surface as the ball is pushed into the specimen.

A problem with the attachment method can be identified from both non-linearity in the load/displacement graph displayed during the test and from the R^2 values reported by the instrument after post-processing.

The following attachment methods were used during this test programme:

- G-clamps for small pipe specimens in the laboratory
- Ratchet lashings for mounting on a pipe (see Figure 1)

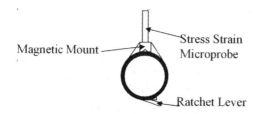

Figure 1. Ratchet strap attachment method.

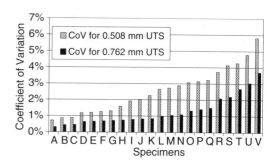

Figure 2. Comparison of coefficient of variation for UTS measurements using 0.508 mm and 0.762 mm indenter sizes.

The first set of tests was completed on small sections of line pipe, which were mounted to the bottom of the instrument using G-clamps. This was to allow measurement under "ideal" mounting conditions in order to eliminate the extent of this effect as far as possible from the tests results.

Subsequent tests were done on complete line pipe specimens using ratchet lashings to attach the instrument. API straps were then cut from these pipe samples and measured with the ABI instrument, attached using the G-clamps. The API straps were then pulled. No significant difference was observed in the ABI results from the two testing arrangements. Ratchet lashings were therefore confirmed as satisfactory for field use. These two methods were used as appropriate for the remainder of the test programme.

4.3 Indenter ball diameter

In order to measure a representative sample of the material, the manufacturer's test procedure calls for the ball to cover at least 5 grains. This is more likely to be achieved with a larger diameter ball. An initial series of tests was therefore completed using the 0.508 mm and the 0.762 mm indenter.

Figure 2 shows the coefficient of variation (CoV = (standard deviation/mean)) calculated for UTS for the first set of test specimens, illustrating the superior performance. The CoV represents the repeatability of the measurement process. Both sets of data are arranged in ascending CoV order, therefore the letters do not relate to the same specimen.

Both the accuracy and precision were found to be superior with the larger indenter; therefore, this was used exclusively for the remainder of the tests.

4.4 Surface preparation

Pits, lumps or debris on the surface alter the load/displacement measured during the test and therefore introduce errors. A surface finish of 63 RMS is therefore recommended by the manufacturer. In practical terms, this means that the surface must be polished using 600 grit paper for the final pass.

A series of tests was completed on unprepared pipe new from the mill, in order to determine the significance of errors. The pipe surface contained mill scale and was considered the best that would be encountered for an in-line tool for example, attempting to take measurements without prior surface preparation. It was found that the correlation coefficient calculated by the instrument for the load/displacement curve dropped from >95% for prepared specimens to 27–80%. Furthermore, the indenter picked up debris, which affected subsequent readings. Error analysis of the yield strength readings showed a significant drop in reported values at the 95% confidence level.

All of the subsequent testing was therefore completed on specimens prepared with 600 grit paper.

4.5 Constants used to calculate yield strength

These tests were completed with:

– the specimen surface polished with 600 grit paper
– the 0.762 mm indenter
– G-clamp or ratchet lashing attachment.

A total of 38 specimens were tested, including the following grades from API 5L.

– GrB
– X42
– X46
– X52
– X56
– X60
– X65
– X70

The yield was calculated first with the standard yield slope of 0.2285. A regression was calculated of the parameter A against the yield value found from the API 5L test and the new formula used to recalculate the yield strength. Both sets of values are shown in Figure 3. It can be seen that although there is little variation between the results at higher strengths, the introduction of a "yield offset" reduces the error for lower strength steels. The standard deviation on the

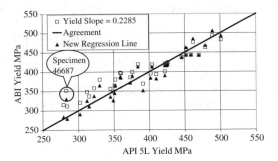

Figure 3. Summary of API 5L and ABI yield strength results.

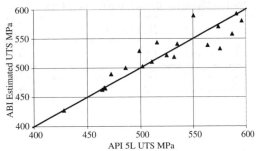

Figure 4. Summary of API 5L and ABI UTS results.

difference between API and ABI yield strength was 20 MPa. The tool is most likely to be needed for older pipelines where records are poor or unavailable, and these are most likely to be of the lower grades.

4.6 Constants used to calculate UTS

Calculation of the constraint factor is not as simple as recalculation of the yield equation parameters. Iteration was therefore used to derive the constraint factor used throughout the programme from initial testing (raw data is stored from the indentation test which may be subsequently reanalysed using different constants).

Following further testing, a regression was done between the API and ABI results to enable an adjustment. The results, which represent the relationship between ABI readings that would be obtained from the field and those from the API straps, are shown in Figure 4. They show a maximum over-prediction of 39 MPa.

UTS is used in some FFP assessment methods, for example DNV-RP-F101 (1999). This is more likely to be used for modern higher strength pipelines however, with known tensile properties.

The standard deviation on the difference between the API and ABI UTS results was again found to be 20 MPa. More tests are planned to reduce the confidence interval on this value.

4.7 Effect of strain history on ABI UTS results

Under conventional tensile testing, work hardening will raise the yield strength but the UTS will remain unaffected. It is important to remember that the ABI test process infers the UTS from compressive deformation of the test piece and it was suggested that strain history may affect the result. A series of tests was therefore completed on flattened and pulled API specimens to determine the significance of this effect under extreme conditions. The results showed a mean error of 38 MPa in over prediction (non-conservative). The maximum over-prediction observed in this sample

was +99 MPa. Again, it is stressed that considerable work hardening had been experienced by these specimens before the ABI test.

A final screening test was done on two API test pieces before and after flattening and the destructive test. The shift in estimated UTS was found to be +19 and +62 MPa, i.e. non-conservative. These tests suggest that the strain history due to pipeline production method, may contribute to the scatter observed in the differences between ABI and API measurements. Further work is necessary to investigate this.

4.8 Through-wall effects

As mentioned previously (section 2.2), the ABI technique measures the local surface properties of the material. It is suggested that some of the errors identified between the ABI and API measurements can be attributed to this effect. The manufacturing process for lower strength steel pipelines (i.e. API GrB and X42) would be expected to produce a slightly higher strength at the surface. For higher strength steels (i.e. X60 and above), manganese segregation to the centre of the pipe wall during cooling can produce a significantly higher strength in the middle. The bulk properties of the pipe as measured by the API test, reflect therefore some integration of this variation.

Specimen 46687 was found to exhibit the highest difference of 80 MPa (28%) with the 0.508 mm indenter; and 47 MPa (16.6%) with the 0.762 mm indenter. In both cases, the repeatability was good with coefficients of variation of 1.0 and 1.8% respectively for a series of tests. This specimen was therefore one of six selected for micro-hardness and micro-examination.

The results of the micro-hardness tests are presented in Table 1, showing an average reduction at the centre of 16%. The photomicrographs showed the correspondingly smaller grains at the surface. While the correlation between hardness and yield strength is known to exhibit a large amount of scatter, the results support the suggestion that part of the error is attributable to through wall strength variation. Furthermore,

Table 1. Through wall hardness for sample 46687.

Location number	HV10 results			
	1	2	3	4
Top surface	150	156	146	150
Pipe wall centre	127	125	123	132
Bottom surface	153	151	150	150

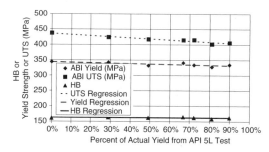

Figure 5. Effect of imposed stress on ABI results.

the difference was reduced with the larger indenter, which penetrated further into the specimen.

4.9 Pre-stressed specimen

It has been reported (Haggag, 1998, Jang et al, 2002) that residual stress affects the tensile results obtained using ABI, though the measured hardness should remain constant. At an imposed stress of 72% SMYS, an apparent reduction of around 14% had been observed by Haggag in limited testing, though the results appeared to be material dependent.

To investigate this effect, a sample pipe complete with end-caps, was pressurised. The yield strength of the pipe was measured in advance using the API method. The pipe was then pressurized up to 90% of this value, with multiple ABI measurements taken at each step. The results are shown in Figure 5.

Although an effect on the measured yield strength was observed, it equated to an apparent reduction of only 3% at 72% of the API measured yield value. The effect was more marked for UTS at 6% for the same pressure. As predicted, there was no correlation between imposed stress and the measured hardness value. It should be noted that the effect results in a conservative error, though more investigation is necessary to quantify the sensitivity to material and other parameters.

5 CONCLUSIONS

This study has shown, that non-destructive field ABI measurements of pipeline tensile properties, give reproducible results with good correlations to API mechanical test results.

The instrument parameters used to achieve this depend on material characteristics. This test program has shown that the use of universal correlation factors can lead to significant errors. A correlation has been determined which minimises the differences between ABI test results and API 5L tensile properties.

PII are engaged in further work to enhance this service and this will be reported in future papers.

ACKNOWLEDGEMENT

The authors would like to thank Mr Gareth Fletcher and Colleagues at Corus Testing Solutions, Teesside, for their help during the course of this test programme.

REFERENCES

American Petroleum Institute, API Specification 5L, 42nd Edition, 2000, Specification for Line Pipe

DNV, RP-F101, 1999, Recommended Practice: Corroded Pipelines

Haggag, F.M. 1993. In-Situ Measurements of Mechanical Properties Using Novel Automated Ball Indentation System, Standard Technical Publication 1204, 27–4, American Society for Testing and Materials

Haggag, F.M. 1998. Computer Controlled Microindenter System, DoD SBIR Phase II Final Report

Haggag, F.M. 1999. Non-destructive Determination of Yield Strength and Stress-Strain Curves of In-Service Transmission Pipelines Using Innovative Stress-Strain Microprobe Technology, ATC/DOT/990901

Jang, J. et al. 2002. Evaluation Of Welding Residual Stresses In Power Plant Facilities By Using A Newly Developed Indentation Technique, OMAE 2002, 23–28 June, Oslo, Norway

Murty, K.L. et al. 1998. Non-destructive Evaluation of Deformation and Fracture Properties of Materials Using Stress-Strain Microprobe, Non-destructive Characteristics of Materials in Aging Systems, Materials Research Society, Pennsylvania, USA

Streisselberger et al., 1991. AG der Dillinger Hüttenwerke United States 49 Code of Federal Regulations [CFR] Part 192, "Transportation of Natural Gas and Other Gas by Pipeline; Minimum Federal Safety Standards"

Health monitoring

Emerging Technologies in Non Destructive Testing, Van Hemelrijck, Anastasopoulos & Melanitis (eds)
© 2004 Swets & Zeitlinger, Lisse, ISBN 90 5809 645 9

Non-destructive testing for health monitoring of adaptive structures

K. Pfleiderer & G. Busse
Institute of Polymer Testing and Polymer Science (IKP), Department of Non-Destructive Testing (ZfP),
University of Stuttgart, Germany

ABSTRACT: Due to the increasing demand for adaptive materials and structures in various fields, new sensors and actuators are being developed that require advanced processing of materials. In this field both damage identification and integrity characterisation are important for a reliable operation of the system. Our contribution first compares the performance of various modern non destructive testing methods for the quality assurance of smart structures. Also the potentials of the methods are discussed regarding different types of defects. We tested adaptive structure models (consisting of glass fibre reinforced composites and piezoelectric actuators) using modal analysis, non-linear vibrometry, lockin-thermography, and air-coupled ultrasound. The results are correlated (data fusion) in order to combine the efficiency of the methods. In the second part we report about a variable coupling between actuator and matrix to simulate both curing and aging of the matrix material (epoxy resin) and also about such a variable coupling between components within the adaptive (or "smart") structure. Subsequently we monitored the transfer function and especially its changes caused by modifying the properties of the coupling in a controlled way. The results were correlated with those of destructive measurements. Some measurement techniques use the actuator as a probe for monitoring structural changes while others use it as a transmitter. Load induced defects (e.g. local disbonds of the materials) are verified non-destructively. As we used thermoplastic material, these defects can be removed by heating and subsequent solidification of the coupling medium. The application-relevant result is that the model structures can be reproducibly modified in order to simulate various conditions and also to monitor the sensitivity of NDE methods responding to these controlled modifications.

1 INTRODUCTION

Technical problems may occur in the manufacturing process of smart structures consisting of a piezo-electric actuator and a glass fiber reinforced epoxy matrix. In order to achieve both a high fiber content and a homogeneous resin distribution, the structure has to be cured under high pressure and temperature. During the process the excessive resin is pressed out of the transfer chamber and the embedded piezo actuator behaves like an obstacle in the flow direction. For embedding the actuator, the structure has to be provided with recesses where substantial stress concentrations can occur. Therefore the fragile piezo ceramic could break easily during the manufacturing process. Cracks do not reduce the efficiency of the actuator in general but the electrical contacts may break so that only a part of the ceramic is controllable. A potential solution is e.g. the insertion of an additional copper net that provides a complete electrical contacting also after such a crack.

If the recesses for the actuators are inaccurately worked, or the actuator slips, further defects can occur, e.g. increased resin concentrations, which affect the mechanical characteristics of the construction unit after hardening. Additionally, the autoclave process can result in an incomplete curing of the resin inside the laminate structure. Both kinds of damage, cracks, and insufficient cross-linking are in the focus of interest.

2 DESCRIPTION OF THE SPECIMEN

To investigate the influence of the embedding on the efficiency of an active structure, a GFRP structure was manufactured whose actuator is surrounded by a thermoplastic layer. The strength of this layer is adjustable by an additionally integrated heating wire. Thus curing by precipitation and aging processes can be simulated.

After the autoclave process one specimen had multiple cracks in the ceramic actuator, probably due to

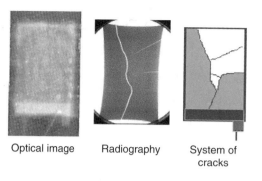

Optical image   Radiography   System of
                                   cracks

Figure 1.   Integrated actuator with cracks separating active
(shaded) and inactive (light) areas.

Figure 2.   Color coded images of the eigenmodes.
Left: Pure bending mode of a symmetrical smart structure.
Right: Symmetry perturbation indicated by torsion mode.

the high temperature and pressure in the autoclave. The
detection of this defect with different methods is the
main goal of the following investigations.

3   MODAL ANALYSIS

Due to the geometry of a testing sample certain eigenmodes are expected. The eigenmodes can be calculated if material properties, boundary conditions, and
geometry are all known.

If the modal structure of a sample is well-known, the
actuators can be positioned at an amplitude maximum
of the standing wave. If the aim is active damping of
oscillations, the information about the position of the
highest amplitude is essential for actuator positioning
(Doebling 1998).

If the actuators are imbedded symmetrically in the
smart structure, only flexural vibration modes are
expected. Figure 2 shows the color coded images of
the local vibration amplitude when the smart structures
is excited by the built-in actuator and inspected by a
scanning laser interferometer. At the left the actuator
is well and symmetrically embedded while the right
example shows superposed torsion modes indicating
a loss of symmetry due to defects.

The result is that the modal behavior of the structures measured between 25 and 2500 cycles per second
is suited to evaluate the performance of smart structures.
A change in the oscillation spectrum can be observed
when the model structure described above (thermoplastic actuator embedding) is heated, as shown by the
spectra taken at different temperatures (Figure 3).

As the spectral changes resulting from the elevated
temperature are difficult to see we expressed them by
the correlation coefficient between each spectrum and
the initial spectrum (Figure 4). The curve displays the
systematic decrease of the correlation coefficient and
its sensitive response to the change of the boundary
conditions.

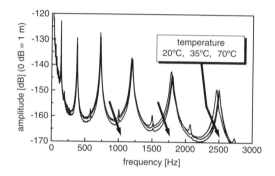

Figure 3.   Oscillation spectra taken at various temperatures.

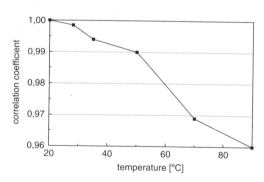

Figure 4.   Correlation coefficient of spectral behavior
affected by the thermoplast temperature.

4   NON-LINEAR VIBROMETY

Linear behavior (response behavior proportional to
the excitation amplitude) is characterized by a linear
transfer function. For elastic waves the linear transfer
function is determined by Hooke's law and the elastic

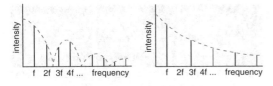

Figure 5. Effect of non-linear behavior on spectral response to mono-frequency excitation f. Left: higher harmonic of a hysteretic non-linearity. Right: higher harmonics of a "clapping" non-linearity.

constant. A crack can transfer only pressure and no tensions perpendicular to its surface "clapping" (Richardson 1979, Hamilton 1986, Solodov 1999). This "mechanical rectification" causes a deviation of the transfer function from the linear process. Also friction influences can generate non-linearity (Zheng 1999). As a consequence integral multiples (= higher harmonics = overtones) of the excitation frequency can be detected using the Fourier transformation. The principle is shown in Figure 5.

For the measurements of non-linearity we used a scanning Laser-Vibrometer (Krohn 2000). In the heated condition both mechanisms of non-linearity can occur and a superposition can result. At this temperature relative motions of the surfaces produced by the cracks are possible, which are prevented in the cold and rigid condition of the thermoplastic material. The actually measured spectral distributions of the higher harmonics (Figure 6) are based on the two processes friction and clapping. At elevated temperatures different frequency spectra are found on and outside the cracks (Figure 7) when the sample is excited at 20 kHz.

On the softened thermoplastic material (by heating) only the second and third harmonics are clearly detectable, which indicates a square and/or cubic non-linearity. Close to the cracks additional harmonics of higher orders can be measured. Due to the high acoustic absorption the elastic waves are extremely spatially confined.

The increasing influence of hysteretic nonlinearity correlated with the heating current around the thermoplastic embedding material of the cracked piezo actuator (see Figure 1) is indicated by the amplitude distributions of the second harmonic on the actuator (Figure 7). The excitation frequency was 20 kHz at only 16 V_{eff}.

At increasing temperature the second harmonic can be localized in a wider area on the actuator, which corresponds approximately to the position of the integrated heating wire laid out in several loops. Though the clapping of the cracks is known to occur at 40 kHz, the contribution of the hysteretic effect is obviously much stronger at the 2nd harmonic.

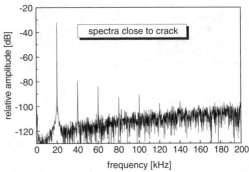

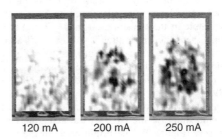

Figure 6. Amplitude of the spectra on (upper) and outside (lower) a crack in a smart structure.

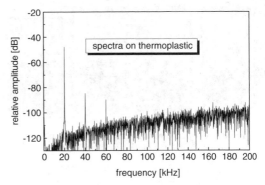

120 mA   200 mA   250 mA

Figure 7. Greyscale coded amplitude distribution of the 2nd harmonic (40 kHz) at various heating currents.

On the other hand the image taken at the fourth harmonic (Figure 8) seems to display some features of the crack pattern thereby indicating that the "clapping" non-linearity of the cracks becomes obvious.

Only cracks transverse to the actuator are visible. In this direction the actuator has its largest strain amplitude, so that here probably the surface movement is at its maximum.

For this smart structure, the variation of patterns observed at different harmonics can be considered as an indicator revealing different kinds of mechanisms and therefore of different defects in the same sample.

Figure 8. Greyscale coded amplitude distribution of the 4th harmonic (80 kHz) at increasing heating currents.

Figure 9. Ultrasound C-Scan in through-transmission mode. Left: pressed out thermoplastic material. Right: well embedded actuator.

5 AIR-COUPLED ULTRASOUND INSPECTION

As liquid couplants like water or gel are not suited for ultrasound inspection of smart structures, we used airborne ultrasound transducers (Hillger 1999) to inspect the samples described above. The signal was coupled into the structure in two different ways. First the ultrasound transducers were positioned on both sides of the sample in the transmission mode. In the second arrangement the transmitter was turned off and the internal actuator was used as a transmitter for ultrasonic oscillations. Thus active and not-active areas of the actuator could be distinguished (Pohl 1998, Mook 1999).

In the through-transmission technique defects of the embedding at the edge of the actuators can be localized (Figure 9). The bright areas in the C-scan indicate thermoplastic material which was pressed from of the actuator bed during the curing process (temperatures up to 120°C/248 degF and pressures up to 7 bar/7 atm).

To determine the efficiency of the structure, as mentioned above, now the actuator was used as a US-transmitter. At room temperature the cracks in the actuator cannot be detected (Figure 10: left image).

At increasing temperature, especially at about 50°C (122 degF), the distinction between active and inactive actuator areas is clearly possible (Stoessel 2000). At high temperatures the thermoplastic embedding of the actuator obviously is no longer able to distribute the elastic wave over the whole actuator bed. The absorption of the softened thermoplastic is dominating so thus the actuator can be evaluated.

| 20°C (68 degF) | 32°C (90 degF) | 55°C (131 degF) |

Figure 10. Ultrasound inspection of smart structure using an actuator as ultrasound transmitter and airborne probe. Influence of temperature on cracked actuator with thermoplastic material around it.

6 SUMMARY

The adjustable absorption of the manufactured smart structure is well suited for the investigation of embedding qualities and for the simulation of curing by precipitation and aging processes. Using modal analysis, non-linear vibrometry, and air-coupled ultrasound inspection one can distinguish between different kinds of defect mechanisms.

REFERENCES

Doebling, S. W., Farrar, C. R., Prime, M. B. 1998. A Summary Review of Vibration-based Damage Identification Methods. *The Shock and Vibration Digest 30*, p. 91–105.
Hamilton, M. F. 1986. "Fundamentals and applications of non-linear acoustics", In: *Non-linear wave propagation in mechanics – AMD-77, The American Society of Mechanical Engineers*, New York ().
Hillger, W., Gebhardt, W., Dietz, M., May, B. 1999. Bildgebende Ultraschallprüfung an CFK-Probekörpern mit Ankopplung über Luft., *DGZfP-Jahrestagung, Berichtsband 63.1*, p. 243–250.
Krohn, N., Stößel, R., Busse, G. 2000. Nonlinear Vibrometry for Quality Assurance. *27th Review of Progress in Quantitative Nondestructive Evaluation (QNDE)*, July 16–21, 2000, Ames, Iowa.
Mook, G. 1999. Zerstörungsfreie Strukturcharakterisierung adaptiver Werkstoffe. *SFB 409 – ADAMES Workshop, Book of Abstracts (1999)*, p. 24.
Pohl, J. Ultrasonic Inspection of Adaptive CFRP-Structures, *Proc. 7th ECNDT (1998)*, p. 111–118.
Richardson, J. M. 1979. Harmonic generation at an unbonded interface – I. Planar interface between semi-infinite elastic media, *Int. J. Engng. Sci., Vol. 17, Pergamon Press*, pp. 73–85.
Solodov, I. Y., Maev, R. 1999. "Overview of opportunities for non-linear acoustic applications in material characterization and NDE", *Proceedings "Emerging Technologies in NDT" 2nd International Conference*, Athens 24–26 Mai, Balkema, Rotterdam, pp. 137–144.
Stößel, R., Krohn, N., Pfleiderer, K., Busse, G. 2000. Air Coupled Ultrasound as a Tool for New NDE Applications. *QNDE 2000, 27th Annual: Review of Progress in NDE*, Aimes (Iowa, USA), 16–21. July 2000.
Zheng, Y., Maev, R., Solodov, I. Y. 1999. "Non-linear acoustic applications for material characterization", *A review in Canadian Journal of Physics, Vol. 77 No. 12*, pp. 927–967.

Emerging Technologies in Non Destructive Testing, Van Hemelrijck, Anastasopoulos & Melanitis (eds)
© 2004 Swets & Zeitlinger, Lisse, ISBN 90 5809 645 9

Validation of damage detection techniques for automatic structural health monitoring during fatigue tests

S. Vanlanduit & P. Guillaume
Dept. Mechanical Engineering, Vrije Universiteit Brussel, Brussels, Belgium

G. Van der Linden
ASCO Industries, Brussels, Belgium

ABSTRACT: In this paper the use of several automatic damage detection techniques will be validated with respect to their use for on-line structural health monitoring during fatigue tests. The overview of techniques considered in the paper contains both low-frequency (1–100 Hz), middle-frequency (10 Hz–1 kHz) and high-frequency (1–10 MHz) techniques. Moreover, both linear and non-linear damage indicators are investigated. From the experimental results in the paper it can be concluded that a combined use of the static stiffness and the transmitted ultrasonic energy leads to the best results. Indeed, in that case both small, medium and large size cracks can be monitored with a simple experimental set up. In addition, valuable information on the size of the crack is available.

1 INTRODUCTION

The research presented in this paper is contained in the framework of the SLAT TRACK project, sponsored by the Flemish Institute for the Improvement of Scientific Research in Industry (IWT) [1]. In this project the damage detection, life prediction and redesign of a slat track – which extends the surface of an airplane wing during takeoff and landing – is considered (see the picture of the slat track in Figure 1). An important task in this project is the performance of experimental fatigue tests and the structural health monitoring during these fatigue tests. Since the slat track has a very high fatigue strength, testing times can typically take several weeks. This testing time still increases when one periodically interrupts the fatigue test to manually examine the structural health. Therefore, in this paper several automatic damage detection techniques were validated with respect their in-line use during fatigue tests.

When studying the crack propagation of the slat track, the rectangular area in the lower part of the cross section (see circle in Figure 1) is the most interesting one (when the propagation extends further than this area, unstable crack growth occurs). For this reason, and also to reduce the testing time, the preliminary validation presented in this paper is performed on a steel beam with comparable dimensions as the lower flange of the track (see Figure 1, left).

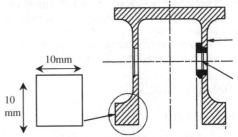

Figure 1. Picture of a slat track (top), cross section of the slat track (bottom-right) and dimensions of the cross section of the beam which is used in the validation of the damage detection techniques.

2 OVERVIEW OF DAMAGE DETECTION TECHNIQUES

The requirements for damage detection techniques with the purpose of detecting and quantifying cracks during fatigue tests differ from those for structural health monitoring during operation or nondestructive testing in the production process of a component.

The major difference is that in the former case the crack location is usually known beforehand (by virtue of numerical calculations, preliminary experiments or experience). Also, more complex instrumentations can be used in laboratory conditions. On the other hand, it is important to be able to accurately monitor the crack during its growth, rather than simply differentiating between intact and damaged states of a component. Last but not least, the damage detection should be automatic. Indeed, in a few weeks time several thousands of inspections are to be performed, which makes manual interaction impossible. During production and operation of the component either less inspections are performed (production) or the inspections are spread over a large time period (several years for the operation of the slat track).

In literature a large amount of references are available with respect to damage detection, both in the vibration engineering and the nondestructive testing community. The idea of using vibration characteristics (i.e. modal parameters) of the structure is attractive because it leads to global damage detection (an extensive overview of several hundreds of research contributions on the subject can be found in Doebling et al (1996)). An important disadvantage of using modal parameters, however, is the low sensitivity (see Vanlanduit et al (2003)). Techniques used in nondestructive testing are usually local but also more sensitive (Cartz, 1995). Unfortunately, some of the methods require too much user interaction (liquid penetration, magnetic particle inspection, visual inspection), while other NDT techniques need bulky instrumentation or large measurement times (radiography, ultrasonic scanning). In this paper, several simple indicators are computed from signals which are measured with the aid of a set of NDT and vibration sensors. Because the indicators are relative values (compared to the initial state of the structure), a comparison of the sensitivity with respect to the crack length is possible.

3 DAMAGE INDICATORS

In the current study various sensors were used:

1. Crack propagation strain gages (CPSG)
2. A load cell
3. An accelerometer
4. An eddy current transducer (EC)
5. Ultrasonic transducers (US)

The CPSG and the load cell measurements result in low frequency data (below 50 Hz), while the accelerometer captures signals upto 1 KHz and the EC and US transducers work in the MHz region.

The measured signals $x(t)$ from each of these sensors are processed using simple signal processing techniques to obtain quantitative and comparable crack indicators:

(a) the relative amplitude (RA) of the i-th frequency component after n fatigue cycles:

$$RA(x_n, i) = \frac{A(x_n, i)}{A(x_1, i)} \text{ with } A(x, i) = |X_i| \quad (1)$$

where X is the FFT of x.

(b) the relative RMS (RRMS) value of the signal after n fatigue cycles:

$$RRMS(x_n) = \frac{RMS(x_n)}{RMS(x_1)} \text{ with } RMS = \sqrt{\frac{1}{N}\sum_{i=1}^{N} x_i^2} \quad (2)$$

(c) the relative total harmonic distortion (RTHD) value of the signal after n fatigue cycles:

$$RTHD(x_n) = \frac{THD(x_n)}{THD(x_1)} \text{ and } THD(x) = \frac{\sqrt{\frac{1}{N/2-1}\sum_{i=2}^{N/2} X_i^2}}{\sqrt{X_1^2}} \quad (3)$$

When the fatigue load signal is not a sine (as is the case here) a more general expression is used:

$$RTHD(x_n) = \frac{THD(x_n)}{THD(x_1)}, \text{ and } THD(x) = \frac{\sqrt{\frac{1}{N/2-j}\sum_{i=j+1}^{N/2} X_i^2}}{\sqrt{\frac{1}{j}\sum_{i=1}^{j} X_i^2}} \quad (4)$$

where j is the number of frequency components in the load signal.

4 EXPERIMENTAL SET UP

4.1 Fatigue tests

The specimen under test is a steel beam with dimensions $400 \times 10 \times 10$ mm. A corner saw cut of 1 mm depth is applied to initiate crack growth more easily. The beam is loaded in cantilevered boundary conditions as shown in Figure 2 (the distance between the clamping and load equals 185 mm). A displacement-controlled experiment is performed with an IST test rig (i.e. a 5 Hz sine with 4 mm amplitude a 8 mm DC offset was applied). Initially (when the beam was still intact), this corresponded to a load between 100 N and 800 N.

4.2 *Damage detection*

To monitor the change in stiffness of the beam, the actuator load cell was used (a 25 kN full scale IST Dynacell, see Figure 2). This is possible because the experiment is displacement controlled (analogously, the built-in actuator's LVDT sensor should be used for force controlled experiments). A successful monitoring of the crack using the load cell would mean that no additional instrumentation would be required, which is an important advantage.

Initially, a loudspeaker was used to excite the beam in the medium frequency range and measure the response with a type 333A32 PCB accelerometer (up to

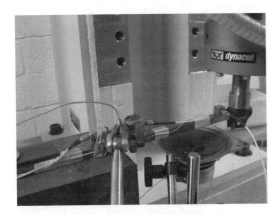

Figure 2.   Setup of the cantilevered beam fatigue test (with instrumentation used for damage detection).

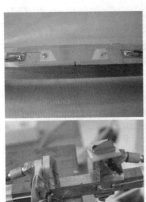

Figure 3.   Detail pictures of the instrumentation used for damage detection.

1 KHz). This would enable us to track the changes of the resonance frequencies in that region. Unfortunately, the radiated acoustic energy was not sufficient to lead to good quality measurements as will be shown in Section 5.

As a reference for the true crack length, Micro-Measurements type TK-09-CPA02 crack propagation strain gages (20 grid lines with 0.25 mm spacing) were used (see Figure 3, top-left). Although the method gave the most reliable monitoring results it is fraught with a reasonably high cost because of the destructive nature of the gages. In addition, the gages can only be used on smooth surfaces (e.g. cracks near welds can not be investigated).

Two high frequent sensors were used in the paper. First, an eddy current transducer (Bently Nevada Proximitor) was placed approximately 1 mm above the surface, near the region where the saw cut was applied (see Figure 3, top-right). Secondly, two Panametrics 10 MHz surface wave transducers (type VIDEOSCAN V537-RM) are used in transmission (see Figure 3, bottom), in conjunction with a pulser/receiver.

The signals from all sensors (except for the ultrasonic transducers) were acquired with a National Instruments data acquisition card. Both the control of the measurements and the post-processing was done in Matlab. The ultrasonic transducers were measured with a HP 100 MHz digital oscilloscope, which was also controlled in Matlab.

5   EXPERIMENTAL RESULTS

The discrete evolution of the crack length could easily be obtained from the voltage over the crack propagation strain gages after exciting it with a 18 Vdc power supply shunting with a 50 ohm resistance and putting a 100 Kohm resistance in series (see the dotted line in Figure 4). By using a linear interpolation, a continuous measurement of the crack length is possible (full line in Figure 4). It should be remarked that this is a measure of the crack length at the side of the beam. This does not mean that the full width of the beam is cracked (indeed, it should be reminded that the crack initiation slot is a corner slot).

The relative damage criteria (which were defined in Section 3) applied to the force signal are shown in Figure 5. Both the relative RMS and relative amplitude decrease significantly: a few percentages during the first four millimeters of the crack propagation process and several orders of magnitude for larger cracks. Together with decrease in amplitude, the nonlinear distortions (THD) decrease initially. In the later stage (cracks >4 mm) the force becomes considerably nonlinear (THD up to more than ten times the initial THD). As a conclusion the force signal indicators are interesting for medium size to large cracks. In addition,

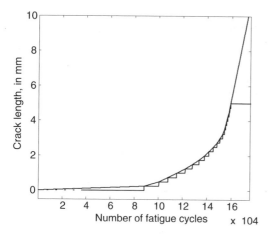

Figure 4. Crack length indicated by the crack propagation strain gage in function of the number of fatigue cycles. Dotted line: CPSG measurement, full line: interpolated.

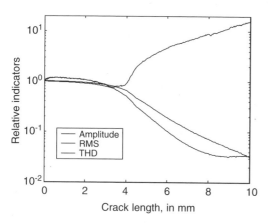

Figure 5. Relative damage indicators for the force signal.

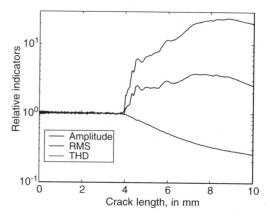

Figure 6. Relative damage indicators for the acceleration signals.

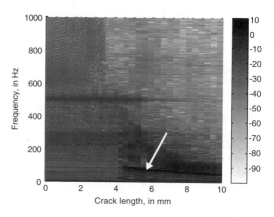

Figure 7. Color map of the frequency dependent amplitude (in dB) of the acceleration signal in function of the crack length. The white arrow indicates the resonance of the beam.

the force (or rather the stiffness which can be computed directly from the force) has a physical meaning (usually a minimally required stiffness is known from the design of the component).

The acceleration signals showed a different evolution (see Figure 6). During the first four millimeters of the crack propagation process, none of the indicators changed significantly. This is the case because the experiment was displacement controlled. For cracks lengths larger than four, the beam was gradually plastically deformed. This resulted in a decrease of amplitude. The RMS value and the THD on the other hand drastically increased. The harmonics of the excitation frequency (which resulted in the increase of RMS and THD) were created because the contact between the actuator and beam was lost due to the

plastic deformation (remark that beam was not clamed to the actuator). Because of this higher frequency excitation the beam could vibrate at its resonance frequency. This is clearly visible by the dark line near 100 Hz in Figure 7. In addition, it can be seen in Figure 8 that the for crack lengths larger than four, the resonance frequency decreases. The change is however not sensitive enough for use as a damage indicator.

The damage indicators for the eddy current sensors shown in Figure 8 give comparable results than those for the load signal (Figure 5). For smaller cracks, the eddy current indicators were even less sensitive.

One possible reason is the fact that the eddy current sensors were positioned in the middle of the beam. Because a corner saw cut was used for crack initiation it took a while until the crack propagated up to the middle of the beam.

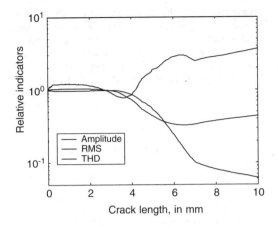

Figure 8.   Relative eddy current damage indicators.

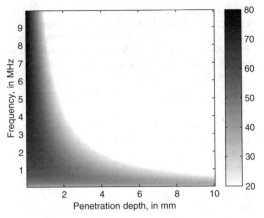

Figure 11.   Color map of the theoretical frequency dependent velocity amplitude (in dB) of the ultrasonic receiver signal in function of the penetration depth.

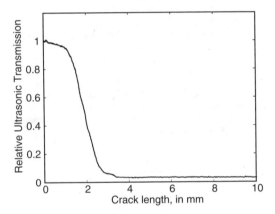

Figure 9.   Relative RMS ultrasonic transmission.

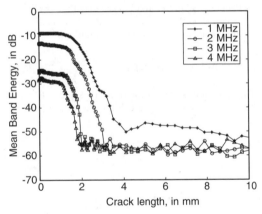

Figure 12.   Transmitted energy for four frequency bands: 1 MHz, 2 MHz, 3 MHz and 4 MHz in function of the crack length.

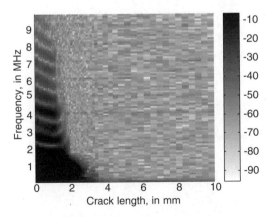

Figure 10.   Color map of the frequency dependent amplitude (in dB) of the ultrasonic receiver signal in function of the crack length.

Finally, the results of the RMS indicator for the ultrasonic transmission experiment are shown in Figure 9. Compared to the other sensors, the results are much more sensitive. Even crack lengths under 1 mm could be detected. However, a straightforward use of the RMS indicator does not allow conclusions further on in the crack propagation stage (almost no energy is transmitted when the crack exceeds 3 mm).

In addition to the high sensitivity of the ultrasonic transmission experiment, it is possible to draw conclusions on the length of the crack by looking at the evolution of the received amplitudes (see Figure 10). Indeed, when comparing the region in the color map in Figure 10 which contains energy (more than −40 dB), with the theoretical penetration depth of Rayleigh waves in steel in Figure 11 (Schmerr (1998), Cheek (2002)) a good correlation can be observed. An alternative representation is given in Figure 12. From this figure, it can be seen that low frequency bands give information on medium sized cracks, while the high frequency bands can be used for the sensitive detection of small cracks.

6 CONCLUSIONS

In this article several damage detection methods for use during fatigue tests were compared using simple signal processing techniques and a variety of sensors. An important conclusion which can be drawn from the validation is that none of the sensors based damage indicators was able to monitor the crack propagation during the full extent of the experiment. Indeed, because most sensors are somewhat complementary, it is advised to use a combination of indicators and sensors for structural health monitoring. An interesting combination is the use of both load amplitude (or equivalently quasi-static stiffness) and ultrasonic transmission. This can be realized with a minimum of instrumentation (because a force cell is built-in in most test rigs). Also, both small, medium and large cracks can be monitored.

ACKNOWLEDGEMENTS

This research is part of the Slat Track project, which has been sponsored by the Flemish Institute for the Improvement of the Scientific and Technological Research in Industry (IWT). The authors also acknowledge the Fund for Scientific Research – Flanders (FWO) Belgium the Flemish government (GOA-OptiMech) and the research council of the Vrije Universiteit Brussel (OZR) for their funding.

REFERENCES

Cartz, L. 1995. *Nondestructive testing*. Materials Park: ASM International.
Cheeke, J. D. N. 2002. *Fundamentals and applications of ultrasonic waves*. Boca Raton: CRC Press.
Doebling, S. W. & Farrar, C. R. & Prime, M. B. & Shevitz, D. W. 1996. Damage identication and health monitoring of structural and mechanical systems from changes in their vibration characteristics: a literature review. Technical Report. LA-13070-MS. Los Alamos National Laboratory.
Schmerr, L. W. 1998. *Fundamentals of ultrasonic nondestructive evaluation: a modeling approach*. London: Plenum Press.
SLAT TRACK project, January 2001–June 2003, Damage detection, life prediction and redesign of critical airplane components, project sponsored by the Flemish Government organization IWT (IWT/000327), partners: ASCO Industries, LMS International, Vrije Universiteit Brussel.
Vanlanduit, S. & Verboven, P. & Guillaume, P. 2003. On-line detection of fatigue cracks using an automatic mode tracking technique. *Jounal of Sound and Vibration*. Accepted for publication.

Acoustic emission for periodic inspection of composite wrapped pressure vessels

M.A. Luzio, R.D. Finlayson, V.F. Godínez-Azcuaga & S.J. Vahaviolos
Physical Acoustics Corporation, Princeton Junction, NJ, USA

Athanasios Anastasopoulos
Envirocoustics, A.B.E.E., Athens, Greece

Philip T. Cole
Physical Acoustics Limited, Cambridge, UK

ABSTRACT: Type 3 composite cylinders are used by the U.S. Navy to store compressed gases. The recertification procedure for these cylinders requires a hydrostatic test and an internal visual inspection every five years. Physical Acoustics Corporation has developed an AE based test method for the periodic inspection of these cylinders. AE baseline tests were performed on over fifty "good" cylinders and damage and/or fatigue cycling was introduced in fourteen cylinders. After damage and cycling the cylinders were tested again, which resulted in a dramatic increase in AE activity compared to that prior to the damage. The most active cylinders were also hydrostatically tested. No cylinders failed this test. However, a subsequent burst test showed a substantially decreased burst pressure for the AE active cylinders. A consistent correlation between burst pressure and AE activity was observed. An accept/reject criterion was developed to differentiate those cylinders with acceptable burst strength from those with unacceptable burst strength.

1 INTRODUCTION

Type 3 composite pressure vessels are fully wrapped carbon-fiber reinforced aluminium lined cylinders. A large number of these composite containers, supplied by different manufacturers, are used by the U.S. Navy for containment of diving mixtures. These cylinders are designed to meet the specifications required by the U.S. Department of Transportation (DOT). This type of cylinder is designed to withstand at least 3.4 times the service pressure before bursting.

Under DOT regulations, the recertification procedure of these Type 3 composite cylinders requires both a hydrostatic test and an internal visual inspection every five years during the allowable 15-year service life of the pressure vessel. After the 15-year period, cylinders are discarded.

The primary problem confronted by the users of these cylinders is the possibility that the composite overwrap will lose strength through impact damage, or fatigue cracks developing in the metal liner as the container undergoes in-service pressure cycles. The hydrostatic test is not an appropriate method for a thorough inspection of these types of cylinders, as it does not provide information required for evaluating the remaining strength of the cylinder. Without this information a prediction of the remaining life of the vessel cannot be made leading to the conclusion that unsafe pressure vessels may be left in service.

A more effective and less costly inspection method is needed to replace the hydrostatic test. Acoustic Emission (AE) has proven to be one of the best non-destructive inspection (NDI) methods for validating the integrity of composite structures. Under a relatively low proof load (normally 110% to 125% of the operating load) a good structure is relatively quiet and therefore passes a predefined accept/reject criterion, while a damaged structure will produce a great deal of AE. Excessive AE is a clear indication of damage in a cylinder.

The work reported in this paper, conducted by Physical Acoustics Corporation (PAC), was aimed at developing new equipment and a new procedure based on AE techniques for testing the Navy Type 3 composite cylinders.

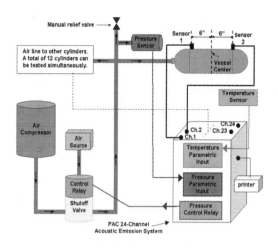

Figure 1. AE system and test setup.

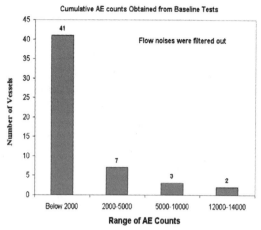

Figure 2. Acoustic Emission results for the 53 new cylinders. Bin distribution represents the range of counts, and bin height corresponds to the total number of cylinders that belong to that category.

2 DESCRIPTION OF THE TEST SYSTEM

The AE system developed for the tests consisted of a 24-channel Digital Signal Processing (DiSP) system using two sensors with built-in pre-amplifiers on each cylinder (peak resonant frequency at 150 kHz) as shown in Figure 1. This system allows for up to 12 pressure vessels to be tested simultaneously. The sensors are placed 12 inches apart from each other, 6 inches on either side of the mid-hoop.

The air used for the pressurization of the vessels was provided by an air compressor. Readings from a pressure transducer, located at the end of the air line, are fed into the pressure parametric input on the AE system. This signal establishes communication with the pressure control relay. A command signal is sent to the control relay (located on the shutoff valve) to turn a secondary air source on or off. The function of the air source is to close and open the shutoff valve, thus controlling the pressure in the vessel by starting or stopping the airflow from the compressor. Another parameter, temperature, was monitored using a temperature sensor connected to the system.

3 TEST PROGRAM

There are two distinct scenarios that the Navy Type 3 cylinders may see while in service. The first scenario represents the case when a pressure vessel is impacted, undergoes extraneous stresses, and obvious damage is identified. In this case the condition of the vessel is readily reported, and an immediate test is performed on the cylinder to assess the extent of the damage. If the test yields results meeting the failure criterion, then the vessel is permanently removed from service and discarded. Otherwise it is returned to service.

The second scenario is representative of the case when damage occurs but, is not identified. In this case a report on the pressure vessel condition is not issued and immediate testing is not performed. The cylinder is allowed to continue in service and will only be tested when the next AE test is scheduled to take place. In order to conduct tests for these two scenarios, a simulation of operating conditions was performed.

3.1 AE inspection of undamaged cylinders

In order to be able to determine if a Type 3 pressure vessel is suitable for continued service, baseline data for good and damaged cylinders were required. This baseline would serve to establish a failure criterion, and was to be based on AE signal features (i.e. AE counts).

To establish this base line, a total of fifty-three Type 3 pressure vessels were tested. All of these cylinders were initially tested in a new condition. Data was acquired using 110% of the service pressure, as required by the ASTM standard. Since the maximum service pressure was given as 3,000 psi, the test pressure used was 3,300 psi. Analysis of the data was then carried out after filtering of flow noise observed during filling of the vessels, and a comparison of AE counts for each pressure vessel was conducted. It was found that the majority of the cylinders had very low AE activity.

Approximately, 77% of the cylinders (41 cylinders) had less than 2,000 counts, 13% (7 cylinders) had a total of 2,000 to 5,000 counts, 6% (3 cylinders) had a total number of counts between 5,000 and 10,000, and finally 4% (2 cylinders) of the vessels yielded between 12,000 to 14,000 counts. Figure 2 shows a distribution of the number of vessels versus AE count

272

range in histogram format. This chart demonstrates that for the test conducted, less than a quarter, approximately 23% (12 cylinders) of the cylinders, gave results that were higher than or equal to 2,000 counts.

3.2 Design of initial failure criterion

After acquiring sufficient information and data on the acoustic behavior of good cylinders, the next step was to select the AE counts to be used for the failure criterion. Initially, one of the Type 3 cylinders (labeled ALT 809-099) was used as a baseline sample. To damage the cylinder, the cylinder was raised to twelve feet and then dropped on concrete impacting on the aft dome. The cylinder was then subjected to AE testing, in accordance with the ASTM standard E2191-02. To reduce flow noise, an air-compressor was used for pressurizing the cylinder, and a slow fill procedure was used.

A close examination of the cumulative counts from the AE test, after damage, showed a total of 29,000 counts. This value is less than the 50,000 cumulative counts criterion for failure as proposed by the DOT-CFFC, but it is much higher than the counts obtained from the fifty-three good cylinders.

The ASTM standard requires that a pressure vessel meet a safety factor of 3.4 times the operating pressure before bursting. Using this multiplying factor, and the design operating pressure of 3,295 psi, as defined by the manufacturer, the minimum acceptable burst pressure was calculated to be 11,203 psi. The damaged vessel, ALT 809-099, was hydrostatically pressurized until burst after the impact damage. It burst at 12,032 psi, which is above the minimum acceptance burst pressure of 11,203 psi. This result exceeds the minimum acceptance burst pressure by 7.4% (829 psi). Since the damaged cylinder burst above the minimum acceptance factor of 3.4, the 29,000 obtained during the AE test could be considered a very conservative figure for an initial accept/reject criterion.

3.3 Verification of initial failure criterion

Additional testing to validate the developed test procedure, and to "test" our initial accept/reject criterion was performed. As previously discussed above, there are two service scenarios that we could have simulated. The first scenario would lead us to subject the cylinders to an immediate hydrostatic burst, after the AE test. Although this approach would yield a more refined accept/reject criteria, it would only have yielded very limited information on the growth of damage under normal operating conditions.

The most realistic approach, to simulate normal operating conditions, could be achieved with a test procedure representative of the second scenario. This scenario consists of cycling the cylinders before burst.

Table 1. Cumulative Acoustic Emission counts before and after damage.

Cylinder ID	Condition	Cumulative AE counts	
		Before	After
ALT 809-078	Good	500	N/A
ALT 809-081	Good	1,300	N/A
ALT 809-009	Instron Indented	1,600	117,000
ALT 809-013	Instron Indented	7,400	118,000
ALT 809-026	Instron Indented	3,100	40,800
ALT 809-015	Dropped	N/A	25,700
ALT 809-020	Dropped	4,300	18,600
ALT 809-079	Dropped	100	16,600

A total of eight new cylinders were selected. Of the eight cylinders, two were left in good condition (no damage), and the remaining six cylinders were damaged. Three of these six cylinders were compressively loaded with an indenter on an Instron machine, and the other three were dropped from a height of fifteen feet impacting on the aft dome. Although the second scenario was being simulated, after damage the cylinders were subjected to a pneumatic proof loading of 3,300 psi in accordance with the test procedure (also as outlined in the ASTM standard). AE data was collected from the proof load test.

The six damaged cylinders produced a greater number of counts when compared to the two good cylinders. Table 1 lists the results of the AE test performed on the eight pressure vessels. In Table 1, it can be seen that the three cylinders indented by the Instron machine yielded a total number of counts far greater than the initial accept/reject criterion of 29,000 counts. In the case of the three dropped cylinders, while there was a large increase in the cumulative AE counts, fell below the initial accept/reject criterion of 29,000 counts. The cylinders were next cycled to simulate the fifteen-year maximum usage period.

Three cylinders were hydrostatically cycled between 1,000 and 3,000 psi, which is the normal pressure range found in service. The three cylinders were selected one from each "group": one indented by the Instron, one damaged by dropping, and the third was a good cylinder. After 780 pressurization cycles, the cylinders were taken to burst. The results from this test would allow a comparison to the initial burst pressure for cylinder ALT 809-099 (the baseline sample) with the burst pressure of these three vessels. Table 2 lists the cumulative counts as well as the burst pressures for all four cylinders.

From the result obtained from cylinder ALT 809-099, it was expected that the good cylinder (ALT 809-081) would burst at a pressure level above 12,032 psi, since no damage was inflicted. The drop-damaged

Table 2. Cumulative AE counts (before cycling) and burst pressure for three cylinders and cylinder ALT 809-099.

Cylinder ID	Condition	Cumulative AE counts	Burst pressure [psi]
ALT 809-081*	Good	1,300	>14,509
ALT 809-009	Indented	117,000	9,369
ALT 809-015	Dropped	25,700	10,624
ALT 809-099	Dropped	29,000	12,032

* Vessel ALT 809-081 did not burst.

cylinder, since it was dropped from fifteen feet instead of twelve feet as before, was expected to burst at a slightly lower pressure.

The undamaged cylinder ALT 809-081 was expected to burst at a very high pressure and it was pressurized to 14,509 psi without bursting. This pressure is well above the minimum acceptable burst pressure of 11,203 psi, and also well above the burst pressure of vessel ALT 809-099. The cylinder previously dropped from fifteen feet (ALT 809-015), burst at a pressure of 10,624 psi on the dome where it had impacted the concrete floor. Also, it was below the burst pressure of cylinder ALT 809-099, as expected. The damaged cylinder, indented by the Instron machine, burst at 9,369 psi in a location on the side-wall where it was pressed.

From the results of these tests, and knowing that the dropped-damaged cylinder (ALT 809-015) produced a total number of counts less than the preliminary accept/reject criterion of 29,000 counts, it was clear that the initial accept/reject criterion was set too high. Based on the 25,700 counts obtained from the dropped-damaged cylinder, a new accept/reject criterion was selected. To compensate for variations usually found in experimental tests, an even more conservative number was taken. The initial 29,000 counts criterion was lowered to 20,000. Figure 3 shows the burst data for the four cylinders tested.

The dashed vertical line shows the revised 20,000 count mark failure limit, and the horizontal dashed line represents the 11,203 psi minimum acceptance burst pressure, 3.4 times the maximum operating service pressure.

3.4 Test pressure increase and verification

Although the normal in service operating pressure for NAVSEA is set at 3,000 psi for in-service cylinders, it may be possible for a cylinder to be pressurized to more than the normal operating pressure, reaching a maximum of 3,300 psi allowed by the safety valve. Since 3,300 psi was our current test pressure, a new test pressure of 110% maximum service pressure needed to be established. According to the ASTM standard, and from our experience in AE testing, the

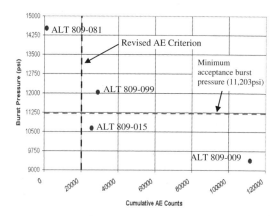

Figure 3. Burst pressure versus cumulative counts for the four cylinders.

Table 3. Results of the five undamaged cylinders used for the 3,630 psi AE test.

Cylinder ID	Number of cycles	Cumulative AE counts Initial (3,300 psi)	Final (3,630 psi)
ALT 809-051	260	200	100
ALT 809-071	260	1,200	5,400
ALT 809-093	160	1,000	300
ALT 809-094	160	1,800	100
ALT 809-098	160	13,700	1,100

testing pressure should be at least 110% of the maximum service pressure. Since the maximum service pressure may reach 3,300 psi, the test pressure would have to be raised to 3,630 psi in order to maintain the 10% over pressure.

With the test pressure raised to 3,630 psi, further testing was needed to determine if this change would alter the AE results in good cylinders. To ensure that the pressure change did not influence the accept/reject criterion developed, five good (undamaged) cylinders were prepared for testing. Before the cylinders were cycled, an AE test at 3,300 psi was performed on all five for later comparison purposes. The results of the tests are shown in Table 3 under the column "Initial" (Cumulative AE counts).

Three of these cylinders underwent pneumatic cycling at the normal service pressure of 3,000 psi for a total of 160 cycles. The other two cylinders were hydrostatically cycled to 3,000 psi, for a total of 260 cycles. The final AE tests were conducted using the revised test pressure of 3,630 psi. Table 3 shows the test results for both groups of cylinders. After the cycling was concluded, all cylinders were AE tested at a pressure of 3,630 psi, instead of the original 3,300 psi.

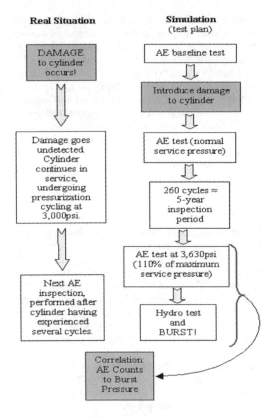

Real Situation	Simulation (test plan)

Real Situation

DAMAGE to cylinder occurs!

Damage goes undetected Cylinder continues in service, undergoing pressurization cycling at 3,000psi.

Next AE inspection, performed after cylinder having experienced several cycles.

Simulation (test plan)

AE baseline test

Introduce damage to cylinder

AE test (normal service pressure)

260 cycles ≈ 5-year inspection period

AE test at 3,630psi (110% of maximum service pressure)

Hydro test and BURST!

Correlation: AE Counts to Burst Pressure

Figure 4. Flow chart showing: a probable real operating situation when damage is not readily detected (left), and the test plan designed to simulate this situation (right).

The results are listed under the column "Final". It can be seen from these values that all cylinders had relatively low AE activity. With the exception of Vessel ALT 809-071, all the cylinders had a lower number of cumulative counts after cycling, with the higher test pressure of 3,630 psi. It was concluded that raising the test pressure from 3,300 psi to 3,630 psi would have an insignificant effect on the AE behavior of undamaged cylinders. Furthermore, this additional increase would not lead to an incorrect evaluation of the integrity of the cylinder.

3.5 Additional cycling and AE testing

Because of the AE dependence on time and pressure, a specific testing method must be strictly followed.

To address the influence of cycling on the AE activity of damaged cylinders using the new test pressure of 3,630 psi, an additional test plan was developed. Figure 4 summarizes the new test plan designed to portray a real life scenario.

In the left column, a representation of what is likely to happen during service when damage is not

Table 4. Description of damage inflicted on the seven cylinders used for the simulation.

Cylinder ID	Damage type
ALT 809-103	Cut: 4 inch length, 1/8 inch depth
ALT 809-014	Impact: Dropped once, 15 ft.
ALT 809-111	Impact: Dropped once, 15 ft.
ALT 809-062	Impact: Dropped twice, 15 ft.
ALT 809-078	Impact: Dropped once, 10 ft.
ALT 809-047	Indented: Instron pressed
ALT 809-016	Indented: Instron pressed

Table 5. AE cumulative counts after the 260 cycles.

Cylinder ID	Cumulative AE counts after cycling (3,630 psi)
ALT 809-103	183,300
ALT 809-014	9,000
ALT 809-111	33,000
ALT 809-062	33,200
ALT 809-078	7,400
ALT 809-047	2,900
ALT 809-016	700

detected, is shown. The right column shows the flow of the test plan simulating this scenario. To carry out this test plan, seven cylinders were damaged. The damage to each cylinder is shown in Table 4.

A cut on cylinder ALT 809-103 was oriented in the longitudinal direction, with the center of the cut at the center of the vessel. The impacted cylinders were dropped onto concrete on the aft domes. The Instron pressed cylinders were indented using a load of 5,500 lb and a tup of 0.55 inches in diameter. Next, an AE test was conducted at the normal operating pressure of 3,000 psi and data was collected to be used for post-test analysis.

The next important step in the test plan, consisted of cycling the cylinders to simulate the second box of the "Real Situation" column (left) in Figure 4. As periodic inspections of Type 3 cylinders are performed every five years, a total of 260 cycles were selected, which is the equivalent number of cycles for one five-year service period. The 260 cycles were performed from 1,000 psi to 3,000 psi. After the completion of the cycling stage, a final AE test was performed at 3,630 psi which is 110% of the maximum service pressure of 3,300 psi. The results of the AE test conducted after cycling are shown in Table 5.

From Table 5, it can be seen that cylinders ALT 809-103, ALT 809-111, and ALT 809-062 had a number of counts greater than the failure criterion of 20,000 counts. The remaining four cylinders had counts below 20,000 counts.

Table 6. Results of the Visual, Hydrostatic, and Acoustic Emission tests performed at 110% service pressure. Note that AE failed the most cylinders based on the number of counts.

| Cylinder ID | Results of the three tests | | | Burst Pressure (psi) |
	Visual	Hydrostatic	AE	
ALT 809-103	Failed	N/A	Failed	9,433
ALT 809-014	Passed	Passed	Passed	12,205
ALT 809-111*	Passed	Passed	Failed	11,464
ALT 809-062	Passed	Passed	Failed	8,252
ALT 809-078	Passed	Passed	Passed	12,878
ALT 809-047	Passed	Passed	Passed	12,343
ALT 809-016	Passed	Passed	Passed	12,673

* Burst during the 1-minute holding period.

To further verify our accept/reject criteria, these seven cylinders were subjected to visual inspection, a hydrostatic volumetric expansion test, and finally a burst test. An important piece of information to be gained from the burst test, is the effect, if any, that cycling has on the detection of damage. When the visual inspection was conducted, only cylinder ALT 809-103, the cylinder with the cut, failed. The other six damaged vessels were hydrostatically tested, as outlined in the DOT-CFFC code. All six cylinders passed the hydrostatic test. Table 6 summarizes the results.

Following these tests, all seven cylinders were burst hydrostatically. Cylinders ALT 809-103 and ALT 809-062 burst below the minimum acceptance burst pressure of 11,203 psi. Cylinder 809-111 burst during the 1-minute hold period at 11,464 psi. According to DOT-CFFC, if a cylinder bursts during the manufacturer's production test at the 1-minute holding stage, the entire lot must be rejected.

As can be seen from Table 6, the pressure vessels that burst below the minimum acceptance level, and the vessel that burst during the 1-minute holding period, are the same vessels that the AE test rejected. The AE test, based on the 20,000 counts criterion, accurately predicted the cylinders that would fail the burst test. The hydrostatic volumetric expansion test, however, did not reject the vessels that burst below the minimum acceptance burst pressure.

3.6 Final validation test

Two additional new cylinders were damaged and tested during the NAVSEA final validation test. Vessels ALT 809-006 and ALT 809-073, were damaged by dropping from 15ft on concrete, the AE tested. The AE activity produced 29,296 and 10,686 counts respectively. Although both new cylinders were dropped from the same height, only one, cylinder ALT 809-006,

failed the 20,000 counts criterion. This indicated that each cylinder sustained a different degree of damage. After the AE test, both cylinders underwent a hydrostatic volumetric expansion test, and subsequently a burst test. While both cylinders passed the hydrostatic test, cylinder ALT 809-006 burst at 9,600 psi, and ALT 809-073 at 11,800 psi. Thus, the AE test was once again confirmed. This test correctly predicted that the 29,296 counts cylinder should burst at a pressure below the minimum acceptance pressure of 11,203 psi. It also correctly predicted that the 10,686 counts cylinder, should burst at a pressure above the minimum acceptance pressure of 11,203 psi.

3.7 Final accept/reject criterion

Based on the results of all tests described, and on the notional logic of Bayes' theorem, an accept/reject (pass/fail) criterion has been set. According to the theorem, four outcomes are possible when a test is conducted. These are labeled as true positives, false positives, false negatives, and true negatives. True positives are the "Correct Prediction of Good Vessels", that is, the vessels that presented counts below the criterion (20,000 counts), and did not burst at pressures below the minimum acceptance burst pressure (11,203 psi). False positives, are the "False Alarms". This class represents the case of a vessel that is rejected due to high counts (above criterion), but does not burst below the minimum acceptance pressure. False negatives are the "Undetected Errors", which would be the case where a vessel is accepted due to low counts (below criterion) but burst below the minimum acceptance pressure. Finally, true negatives are the "Correct Prediction of Bad Vessels". In this case, the cylinder was rejected correctly due to high counts (above criterion), and burst at a pressure above or at the minimum acceptance pressure.

Figure 5 shows the final acceptance/rejection results. The upper-left area of Figure 5 contains six points representing six of the vessels that exhibited very low counts, below the 20,000, that burst above the minimum acceptance level of 11,203 psi. The lower-right area of the graph, denotes the area where cylinders were rejected correctly. These cylinders produced a high number of counts, and burst at pressures lower than 11,203 psi. Five cylinders were classified as meeting these criteria.

In the upper-right area, cylinders had high counts (above 20,000 criterion) but did not burst below the minimum acceptance level. Only two of the cylinders tested fell in this area. Finally, the fourth area on the lower-left, cylinders would have low counts, but also burst below the minimum of 11,203 psi. This area would be representative of damaged cylinders not detected by the AE method. No cylinders tested in this program fall into this category. This shows that the

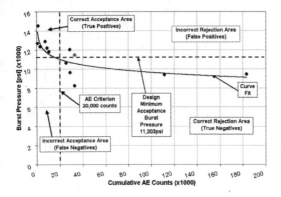

Figure 5. Final distribution of areas for correct and incorrect accept/reject decisions. Correlation between AE counts and burst pressure for 13 cylinders determined these areas.

AE test did not pass the vessels that were unsuitable for use, unlike the hydrostatic volumetric expansion test. Overall, these results show that the accept/reject criterion is both accurate and suitably conservative.

4 CONCLUSIONS

The results obtained in this program show that the AE test was able to detect damaged cylinders that burst below the minimum acceptance pressure, while the hydrostatic test did not identify these cylinders correctly.

The test procedure proposed by PAC suggests that a test pressure of 3,630 psi, which represents 110% of the maximum service pressure, is ideal for performing inspection of the vessels used by the Navy. By pressurizing the cylinders to 110% of the maximum service pressure, any previous service cycling effects are surpassed by the effects of the higher pressure.

The 20,000 AE counts limit for the acceptance or rejection of Type 3 composite cylinders is a conservative number that prevents unsuitable cylinders from remaining in service.

NDT of civil engineering structures

Emerging Technologies in Non Destructive Testing, Van Hemelrijck, Anastasopoulos & Melanitis (eds)
© 2004 Swets & Zeitlinger, Lisse, ISBN 90 5809 645 9

Condition assessment of concrete half joint bridges

J.R. Watson & P.T. Cole
Physical Acoustics Limited, Cambridge, England

A. Anastasopoulos
Envirocoustics ABEE, Athens, Greece

ABSTRACT: The deterioration of concrete bridges in the UK is largely due to chloride attack of reinforcement steel, which reduces the tensile strength of the structure. Civil Engineers have the task of quantifying damage within structures in order to prioritise maintenance. Acoustic Emission (AE) has become an established technique in many industries for assessing the extent of damage in a variety of different structures and materials. Trials using procedures and evaluation methods, developed by Physical Acoustics Limited supported by research carried out at Cardiff University, were carried out on the concrete half joints of a motorway bridge on the M6 motorway, England. This paper discusses the use of AE as a condition assessment tool for concrete structures and presents the findings of the M6 monitoring as a case study.

1   INTRODUCTION

Within the UK trunk road and motorway network there are approximately 10,000 bridges, of these over eighty percent are predominantly concrete. The majority of these bridges were constructed during the late 1960s and 1970s during the rapid expansion of the UK's road network, and incorporated many new design features, which increased the speed of construction. During this period reinforced concrete was considered to be a material that would require little or no maintenance over their design life. In hindsight however this is far from true. Today the UK's engineers have inherited deteriorating concrete structures (some rapidly deteriorating), which have suffered from prolonged chloride attack, exaggerated by poor construction such as insufficient cover and poor water proofing. This has reduced the capacity of concrete bridges to resist increasing bridge loading, and in some cases has lead to bridges failing structural assessments.

One particular concrete design detail that has been found to be concern is the half joint. The design consists of a supporting nib, which can either be continuous e.g. supporting slabs or multiple smaller ones e.g. supporting beams. The half joint allows the use of precast slabs or beams, which are craned into position resting on bearings on the half joint nib. The combined load from the beam and live load is transferred into the nib which carries much of the shear and tensile loading

by means of high density steel reinforcement, which may include post tensioning.

Defective expansion and articulation joints can allow the ingress of water carrying chlorides from road salts, which percolate and collect in the nib. Over time the build up of chlorides can lead to deterioration and reduction of the steel section, sometimes in discrete locations, and cause a subsequent loss of shear and tension resistance in structures. One of the first indications of this occurring can be the spalling of cover concrete and structural cracking within the nib. It is of great importance that bridge engineers are able to identify the presence of this type of damage as early as possible to allow suitable remedial work and management of the structure. It is also desirable to accurately assess the condition of individual half joints to identify areas of damage and to rank these in terms of severity, allowing the prioritisation of actions and maximise cost efficiency.

Following on from the development of successful condition assessment and damage location techniques for steel bridges, Physical Acoustics Limited were approached by consultants Atkins and the UK Highways Agency, to carry out a trial to evaluate the potential of AE as an assessment method for half joints. A bridge in Cumbria was suggested by UK HA Area 19s Maintaining Agent's, AmeyMouchel, as an excellent subject for a limited trial. Borrowbeck bridge is an under bridge just south of Junction 38 on the M6 motorway in the North West of England,

Figure 1. Borrowbeck bridge.

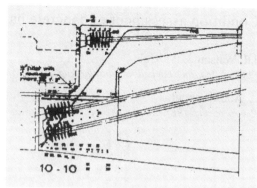

Figure 3. Post-tensioning within the half joint.

Figure 2. Half joint on Borrowbeck bridge.

Table 1. Visual condition of half joints tested.

Half joint location	Suspected condition
B1	Good
B6	Good
B7	Good
E1	Cracked, visibly corroded
E6	Cracked
E7	Cracked
F6	Cracked
F7	Cracked

which had been extensively scaffolded to allow a detailed inspection of its northern/southern hammerheads and half joints. The bridge spans a gully from a mountain stream and an access road to a farm and is shown in Figures 1 and 2.

The bridge consists of two spans to the abutments from the north and south hammerheads with a central drop in span. All span supports are half joint nibs which each support two precast beams, either concrete box beams or pre-tensioned I beams. The northern and southern hammerheads are post-tensioned concrete box girders that have the cable anchorage's located in the nib of the half joint, shown in Figure 3. This contributes significant compression strength to the half joint to counter shear and tensile forces.

As part of the assessment of the structure extensive chloride samples were taken from the soffit and half joints. Analysis of the samples indicated chloride levels up to 2.5% of the concrete cement content, with levels of 0.32% to 2.2% at 80 mm depth. Many of these areas coincided with visual evidence of damage such as delamination, spalling and surface cracks. From

this information AmeyMouchel identified eight half joints nibs for assessment by AE monitoring, considered to be in a variety of conditions ranging from good, to those having evidence of damage. Details of these joints are shown in Table 1.

2 AE CONDITION MONITORING STRATEGIES

The procedure developed by Physical Acoustics Limited for assessment of half joints follows on from the extensive work of Physical Acoustics Nippon and Kumamoto University in Japan (Ohtsu 1987, Ohtsu & Shigeishi 1993, Yuyama & Ohtsu 1999), incorporating new research into concrete fracture signal behaviour and damage energy source analysis carried out at Cardiff University. The procedure utilises two monitoring strategies; Zonal and 3-dimensional Local monitoring. Both strategies were complimented by structural displacement and ambient temperature monitoring, which allows correlation of AE activity from damage sources with load conditions and environmental changes.

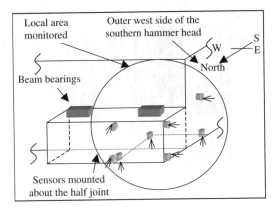

Figure 4. Local monitoring array on half joint E1.

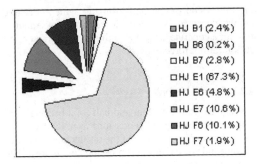

HJ B1 (2.4%)
HJ B6 (0.2%)
HJ B7 (2.8%)
HJ E1 (67.3%)
HJ E6 (4.8%)
HJ E7 (10.6%)
HJ F6 (10.1%)
HJ F7 (1.9%)

Figure 5. Percentage of cumulative energy released from each half joint nib.

2.1 Zonal monitoring

The first monitoring strategy used is Zonal assessment. This comprises of a single sensor mounted on the half joint nib soffit that detects signals originating from active damage sources within the joint. It has been determined that this sensor is capable of detecting a 0.5 mm pencil lead break on the surface of the structure at distances greater than 6 m. This gives the Physical Acoustics AE system sufficient sensitivity to detect signals from microfracture of concrete, with energies in the region of 10^{-12} Joules, $\sim$10,000,000 times smaller than post-tension wire fracture. The individual Zonal monitoring sensor is able to assess a large volume of structure during normal traffic loading. Analysis of this data after the removal of any extraneous noise determines the significance and intensity of damage mechanisms within the joint and the conditions under which they are caused. Comparison of the data from individual sensors allows condition ranking of areas of slab joint or the nibs of beam half joints. This facilitates the prioritisation of structures and areas for further AE monitoring, if required.

2.2 3D local monitoring

The second strategy employed in half joint assessment is 3-dimensional monitoring. Sensors locations about the half joint E1 are shown in Figure 4. This strategy seeks to acquire detailed information about unique sources within the joint, which includes source intensity and severity. This was achieved through correlation with controlled crack growth experiments and attenuation modelling at Cardiff University. The damage and growth conditions together with approximate location are displayed graphically in either 2 or 3 dimensions.

3 AE ASSESSMENT OF BORROWBECK BRIDGE HALF JOINTS

Testing was carried out according to Physical Acoustics Half Joint Monitoring procedure, according to PAL ISO 9002 Quality System. A Physical Acoustics "DiSP" AE testing system was used with special sensors optimised for detecting microfracture in concrete. The displacement of the half joint soffit was monitored throughout the test and logged with AE data by the DiSP system together with temperature. This allowed correlation of environmental conditions with AE data.

Each of the eight joints identified were zonally monitored using one AE data sensor mounted on the soffit of the half joint. Soffit mounted sensors make the procedure and results from Borrowbeck transferable to other types of structures such as slab half joints.

The sensitivity of the system was set to detect signals with 1/10,000th the energy of an 0.5 mm pencil lead break at 0.5 m distance. The pencil break or "Hsu-Nielsen" source is the field calibration standard, as per BS EN 1330-9:2000 (British Standards 2000). Based on Cardiff research, this level of sensitivity exceeds that required for locating the early stages of concrete microfracture. All monitoring was carried under normal traffic flow. Following Zonal monitoring of the eight half joints, AE data was analysed in order to identify the most damaged joint. Eight sensors were mounted on this half joint, 3 sensors on the soffit, 3 on the external side of the joint and 2 on the vertical end face of the joint to the right hand side of the outer beam, as shown in Figure 5. Sensors were mounted as previous, calibrated and location performance checked against artificial surface sources.

4 MONITORING RESULTS

4.1 Structural calibration results

Study of signal attenuation found that the "Hsu-Nielsen" field calibration source could be detected at the other side of the half-joint nib with a sizeable amplitude, which was considered to be low attenuation for a concrete structure of this size, geometry and structural reinforcement.

4.2 *Zonal monitoring results*

AE signals from the beam and half joint bearing were filtered out, leaving only emissions from the half joint itself. A summary of some of the results after data processing and noise removal is presented in Figure 5. The half joint location supporting beam E1 was found to be the most active, accounting for 67.3% of all the signal energy from all monitored half joints. Its position is under the outer beam on the North bound carriageway. This correlated with high chloride levels, up to 2.5%, identified from concrete samples taken from this half joint. Half joint E1 had suffered extensive visible spalling due to corrosion, and was considered to be one of, if not the most degraded half joint on the structure. Half joints at the south end of the southern hammer head which support beams B1, B6 and B7, were identified by the Maintaining Agent as being of sound condition. These were found to emit very low energy signals accounting for less than 6% of total emission detected, suggesting very low levels of damage present or occurring during the monitoring period.

The emission from the most damaged half joint nib, E1, was directly related to displacement (load). As a vehicle passes over the rear of the table the displacement becomes positive due to the cantilever effect of the table over the pillar, reversing when past the pillar support. The stress then increases rapidly on the half joint as the vehicle passes over it and on to the beam, and emission primarily occurs as stress increases at this point. Some emission occasionally occurs on unloading, this is normal behaviour known as crack closure noise. Signal analysis of data from half-joint E1 (and others) indicates that the signals are from micro-fracture of concrete and secondary crack closure emissions. This is supported by Cardiff University research.

Data from half joint E1 was analysed to determine the location of active damage sources present. The location of emission was calculated by triangulation, from the time arrival of signals, a similar method to earthquake epicentre location. The 3-dimensional location effectively "looks inside" concrete structures, in this case the majority of emission appeared to originate from two locations, shown in Figure 6. The first is a focused source, located approximately 1.6 m from the end face within the half joint nib in an area of high shear stress. The second is located at the outer surface of the nib which was near an area of that had suffered extensive cover spalling due to corrosion of the reinforcement shown in Figure 7.

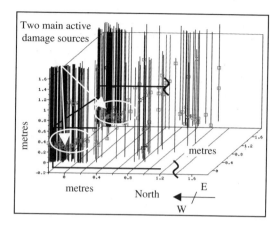

Figure 6. 3-D source location of active damage within the nib of half joint E1.

Figure 7. Spalled concrete near the second source.

Location accuracy depends upon the position of the source inside the array, and the signal wavelength, in this case it is expected to be within ± 0.1 m.

Through correlation of displacement of this half joint nib and the AE activity, it was possible to confirm that most of this activity occurred at maximum load, suggesting that they are indeed damage mechanisms.

The characteristics of the located emissions were found to have distributed energy levels, averaging 10^{-13} joules, peaking at 10^{-11} joules. This indicates active micro-fracture concrete.

5 CONCLUSIONS

The Physical Acoustics Half Joint monitoring procedure, as used on Borrowbeck Bridge, has proven to be practical for use on motorway bridge half joints under normal traffic conditions, and has demonstrated its ability to identify and rank their condition.

3-dimensional monitoring has provided detailed source analysis, together with location information on the most badly deteriorated joint, achieved with normal traffic loading.

ACKNOWLEDGEMENTS

The authors would like to acknowledge and thank the following; UK Highways Agency, UK HA Area 19 Maintaining Agents AmeyMouchel Joint Venture, especially Mr. H. Roper and Mr. D. Pollard for all their valuable assistance in this trial. Physical Acoustics would like also to thank Dr. A. W. Davies, Dr. K. Holford and Mr. T. Bradshaw at Cardiff University's School of Engineering for their continuing assistance in development of structural monitoring techniques.

REFERENCES

British Standard. 2000. Non-destructive testing Terminology Part 9: Terms used in acoustic emission testing, *BS EN 1330–9:2000.*

Ohtsu M. & Shigeishi M. 1993. Identification of AE Sources By Using SiGMA-2D Moment Tensor Analysis. *Journal of Acoustic Emission Volume 11, No. 4.*

Ohtsu M. 1987. Acoustic Emission Characteristics in Concrete and Diagnostic Applications. *Journal of Acoustic Emission, Volume 6, No. 2.*

Yuyama S., Ohtsu M. 1999. A Proposed Standard for Evaluating Structural Integrity of Reinforced Concrete Beams by Acoustic Emission. Acoustic Emission: *Standards and Technology Update, American Society for Testing and Materials.*

DIC for damage detection at the surface of a pre-stressed concrete beam under static loading

D. Lecompte & J. Vantomme
Royal military Academy, Department of materials and construction, Brussels, Belgium

ABSTRACT: The purpose of using the digital image correlation technique (DIC) in the present project is to evaluate if, according to increasing static loading of a reinforced concrete beam, higher strain concentrations, in regions where cracks occur, will be detected. Furthermore it is examined if after cracking any changes in the strain distribution of the region between two cracks appear and if any information about the dilatation of existing cracks during the static loading can be extracted from the DIC measurements. The paper states that it is possible to follow the evolution of cracks and furthermore it shows that it is difficult to predict crackforming and that the obtained results should be treated with extreme care.

1 INTRODUCTION

The present paper is defined in the framework of the FWO-Vlaanderen-project Nr G.0266.01, where the objective is to evaluate the behaviour of a realistically sized beam, during static and dynamic loading, using a 4-point bending test. However, the classical experimental validation techniques (strain gauges, extensometers, ...), don't allow a correct apprehension of the local damage mechanisms leading to the rupture of the structure, due to the heterogeneous nature of concrete. A more relevant modelling, relative to these mechanisms, can be assisted by a full field measuring technique focussed on a given region of interest of the loaded specimen. Models to accurately predict the influence of a loss of stiffness and strength or to forecast a dynamical effect due to failure, under certain loading conditions are needed, but this is only possible through detailed and valid measuring results. A lot of questions are still in search of an answer. One has to know for example how the stressed concrete behaves between two cracks and what the influence of these regions could mean with respect to the general behaviour of the beam (tension stiffness), or why some cracks are permanent and others close again during unloading. The question is: is it possible to predict crack forming before a macroscopic crack reveals itself and can the crack-evolution during the process be monitored. It is in this framework that the present study is defined: to take a closer, yet full field, look at the behaviour of a pre-stressed concrete beam under a given static loading cycle. The chosen approach is the application of a digital image correlation technique, based on a given speckle pattern, exhibited by the specimen or obtained by a paintspray.

This paper presents a theoretical overview of the image correlation technique, discusses the measurement principle and the obtained results.

2 DIGITAL IMAGE CORRELATION

2.1 Introduction

The digital image correlation technique is an optical-numerical measuring technique, which offers the possibility to determine displacement and deformation fields at the surface of objects under any kind of loading, based on a comparison between images taken at different loading steps. This technique has been used in different domains and many of its applications have been reported (De Roover et al. 2001, Geers & Schimek 1999, Doumalin et al. 1999, Tillie 1996, Vendroux & Knauss 1998). It allows to study quantitatively and qualitatively, the mechanical behaviour of materials under certain loading conditions. In comparison with strain gauges and extensometers, which measure only local and unidirectional strains, this technique presents two main benefits. On the one hand, it offers the possibility to obtain non-contact measurements, which therefore are not influenced by any mounting of the system on the object and on the other hand, it permits to generate local as well as global displacement and deformation quantities of the object under loading. This means that, because of the full-field information, useful results can also be expected for more heterogeneous materials.

2.2 Strain distribution

It is possible to describe the deformation of the object using the strain distribution. A measure of the deformation of a line segment can be defined as follows:

$$\lambda = \frac{l + \Delta l}{l} \tag{1}$$

where λ is the stretch ratio defined as the relative elongation of an infinitesimal line segment. The strain value ε can then be defined as a function of λ. One of the possible expressions is the following:

$$\varepsilon = \lambda - 1 \tag{2}$$

This function of λ defines the technical strain and is used in the present work to determine the strain distribution, where the obtained values are expressed in percent. It is the definition of a one-dimensional strain measure, but it can easily be extended to the two-dimensional case if we suppose that the gradient tensor "F" defines the transformation of the line element dX into the line element dx. This yields the following equation:

$$dx = F.dX \tag{3}$$

As the deformation gradient defines the rotation as well as the stretch of the line segment, a decomposition into a rotation and a stretch tensor is appropriate:

$$F = R.U \tag{4}$$

For the calculation of the deformation gradient tensor from a two-dimensional displacement field, the two-dimensional co-ordinates of each object point must be known in both its undeformed and deformed state. It is mandatory that for the calculation of the deformation tensor for a point, a homogeneous state of strain has to be assumed for the set of neighbouring points. The relation between the co-ordinates of the deformed points p_v and the co-ordinates of the undeformed points p_u is the following:

$$p_v = u + F.p_u \tag{5}$$

It describes a linear system of equations where u is the translation vector and the unknowns are the four parameters of the deformation gradient tensor F. The system becomes over-determined if more than three points are chosen. A Gaussian least squares algorithm

is then used to solve the system. The translation vector "u" describes the translation of the point for which the deformation tensor is calculated. Once this is done, the strains can be derived from it:

$$U = \begin{pmatrix} U_{11} & U_{12} \\ U_{21} & U_{22} \end{pmatrix} = \begin{pmatrix} 1 + \varepsilon_x & \varepsilon_{xy} \\ \varepsilon_{xy} & 1 + \varepsilon_y \end{pmatrix} \tag{6}$$

The strains ε_x and ε_y depend on the used co-ordinate system. To avoid this inconvenience, a major and minor strain can be calculated, which are independent of the governing co-ordinate system. The stretch matrix U can be transformed into its diagonal form and its corresponding eigenvalues are given by the following expression:

$$\lambda_{1,2} = 1 + \frac{\varepsilon_x + \varepsilon_y}{2} \pm \sqrt{\left(\frac{\varepsilon_x + \varepsilon_y}{2}\right)^2 - \left(\varepsilon_x \varepsilon_y - \varepsilon_{xy}^2\right)} \tag{7}$$

Those eigenvalues are the actual stretch ratios and can be transformed into corresponding strain values ε_1 and ε_2, depending on the choice of the transformation, which in our case is the technical strain. The corresponding eigenvectors determine the two directions of major and minor strain. These values are used for the evaluation of the strains in the present study.

3 EXPERIMENTAL PROCEDURE

3.1 Testing procedure and camera settings

The experimental principle consisted in loading the pre-stressed concrete beam up to different loading values and subsequently unloading it. After the static loading and unloading procedure, a dynamic analysis of the beam – which is not discussed here – was performed. Three tests had already been performed where maximum loads of respectively 70 kN, 135 kN and 250 kN were imposed per jack. This paper is focused on the exploitation of the images that were taken during the fourth test in a series of tests until rupture of the beam. It was tempted to retrieve information about cracking during the first three tests, but a choice had to be made with respect to the relation between the image size (a size big enough to enhance the probability of having a crack appearing in the chosen image field) and the obtained resolution. Unfortunately, no crack became visible in the selected zone. Therefore, in the fourth test a region of interest was chosen around an existing crack. The pictures will show black crack marking lines near the cracks that originated in the previous loading procedures. These markings are a part of the testing process and can thus not be avoided

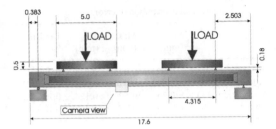

Figure 1. Experimental set-up (dimensions in m).

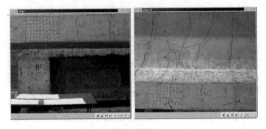

Figure 2. Image field of the camera and calculation mask in the undeformed (left) and deformed (right) state.

while taking the necessary pictures. These lines have a disturbing influence on the treatment of the images. Figure 1 shows the experimental set-up.

The apparatus for the Digital Image Correlation analysis consists of a PC computer, a VOSSKUHLER CCD-1300 camera with a resolution of 1280 × 1024 pixels by 12-bits, fitted with a high power Schneider 17 mm camera lens. The camera is fixed on a tripod to assure the stability, which is necessary for the precision of the measurements. Greyscale images (12-bits resolution converted into 8-bits resolution) are real-time captured through a framegrabber and processed by a Dual Socket Pentium III Computer with the ARAMIS 2D Software (GOM – Gesellschaft für Optische Meßtechnik mbH).

The surface of the specimen had only to be slightly painted thanks to the important contrast pattern of concrete. It is equally important that the pattern doesn't exhibit large areas of constant brightness, but concrete once again possesses this characteristic. The objective of the little paint sprayed onto the surface of the specimen, is to allow a straightforward determination of starting points, used by the program to start the correlation algorithm.

The shutter time is adjusted in such a way that the distribution of grey levels is spread out as large as possible. It was chosen the smallest possible in order to increase the sharpness i.e. to avoid noise, caused by the vertical displacement of the beam during the scanning of the image.

While focussing on the specimen, the diaphragm of the front lens is opened to narrow the depth of field and to allow a precise object focus. It is again closed afterwards to increase the depth of field.

The spatial resolution is obtained by determination of the number of pixels corresponding to a measured distance indicated on the object surface. Its value equals 0.4425 mm/pixel. This rather large value is due to size of the domain that had to be chosen in order to be able to follow the complete vertical displacement of the zone of interest. This is explained in Figure 2 where the undeformed state and the maximum deflection of the beam is shown and where it is obvious why this large scaling factor was obtained.

The CCD camera is positioned at the side of the concrete beam, perpendicular to its longitudinal axis and is focused on the bottom flange. The focal length is approximately 200 cm, in order to obtain the analysis of a domain with dimensions 13.5 × 9.8 cm during the whole vertical displacement path due to the flexural loading procedure. The specified zone was chosen in order to be able to follow an existing crack and in addition to increase the possibility of detecting the formation of a new crack. A calculation mask (Fig. 2) is therefore applied to the images taken.

Pictures are taken every 20 kN until 200 kN. From there on, image acquisition is performed every 10 kN until a load of 295 kN (maximum load per hydraulic jack). The correlation is done by choosing the first and undeformed image as the reference image for the matching of all the following images. All the results have to be seen with respect to the reference image.

In the present study a facet size of 13 × 13 pixels is chosen with an overlapping between the facets of 6 pixels to perform the image correlation. It was impossible to choose smaller facet sizes, since the matching of all of the picked startpoints between the stored images became impossible. When facets become too small, the correlation algorithm has problems finding corresponding ones. This can be explained by the fact that ascending facet sizes are related to an ascending uniqueness of the facets.

3.2 *2D versus 3D*

The option offered using two cameras, is the determination of the out of plane movements. Some reasons why the 3D measurement option is not preferred over the single camera technique are listed below.

It was considered to place the camera's underneath the tested beam in order to visualise a larger region of interest without having to take into account the vertical displacement of the beam, since this displacement can be discriminated by the software. However, two problems arise when this option is chosen. The first one is the presence of dust and the possible projections of concrete due to the loading of the beam. Both actions are detrimental with respect to the functioning of the camera's and the lenses. The second one is of a different order. It is obviously possible to measure out of plane displacements, however one should take into account that this kind of displacement affects the initial focussing. This measurement technique is thus advised, only if the displacements are within a certain region and do not affect the original sharpness of the images. Knowing that the vertical displacement of the centre of the beam is estimated at ± 40 cm, this could yield some correlation problems.

Another argument for omitting the use of two cameras is the essential and not always straightforward calibration procedure. In the present testing conditions, one could find difficulty in calibrating the cameras in a successful manner. For a more detailed description of the calibration procedure, the ARAMIS manual is referred to.

The final reason to use 2D instead of 3D measurements is that the vertical displacement of the beam practically takes place perpendicular to the optical axis. There will virtually be no matter of out of plane translations.

The enumerated arguments lead to the choice of using only one camera to picture the deformation and displacement of the investigated part of the concrete beam.

4   MEASUREMENT ACCURACY

Because many factors of both hardware and software affect the reliability and accuracy of the measured displacement and strain maps, there is a need to assess the quality of the measurements; the errors need to be quantified. The problem encountered while using the camera set-up in an unusual environment is that the incoming light can not be controlled as in laboratory conditions and the possible vibrations of the measuring platform are not known. The simplest procedure to assess the global error (no discrimination between the factors of influence) on the measurements in this circumstance, is to compare two different images taken at the same moment in the undeformed state of the specimen. These results will show the total existing minimum noise for the different measurement outputs. A more general approach of the error estimation is suggested by Smith et al. (1998) and is applied by Geers & Schimek (1999) to the ARAMIS Software.

In order to obtain a statistical distribution of the errors, a rectangle covering the largest part of the image (the borders were left out) is selected in which the mean value and the standard deviation is calculated.

One can notice that the obtained values are relatively important (Table 1). This is probably due to the higher value of the spatial resolution (0.4425 mm/pixel). Since the estimated accuracy of the correlation technique lies between 0.1 and 0.01 pixel, the results

Table 1.   Statistical distribution of errors.

	Min (mm)	Max (mm)	Mean (mm)	Deviation %
x-displacement	−0.022	0.046	0.011	0.009
y-displacement	−0.043	0.044	0.002	0.007
Major strain (%)	−0.25	0.59	0.12	0.12
Minor strain (%)	−0.58	0.28	−0.12	0.12

become reasonable. They have nevertheless to be taken into account for the evaluation of further results. However, this noise does not prevent typical patterns, related to certain deformations, to appear in the graphical results.

5   EXPERIMENTAL RESULTS

5.1   Evaluation of y-displacement

When evaluating the results with respect to the displacements of the zone of interest projected onto the y-axis, it can be seen that a uniform vertical displacement doesn't take place. Other than the rigid body movement, a rotation of the zone can be noticed as well. The only indication of existing cracks is revealed by a certain disturbance of the displacement pattern. An explication can be found in the fact that minor deformations within the selected zone are completely overpowered by the large vertical displacements of the beam, and furthermore because of the direction in which the crack opens during deformation (principally along the x-axis)

5.2   Evaluation of x-displacement

It is expected for the displacements projected onto the x-axis, to reveal more information concerning the influence of the cracks on the measured values of displacement. The existing crack is responsible for the different displacement values of the left and right-hand side of that same crack.

A possible procedure for evaluating the evolution of the crack opening, is to consider the x-displacements of two different points, situated near and at both sides of the crack. This is shown in figure 3, where a difference in x-displacement between the two points, having the same y-co-ordinate, is noticed. The shape of the curve, which shows a change in direction of the x-displacement at load-step 12, can be explained knowing the following: at the given load step, the supports of the beam were alternated, in favour of the centring of the jacks. Where in the beginning of the loading, the rolling support is situated at the left end of the beam and the fixed one at the right end, they are switched at load-step 12. It demonstrates, however, that the

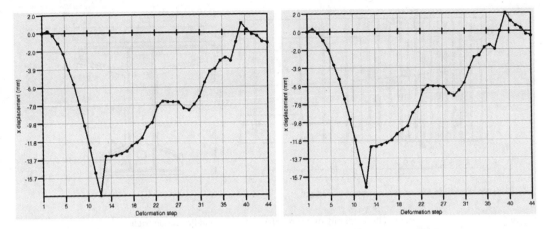

Figure 3.   x-displacement of point left (left) and right (right) of crack during loading and unloading.

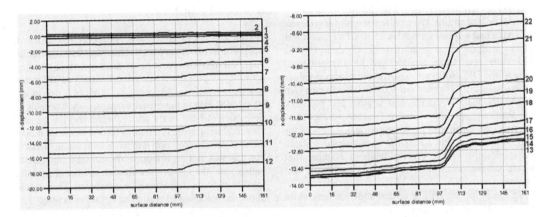

Figure 4.   x-displacement in a given section through deformation states 1–12 (left) and 13–22 (right).

displacements within the studied zone are corrupted with the displacement values of the rather important rigid body movement. This complicates the exploitation of the obtained results.

Another possibility of clarifying a jump between points at both sides of a crack, is to define a section parallel to the x-axis at a certain deformation step or throughout all the deformation steps. This principle is made clear in Figure 4. The jump, made by the line segments at a certain distance from the border of the selected zone, indicates a difference in x-displacements. This difference can be used to calculate the widening or narrowing of the crack, given a certain reference state. It means that for a new fissure the opening at any deformation state can be calculated but for an existing crack only the changes in opening can be calculated. It is possible to consider different sections at different places within the zone of interest.

5.3   Evaluation of major and minor strain

The calculated strain values, based on the determined displacements should yield the most interesting information about the behaviour of the selected zone. The strains to be analysed are obviously the major and minor strains with their corresponding directions. Due to its rotation, the piece is no longer aligned with the original co-ordinate system, hence there is no longer interest in evaluating ε_x or ε_y. Not only the values are important, also the related directions of the strain have to be looked at. In fact, they have to be evaluated simultaneously to draw any conclusions from their values.

When looking at the overall pattern of the major and minor strain one can distinct two different directions along which the corresponding strain values increase (or decrease). Those directions coincide very well with the corresponding principal directions.

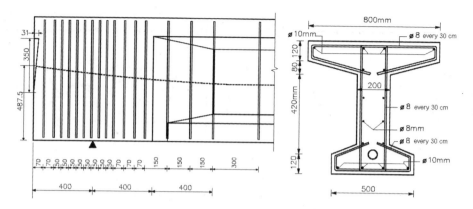

Figure 5.   Reinforcement present in the extremity of beam (left), and in the middle section (right).

First of all it must be clear that the measured strains can not be real strains. If a tensile strength for concrete of 3 MPa and a Youngs modulus of 30 GPa is accepted, cracks will appear at a strain value of about 0.01%. This value is ten times smaller than the obtained accuracy of the DIC-Software. The measured strain values are significantly higher than 0.01%. This means that ARAMIS doesn't provide real strains, but probably measures the opening of micro fissures, formed during previous loading steps. An additional argument for this deduction is that ARAMIS also calculates strain values for the visible cracks, while it is clear that where cracks occur it is impossible to obtain real strains.

In the following will be talked about strains while referring to micro fissures.

To explain the obtained results relative to the principal strains, one has to study the existing reinforcement present in the beam. Figure 5 shows the location of the armouring. It is possible that the post-tensed cable has an influence on the strain distribution in the considered region, where it is located at the upper part and presents a small inclination, but this cable is positioned in the center (horizontally) of the beam section. Therefore, reasonable doubts exist with regard to its influence on the surface strains.

However, not only the post-tensed cable may have a certain influence, the longitudinal and transversal reinforcements certainly have their effect on the selected zone as well. Especially the longitudinal reinforcement of which is known that it's positioned at the bottom of the lower flange and with the rods at the extremity (left and right) of the section, located at ± 2 cm of the vertical surface (Fig. 5 right). A feasible explanation for the lower strain values at the bottom part of the selected zone is that the reinforcement bars partially prevent the existing micro fissures to open more.

6   CONCLUSIONS

A first and important remark is that the resolution of the pictures is not very high. This has a negative influence on the deformation measurements and their accuracy, especially when the deformations are rather low, as it is the case for concrete. It could however not be avoided, due to the experimental set-up, where large deflections of the beam are enforced and the camera view had to be wide in order to follow its entire movement. A possible solution is to work with smaller beams, but this was not an option in the present study or to consolidate the camera set-up with the beam in order to eliminate its rigid body movements. This would allow the selection of a smaller region of interest and increase the obtained resolution.

The vertical displacement of the considered region can, without any problem be measured. It was however not the objective of this study. The related results didn't present much information on the cracks either. As expected, the opening of the cracks was more marked in the direction of the x-axis. Small rigid body movements, like the encountered rotation of the inspected zone, can be detected easily.

Information about the cracks, and their widening and narrowing can be deduced from the results on the x-displacements. As shown before, it is possible to calculate the opening of a new crack or the evolution of an existing one.

The strain measurements allow the detection of cracks at an early stage. Small changes in the strain pattern indicate the beginning of a crack, which is at that moment not necessarily visible with the naked eye.

However, an important reflection is that what is seen and measured by the cameras, only indicates what happens on the surface of the concrete beam. It is probable that cracks are initiated within the concrete section and only show up at the surface after having followed a certain crack-path. It is clear that based on the results obtained by ARAMIS this crack-path can not be detected.

REFERENCES

Aramis Users Manual 2000. Gom (GOM – Gesellschaft für Optische Meßtechnik mbH), Gom mbH.

Avril, S., Vautrin, A., Hamelin P. & Surrel, Y. 2001. Caractérisation par une méthode de grille de la fissuration de poutres en béton armé réparées par matériaux composites. *Colloque sur la photomecanique 2001*: 175–182.

Bergmann, D. & Ritter, R. 1996. 3D deformation measurement in small areas based on grating method and photogrammetry. In C. Gorecki (Hrsg.), *Optical Inspection and Micromeasurements Proc. SPIE* 2782: 212–223.

Bulhak, J. & Surrel, Y. 2001. Mesure de déplacements et de déformations: quelle résolution spatiale? *Colloque sur la photomecanique 2001*: 1–8.

De Roover, C., Vantomme, J., Wastiels, J., & Croes, K. 2001. DIC for deformation assesment: A case study. Royal Military Academy, Belgium

Doumalin, P., Bornert, M., & Caldemaison, D. 1999. Micro-extensometry by Image Correlation Applied to Micromechanical Studies Using the Scanning Electron Microscopy. In *International Conference on advanced technology in experimental mechanics 1999* I: 81–86.

Geers, M. & Schimek, E. 1999. Experimentele analyse van schade en breuk in beton door middel van contactloze veldmetingen. Royal Military Academy, Belgium.

Smith, B.W., Li, X. & Tong, W. 1998. Error assessment for strain mapping by digital image correlation. *Experimental techniques* 22: 19–22.

Tillie, L.W.M. 1996. Verplaatsingsmetingen m.b.v. beeld-corre- latietechnieken. Faculteit der werktuigkunde, TU Eindhoven,WFW rapportnummer 96.153.

Vendroux, G. & Knauss, W.G. 1998 Submicron deformation field measurements: Part 2. Improved digital image correlation, *Experimental mechanics* 38 (2): 86–92.

Emerging Technologies in Non Destructive Testing, Van Hemelrijck, Anastasopoulos & Melanitis (eds)
© 2004 Swets & Zeitlinger, Lisse, ISBN 90 5809 645 9

A combined experimental–numerical study of tensile behaviour of limestone

K. De Proft & W.P. De Wilde
Vrije Universiteit Brussel, Department Mechanics of Materials and Constructions,
Brussel, Belgium

G.N. Wells & L.J. Sluys
Delft University of Technology, Faculty of Civil Engineering and Geosciences, Delft,
The Netherlands

ABSTRACT: In this paper, a combined experimental-computational study of double-edge notched stone specimens subjected to tensile loading is presented. In the experimental part, the load-deformation response and the local displacement field are recorded. The evolution of the crack path, during loading, is followed with an Electronic Speckle Pattern Interferometer (ESPI). This device allows monitoring of the structure, without contact. Both experimental results are used to validate the numerical model for the description of fracture. The model uses displacement discontinuities to model cracks, which are implemented using the partition of unity property of finite element shape functions. In the discontinuity, a plasticity based cohesive zone model is used. Numerical simulations are compared with experimental observations.

1 INTRODUCTION

Currently, a variety of computational techniques exist to describe fracture of quasi-brittle materials. These numerical models must be able to simulate the behaviour of brittle materials for different loading conditions. Therefore, experimental data are very important.

Firstly, experimental data is needed in order to determine whether the proposed numerical models are capable of simulating the observed behaviour. Secondly, experimental data are also necessary to obtain a certain number of model parameters included in the numerical model. Conversely, numerical models can also be used to improve the experimental design.

Obviously, the link between experiments and computational tools is extremely important. In this paper, a combined experimental-computational study of double-edge notched (DEN) stone specimens, subjected to tensile loading is presented. In the first section, the experimental set-up is presented followed by a discussion on the experimental results in the second section. Both global, *i.e.* load-deformation response, as well as local, *i.e.* displacement field around the crack tip, information is recorded. Then, the cohesive zone model based on partition of unity is used to model the DEN tests. A plasticity based cohesive

zone law is adopted. In the final section, numerical results are compared with experimental values.

2 EXPERIMENTAL SET-UP

For the experiments, a natural stone called 'Massangis' is used. All specimens are 120 mm high and 50 mm wide. The thickness is 11 mm. Notches of 7 mm deep and 1 mm wide are sawn in the middle of both sides of the specimen. The geometry of the specimen is shown in figure 1. Two Linear Variable Differential Transducers (LVDT) are used for the measurement of the deformation. The LVDT's are placed over the notches on each side of the specimen, as indicated in figure 1. The vertical measuring range of the LVDT's is 20 mm. Due to the notches, the crack will be located within the range of the LVDT's. When a macro crack starts to grow, the deformations tend to localize in the cracked area. Other parts of the specimen will unload. When the crack is not in the range of the LVDT's or when the measuring range of the LVDT's is too large, a snap back will occur making the measurement of the post peak behaviour impossible. In the other case, when the crack is situated in the range of the LVDT's, the measured deformation increases gradually. The average signal of the LVDT's can then be used as the control signal for the test. The average signal is also used in the load-deformation curves.

The experiments are performed with an INSTRON 4505 testing bench. The specimens are directly glued to the loading platens, so that the boundaries of the specimen cannot rotate.

Tensile loading is applied by a uniform vertical displacement of the boundary. All tests were performed under displacement control at a rate of 0.3 µm/s.

An ESPI (Electronic Speckle Pattern Interferometer) device is used to record the local displacement field at different load steps. The specimen is illuminated by laser light and speckles appear on the lighted surface. A CCD camera captures the reflected light. The observed speckle pattern includes information about the deformation of the specimen. By subtracting different speckle patterns, interference fringes are formed. These fringes contain information about the displacement of the studied specimen. Unlike strain gauges, there is no contact with the studied specimen and the strain field, which can be computed with the software, of a section of the specimen can be studied. A user-defined border restricts the measuring area of the ESPI. Within this border, a reference point is defined. This reference point is assumed not to move and the displacements of all material points situated inside the border are referred to the reference point. In order to compare with numerical results, five paths are defined along which the displacements are monitored at several load steps. The different load paths and the position of the reference point are shown in figure 2. Subtracting the displacements along path 1 from the displacements along path 2 results in the deformation between those two paths.

3 EXPERIMENTAL RESULTS

A typical load-deformation curve is shown in figure 3. The behaviour is nearly linear elastic until the peak load.

After the peak load, a sharp drop is observed indicating a very brittle response of the material.

In this stage, deformation is localized in a single macro crack and the behaviour is highly non-linear. The black dots represent load levels where snapshots of the displacement field are taken. The deformation profiles, obtained by subtracting path 1 from path 2, are shown in figure 4.

For load level I, deformations are more or less uniform. For all other load levels, the obtained deformations are highly non-uniform indicating crack growth from the left to the right notch. The location of the crack tip during loading can be followed with the

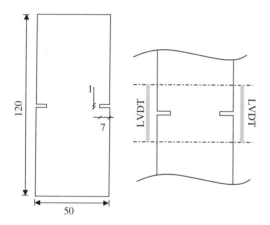

Figure 1. Geometry of the specimen (all dimensions in mm) and placement of LVDT.

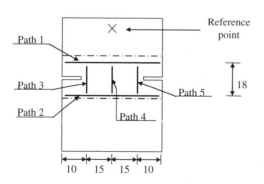

Figure 2. Position for reference point and paths for ESPI measurement.

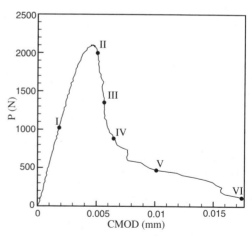

Figure 3. Representative load-deformation curve for a monotonic loaded specimen.

296

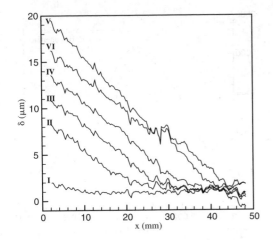

Figure 4. Deformations measured between path 1 and path 2.

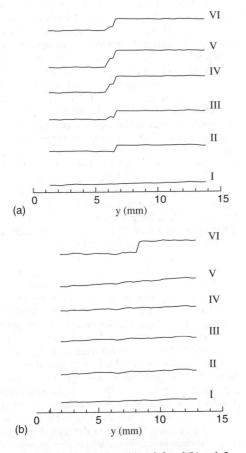

Figure 5. Displacements along (a) path 3 and (b) path 5.

deformations. Near load level VI, the crack has crossed the specimen. Figure 5a and figure 5b give the displacements along path 3 and path 5 respectively. As can be seen, at load level II, the crack started at the left notch resulting in a jump in the displacement field. At the right side of the specimen, the displacements are still continuous. The crack has reached the right side at load level VI.

4 NUMERICAL MODEL

The experimental results have been compared with numerical simulations. Cracks are modelled as displacement discontinuities and are implemented within the finite element context using the partition of unity property of the finite element shape functions. When a crack crosses a finite element, nodes are locally enhanced by additional degrees of freedom with the Heaviside step function as an enhancement basis (Wells & Sluys, 2001, Moës et al., 1999). The displacement field, for a body crossed by m non-intersecting discontinuities, is obtained by:

$$\mathbf{u} = \mathbf{Na} + \sum_{i=1}^{m} H_{\Gamma_i} \mathbf{Nb}_i \qquad (1)$$

where $\mathbf{N}$ are the finite element shape functions, $H_{\Gamma i}$ is the Heaviside step function centred on the discontinuity, $\mathbf{a}$ are the regular degrees of freedom and $\mathbf{b}$ are the enhanced degrees of freedom. The governing finite element equations can be obtained as:

$$\begin{bmatrix} \mathbf{K}_{aa} & \mathbf{K}_{ab_1} & \cdots & \mathbf{K}_{ab_m} \\ \mathbf{K}_{b_1a} & \mathbf{K}_{b_1b_1} & \cdots & \mathbf{K}_{b_1b_m} \\ \vdots & \vdots & \vdots & \vdots \\ \mathbf{K}_{b_ma} & \mathbf{K}_{b_mb_1} & \cdots & \mathbf{K}_{b_mb_m} \end{bmatrix} \begin{Bmatrix} d\mathbf{a} \\ d\mathbf{b}_1 \\ \vdots \\ d\mathbf{b}_m \end{Bmatrix} = \begin{Bmatrix} \mathbf{f}_a^{ext,t+\Delta t} \\ 0 \\ \vdots \\ 0 \end{Bmatrix} - \begin{Bmatrix} \mathbf{f}_a^{int,t} \\ \mathbf{f}_{b_1}^{int,t} \\ \vdots \\ \mathbf{f}_{b_m}^{int,t} \end{Bmatrix} \qquad (2)$$

where

$$\mathbf{K}_{aa} = \int_{\Omega} \mathbf{B}^T \mathbf{C}^e \mathbf{B} \, d\Omega$$

$$\mathbf{K}_{ab_j} = \int_{\Omega} H_{\Gamma_j} \mathbf{B}^T \mathbf{C}^e \mathbf{B} \, d\Omega$$

$$\mathbf{K}_{ab_j} = \int_{\Omega} H_{\Gamma_j} \mathbf{B}^T \mathbf{C}^e \mathbf{B} \, d\Omega \qquad (3)$$

$$\mathbf{K}_{b_jb_j} = \int_{\Omega} H_{\Gamma_i} H_{\Gamma_j} \mathbf{B}^T \mathbf{C}^e \mathbf{B} \, d\Omega$$

$$\mathbf{K}_{b_jb_j} = \int_{\Omega} H_{\Gamma_j} \mathbf{B}^T \mathbf{C}^e \mathbf{B} \, d\Omega + \int_{\Gamma_j} \mathbf{N}^T \mathbf{D} \mathbf{N} d\Gamma_j$$

where $\mathbf{C}^e$ is the continuum elastic material tensor, $\mathbf{D}$ is the material tangent for the discontinuity. It is assumed that the considered element is crossed by discontinuity j and influenced by discontinuity i. Detailed information can be found in De Proft (2003).

The continuum is assumed to remain elastic, while a plasticity based cohesive zone model describes the behaviour at the discontinuity. The adopted plasticity model was proposed by Carol et al. (1997) and uses a hyperbolic yield function:

$$f = T_t^2 - (c - T_n \tan \phi)^2 + (c - f_t \tan \phi)^2 \qquad (4)$$

where $\mathbf{T} = \{T_n, T_t\}$ are the normal and tangential component of the traction vector, c is the cohesion, f_t is the tensile strength and ϕ is the internal friction angle of the material. For tension, an associative flow rule is adopted. The cohesion and the tensile strength are a function of the energy dissipated during fracture. For tension, the energy dissipated during fracture is defined as:

$$dW_{cr} = T_n d\Delta_n^{pl} + T_t d\Delta_t^{pl} \qquad (5a)$$

$$W_{cr} = \int_0^t dW_{cr} \qquad (5b)$$

where $\Delta^{pl} = \{\Delta_n^{pl}, \Delta_t^{pl}\}$ is the normal and tangential component of the plastic separation.

With equation (5), the tensile strength and cohesion change according to:

$$f_t = f_{t0}\left(1 - \frac{W_{cr}}{G_{fI}}\right) \qquad (6a)$$

$$c = c_0\left(1 - \frac{W_{cr}}{G_{fII}}\right) \qquad (6b)$$

with c_0 and f_{t0} the initial value of cohesion and tensile strength, G_{fI} the mode-I fracture energy and G_{fII} the mode-II fracture energy.

The tangential stiffness and the stress update is obtained according to classical elasto-plasticity.

5 NUMERICAL SIMULATIONS VS EXPERIMENTAL RESULTS

In this section, the numerical model (cohesive zone model with a plasticity based cohesive law) is used to simulate the DEN tensile tests. The deformation profiles, shown in figure 4, clearly showed that non-symmetric crack growth occurred during the tests. The numerical model, presented in the previous section, should be enriched in order to capture this

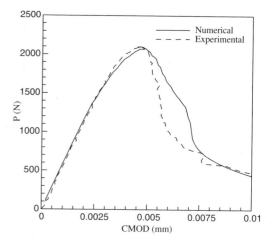

Figure 6. Load-deformation curve for non-symmetric crack growth due to a weaker element.

non-symmetric crack growth. Two types of enrichment are taken into account: (i) local weakening of the material at one notch or (ii) introduction of a bending component. Both types are further explored.

Figure 6 compares the experimental load-deformation curve with the simulated curve. A weaker region near the left notch is assumed. The adopted model parameters are: $f_t = 6.7$ MPa, $c = 20$ MPa, $\phi = 26.35°$, $G_{fI} = 0.03$ N/mm and $G_{fII} = 0.1$ N/mm. Obviously, the peak load is captured correctly. The post peak behaviour differs since the computed response is more ductile.

Figure 7 shows the deformation profiles, measured between path 1 and path 2, for different load levels. In the beginning, the computed deformation profiles follow the experimental observed curves well. Only for load level IV, the computed and measured deformation profiles differ. Moreover, the computed profile shows a more uniform distribution.

Another way to incorporate non-symmetric crack growth is by introducing a bending component in the set-up. In this case, the load is applied with a small eccentricity. The adopted model parameters are: $f_t = 7.3$ MPa, $c = 20$ MPa, $\phi = 26.35°$, $G_{fI} = 0.035$ N/mm and $G_{fII} = 0.1$ N/mm. The obtained load-deformation curve and the deformation profiles are shown in figure 8 and figure 9, respectively. Examining figure 8, it can be seen that the peak load is reproduced very well. Again the post peak shows less agreement with the experimental observation. The computed deformation profiles show a very good agreement, as indicated in figure 9. The non-symmetry is captured in all load levels. Again, after the drop in the load-deformation curve, the computed deformation profiles become more uniform as is shown in figure 10.

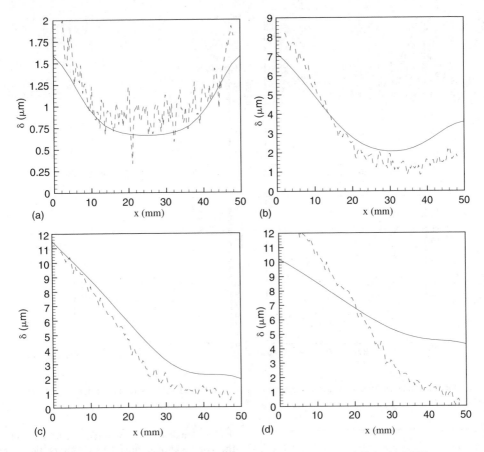

Figure 7.   Deformation profiles at (a) load level I, (b) load level II, (c) load level III and (d) load level IV.

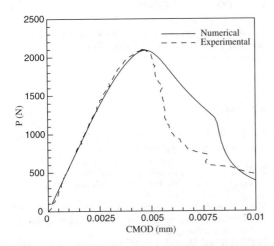

Figure 8.   Load-deformation curve for DEN test with bending.

6   CONCLUSIONS

In this paper, a combined experimental-computational study of tensile behaviour of limestone is presented. During the experiment, both global and local measurements were performed. Globally, the load-deformation curve was recorded. Locally, the displacement field around the crack tip was measured. It was shown that the use of the ESPI technique added important information. Measurements showed that the obtained deformations are non-symmetric. Moreover, the ESPI is very useful since there is no contact with the specimen, and consequently, the measurement is not interfered.

For the numerical simulations, the cohesive zone model based on partitions of unity was used in combination with a plasticity based model. In order to correctly describe the experiments, the model must be enhanced. Therefore, a weaker region or an additional bending component was added. It was shown that

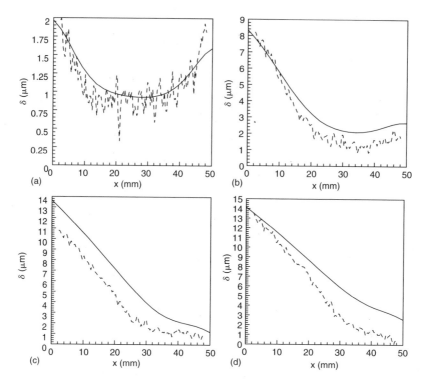

Figure 9. Deformation profiles obtained with eccentric loading for (a) load level I, (b) load level II, (c) load level III and (d) load level IV.

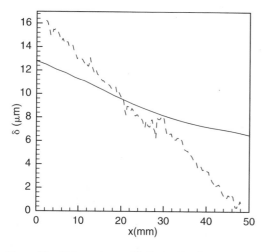

Figure 10. Deformation profile for load level V.

adding bending resulted in the best agreement with the experiments.

Obviously, the results show that the comparison of numerical simulations with experimental data should be carried out with great care. A fit of the simulations to the global data is not sufficient to conclude that the computational model can capture the material behaviour.

ACKNOWLEDGEMENTS

Financial support from the FWO-vlaanderen (Fonds voor Wetenschapelijk Onderzoek, Fund for scientific research - Flanders) is gratefully acknowledged.

REFERENCES

Carol, I., Prat, P.C. and Lopez, C.M. 1997. Normal/shear cracking model: application to discrete crack analysis. *Journal of Engineering Mechanics*, **123**, 765–773.

De Proft, K. 2003. A combined experimental-computational study to discrete fracture of brittle materials. Phd thesis, Vrije Universiteit Brussel.

Moes, N., Dolbow, J. and Belytschko, T. 1999. A finite element method of crack growth without remeshing. *Int. J. Num. Meth. Eng.*, **46**, 131–150.

Wells, G.N. and Sluys, L.J. 2001. A new method for modelling cohesive cracks using finite elements. *Int. J. Num. Meth. Eng.*, **50**(12), 2667–2682.

Emerging Technologies in Non Destructive Testing, Van Hemelrijck, Anastasopoulos & Melanitis (eds)
© *2004 Swets & Zeitlinger, Lisse, ISBN 90 5809 645 9*

Non-destructive inspection method for evaluating conditions of reinforced concrete structures after extreme influence

G. Muravin, L. Lezvinsky & B. Muravin
Margan Physical Diagnostics Ltd., Israel

ABSTRACT: The article describes complex of non-destructive test methods that the authors have designed, developed and used for inspection of reinforced concrete structures (buildings, bridges, tunnels, metro stations, foundations) that have been affected by extreme influence such as fire, earthquake, explosion, landslide activity, etc. It considers original methods for flaws revealing, identifying their type and assessing their danger level according to design and/or fracture mechanics criteria.

1   INTRODUCTION

Fracturing and failures in reinforced concrete structures affected by extreme conditions have increased significantly during the second part of the twentieth century, in spite of improved construction procedures and the high quality of materials used. The result has been a dangerous expansion in the number of fatal disasters, catastrophes, and their harmful social and economic consequences. Fire alone has led to the loss of more than 1% of national budget of developing countries. Interest in the ability of NDI techniques to reveal flaws in damaged reinforced concrete structures and evaluate their condition is, therefore, obvious.

Although NDI methods such as sclerometric test, UT, radiography, radar and stress wave are in extensive use, the information, they provide is limited and more suitable for qualitative rather than quantitative evaluations of a structure.

Traditional NDI techniques do not allow evaluating precisely:

- The locations of high stress-concentrations zones, micro- and macro-cracks propagation or fracturing of bonds with reinforcement.
- The overall condition of the concrete properties such as compressive, tensile and yield strength, elastic modulus, fracture toughness and its uniform distribution throughout the structure.
- The undamaged parts of the construction and differentiate between them and the damaged parts.
- Changes in the magnitude and orientation of the main stresses in the structure and the real stress distribution through the structure.
- Disconnections between a concrete structure and the rocks.

- The position of leaks and potential leaks developing from damage to the structure by overstressing.
- The decreased load carrying capacity of the structure and the danger level of flaws in fracture mechanics criterion.

Currently, there are no international standards or specific authoritative documents specifically concerned with the non-destructive evaluation of reinforced concrete structures subjected to extreme conditions. As a result, although structural elements may show no outward signs of degradation, designers generally assume that elements may contain future hazards and, therefore, apply conservative assessment of the structures' condition. Consequently, many reinforced concrete structures have been unnecessary demolished.

Existing situation has motivated us to propose and apply in practice a set of necessary and sufficient indications for diagnosing reinforced concrete structures that have been subjected to extreme conditions. This article describes the capabilities, advantages and limitations of our Quantitative Acoustic Emission (QAE) NDI methodology and its great potential for revealing, diagnosing and preventing unexpected failures that can occur in reinforced concrete structures.

2   REVEALING SUSPECTED ZONES, IDENTIFYING FLAWS AND VERIFYING OF DAMAGED AREA LIMITS

2.1

Fire, earthquake, explosion, landslide activity and other extreme conditions usually create stress-concentrations, nucleation and development of micro- and macro-cracks, and fracturing of bonds with reinforcement. Therefore, irrespective of the kind of disaster, it is necessary to locate zones where dynamic changes have occurred in the material's structural integrity.

Complicated processes of interaction between cracks and inclusions, cracks and voids, cracks and reinforcement, different mechanisms of blocking and unblocking micro-cracks, each act of flaws motion, increasing or decreasing speed of flaws, change direction of its propagation lead to irreversible dissipation of energy and to appearance of burst AE pulses that can release accumulated energy for very short time, between 10^{-9} and 5×10^{-6} seconds. The appearance of burst AE was observed during the examination of buildings, bridges, tun-nels, metro stations, foundations, etc. after extreme influence.

The problem does not only concern establishing the fact of AE appearance. It is also necessary to find methods and means for recording and recognizing the AE associated with each specific flaw type and then to establish its danger level. This is complicated and requires segregating different processes, realizing their image recognition according to AE data.

There are several ways to determine the location of flaws. In one of the methods, a high-threshold filtration of AE signals with frequencies above 100 kHz was used. As a result, signals carrying information about changes in the concrete structural integrity formation were filtered out. In addition, a substantially long time was required in order to acquire enough AE signals for proper analysis and conclusions about the development of flaws.

Another way to reveal defect location consists of recording AE with the high sensitivity. This solution is associated with major problems connected with the use of an overly highly sensitive measuring system. The reliability of the traditional AE linear location method decreases significantly as the quantity of recorded signals increases simultaneously at different sensor locations. Currently available AE systems are not designed for this; hence errors of AE source location could be about 30%.

We use the last option in combination with selecting events that have energy and frequency bands specific for flaws development to provide their reliable location.

To this aim, the three dimensional graphs of probability density and ellipses of dispersion "energy-average frequency" of AE signals for all zones suspected of flaw development were calculated and were analyzed. Then located AE sources were classified according to the difference in AE energy, average frequency, coefficient correlation and other informative indications and combined the data into ellipses of dispersion (Muravin, 2000a).

2.2

Let us consider as an example, the analysis of AE in the foundation of building that had been subjected to landslide activity and wave shocks from frequent explosions (Fig. 1). Here were revealed several zones where analysis established the presence of not less than three interacting sources of AE of different natures.

These AE signals were combined into ellipses of dispersion, which were compared with data recorded during the laboratory examination of standard specimens with known defects (AE Fingerprints) of:

– Different stages of concrete deformation, which was in normal condition.

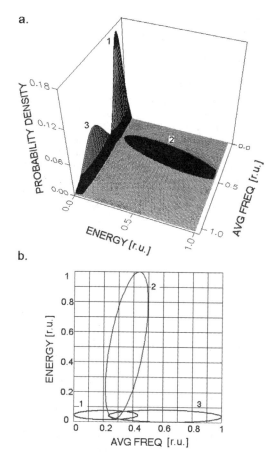

Figure 1.   Three interacting AE sources associated with: 1 – Concrete structural integrity formation. 2 – Fracturing of bond connections between reinforcement and concrete. 3 – Corrosion product development at the cable surface, no micro-cracks in the wires of prestressed cables.

– Different stages of concrete deformation, which had undergone dehydration of the cement stone, alkali-aggregate reaction and other influence that led to concrete degradation and decreasing its fracture toughness.
– Debonding from rebar.
– Electrochemical corrosion development in reinforcement.
– Stress corrosion cracking in reinforcement, etc.

Analysis showed that the AE signals combined in ellipse 1 were observed in all investigated zones along the foundation and made up more than 95% of the total number of recorded AE signals. Similar signals were detected in core specimens taken from foundation. It was, therefore, concluded that the AE signals of group 1 represent noise associated with concrete structural integrity formation.

It was pointed out that signals of group 2 were associated with fracturing of the bond connections between the reinforcement and the concrete, which we observed not once in the laboratory and in field conditions (Muravin, 2000a).

Signals of group 3, which had the highest frequency, were classified as associated with stress corrosion cracking. This was done based on the method described in our patent (Muravin et al. 1990), which enables the recognition and evaluation of the danger level of stress corrosion development in prestressed cables. Current analysis of AE records did not reveal zones where stress corrosion development had lead to the appearance of micro-cracks in the wires of prestressed cables.

Overall inspection of the foundation enabled the location of suspected zones, determination of flaw types and the differentiation between undamaged and damaged parts of the structure. Such operations are common during inspection of variety reinforced concrete structures. The difference consists in identifying flaws. Therefore, due to the wide range of possible failures specifically following fire and/or radiation, our database contained information about changes of the physical-mechanical state of concrete and AE properties associated with:

– The degree of damage that can occur in a concrete due to the evaporation of the water and the breakdown of hydration.
– The recovery rate of concrete structural integrity after radiation and fire and it is acceptable as well as critical deviations from the initial values.

This data-base have been used whenever it was necessary to evaluate reinforced concrete structures after fire or age-related incidents of concrete and reinforced concrete structures degradation that was revealed in different nuclear plants (Muravin et al. 1990).

2.3 Acoustic emission and concrete structural and mechanical properties. Detecting uniformity of concrete strength

In our experience and known statistics, faults in structures may come from using low quality materials or from local overstresses. In both cases, standards require checking the concrete strength at two random points for every square meter and calculating the concrete strength distribution throughout the entire structure. The time involved makes it almost impossible to handle the number of test points on a superstructure by traditional NDI techniques. On the other hand, using the AE method would reduce the inspection time by a factor of about one hundred.

Our laboratory and field tests have shown that continuous evolution of concrete structural integrity, the formation of hydrates, and crushing of crystalline cells result in AE. Parameters of AE associated with benchmarks of micro-cracking, R_c^0; failure load, R_c^b; and crack growth resistance, G; as well as ellipses of dispersion "energy-average frequency" of AE signals are invariable and differing for different types of concrete and external conditions (Fig. 2; Muravin et al. 2000b). The higher values of compression strength correspond to lower values of fracture toughness and to the higher stresses of micro-cracking. Correspondingly, as the strength of the concrete decreases, the stresses of micro-cracking also decrease. Each point on the $E = f(R_c^0, R_c^b, G)$ surface represents a dispersion ellipse of AE characteristics for different kinds of concrete. The dispersion ellipses depicting a narrower dispersion of AE signals flow average energy

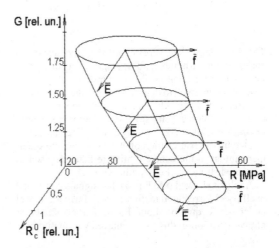

Figure 2. Ellipses of dispersion "energy-average frequency" of AE signals for different types of concrete versus benchmarks of micro-cracking, R_c^0; breaking load, R_c^b; and crack growth resistance, G.

and count rate correspond to concrete specimens with high breaking loads and high stresses of micro-cracking. The specimens with higher strength have higher AE energy and frequency.

This information has been used successfully for inspection of industrial structures, tunnels, bridges after fire and radiation (Muravin et al. 1999, 2000b). When measurement established the presence of AE signals that correspond to the formation of structural integrity of specific concrete (indicative parameters stored in databank) it was assumed that degradation has not influenced the hydration process significantly. The absence of signals indicated irreversible damage to the concrete. The appearance of the AE signals specific to micro- and macro-cracking allowed pin-pointing structural elements where there were local-ized overstressed zones, growing cracks, and extreme deformations.

We were able to record the data about one hundred meters of bridge span per hour and establish zones where the concrete strength differs from average value more than 10%. Statistical sample inspections of con-crete strength distribution using Schmidt hammer tests, UT, fracturing of core specimens (24 per structure), and evaluating results by the Fisher and the Student criteria of uniformity confirmed that the strength of the concrete and the coefficient of variation of con-crete strength were according to design criteria in the zones evaluated as up to standard by the AE inspec-tion. The coefficient of variation of concrete strength was significantly bigger (about 23% as opposed to of 9%) in the suspected zones.

3 FLAW DANGER LEVEL ASSESSMENT

3.1

The AE signals that were recorded in the suspected zones of the structure and segregated according to flaw type were combined into ellipses of dispersion "energy-count rate" to determine the stage of concrete deformation. Then AE data of fracturing processes (Fig. 3, colored ellipse) were compared with laboratory-made "AE Fingerprints". The last were created by the examination of standard concrete speci-men similar to the concrete mixture that was used in examined structure (Fig. 3, ellipses 1–4).

Analysis showed (see Fig. 3) that signals combined in ellipse 5 correspond to micro-crack formation level under stress equal of about 0.3–0.5 Rfract. The error of classification in this case achieved was about 5%.

3.2

The dynamic-acoustic method that we created (Muravin, 2000a) has been frequently and successfully

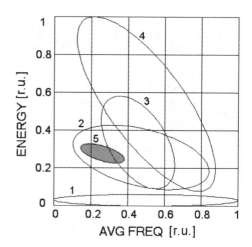

Figure 3. The ellipses of dispersion of the AE pulses "energy-average frequency" used to classify the damage development. 1 – Closing of primary micro cracks and pores. 2 – Micro-crack formation. 3 – Nonlinear creep development. 4 – Main crack development. 5 – Investigated process.

used for stress measuring in reinforced concrete structures. Elements were loaded by shooting dowel using a building pistol. The dowel penetrated into the reinforced concrete structure and created overstresses and micro cracks. The AE parameters were recorded synchronously and then compared with data from laboratory investigations of specimens with similar properties and examined under different pressures by calibrated shock.

By this way it was:

– Determined the current stress level in reinforced elements.
– Revealed the positions of zones where the local val-ues of the stresses exceed maximal design stresses.
– Evaluated the absolute values of non-uniformity of stress redistribution in the pre-stressed elements.
– Established the rupture of the pre-stressed cables, wires, and pre-stressed tendons.

The method achieved an accuracy of eighty five percent while a coefficient of concrete strength uni-formity was equivalent to ten percent.

4 CONCLUSIONS

It was demonstrated the possibilities of our Quantitative AE method to realize:

– The revealing suspected zones, identification of flaws and verify damage area limits.

- The overall examination of reinforced concrete structures properties and uniformity of its distribution through the structure.
- Measuring the real stress distribution through the structure.
- Evaluation of the danger level of flaws in fracture mechanics criterion.

REFERENCES

Muravin, G.B., Lezvinskaya, L.M., Makarova, N.O., Volkov S.I. 1990. Acoustic Emission Method of Determination the Subcritical Cracks in Materials Caused by Corrosion under Stress *A.S. 1744640. Claim N 4877352/28/105773. Claim date 25.10.1990.*

Muravin, G., Lezvinsky, L., Muravin, B. 1999. Evaluating the Condition of Reinforced Concrete Structures after Fire. Concrete Repair, Rehabilitation and Protection. *Proceedings of the International Conference held at the University of Dundee, Dundee, 6–10 September 1999*: 709–722.

Muravin, G. 2000a. *Inspection, Diagnostics and Monitoring of Construction Materials and Structures by the Acoustic Emission Method,* London: Minerva.

Muravin, G., Lezvinsky, L., Muravin, B. 2000b. Evaluation of Reinforced Concrete Structures Degradation, *Workshop Proceedings. Nuclear Energy Agency. Committee on the Safety of Nuclear Installations. Workshop on the Instrumentation and Monitoring of Concrete Structures. Brussels, 22–23 March 2000*: 257–269.

Special techniques

Emerging Technologies in Non Destructive Testing, Van Hemelrijck, Anastasopoulos & Melanitis (eds)
© 2004 Swets & Zeitlinger, Lisse, ISBN 90 5809 645 9

Real-time photoelastic stress analysis – a new dynamic photoelastic method

Michel Honlet
Honlet Optical Systems GmbH, Neu-Ulm, Germany

Geoff C. Calvert
University of Warwick, Great-Britain

Jon R. Lesniak & Bradley R. Boyce
Stress Photonics Inc., Madison (WI), USA

ABSTRACT: The combination of a set of innovations has brought photoelastic stress analysis up-to-date and now offer the opportunity to carry out, in a non-destructive way, *real* time stress monitoring of components or structures. Having an improved sensitivity, this technique allows to monitor changing stress patterns as they happen. Also, much thinner coatings can be used. These can easily and quickly be spread onto the structure and are tinted to allow automatic measurement of coating thickness. The use of cleverly designed optics means that every image acquired by the CCD camera contains all the information necessary to display a calibrated strain pattern, enabling high-speed and live dynamic imaging of stresses. Some examples show the potential to use this non-destructive technique in an industrial environment for different applications.

1 INTRODUCTION

Over the last 10–15 years various automatic photoelastic systems have been developed and used to visualize strain and stress pattern on materials. This optical technique allows to achieve non-destructive tests on real components over a long period of time. The full-field data is important to improve fatigue predictions of finite element calculations models since high strain gradients increase the risk of early failure. Up to now photoelasticity has been used mostly for static or quasi-static applications, both in engineering test laboratories as well as in real industrial environment. But until the advent in 1999 of novel coating techniques for the transparent coating, photoelasticity included a long preparation job, as the rigid plates of the birefringent coating had to be formed manually to match the object surface contour before being glued to it. This extensive work is one of the major reasons why this technique was on its best way to disappear.

Nowadays, using easy to apply photorefractive coatings, it has become a relatively quick and simple task to prepare a complex shaped part and monitor remotely the stress distribution under a variety of loading conditions, including assembly stresses. In the past, such a non-destructive evaluation required either static incremental loading or inherent residual stresses (as in glass). The new generation of digital photoelasticity, called Grey Field Polariscope (GFP) and which also appeared in 1999, introduced a much more sensitive fringe analysis: instead of automatic counting and evaluation of isochromatic lines, the GFP recording and processing method analyzes only grey levels of light in the subfringe range.

It is mainly the need for a non-destructive tool for quality control the glass industry that drove forward the development and capabilities of this new generation, as shown in the examples below. The outcome is a "real" time analysis and for the first time a true *dynamic*, quantitative photoelastic stress analysis has become possible.

2 GETTING STARTED

Basically, each modern photoelastic analysis system works using circular polarized light sources and a CCD camera. The first so-called digital system uses a constantly rotating analyzer through which, during each revolution, a sequence of pictures is recorded, thus allowing a quasi-static analysis.

The greatest advance with the GFP is its ability to accurately measure fractions of a fringe order, so that it is not necessary to produce multiple fringes. Due to

Figure 1. Epoxy based tinted coating is easy to apply. Using color extraction routines the coating thickness can easily be measured for the whole covered area and used for calibration.

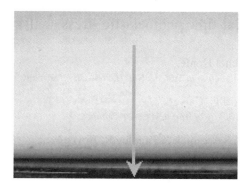

Figure 2. Analysis of a section from a car windshield. The tensile stresses from the central zone turn into a strong gradient of compressive stresses towards the edge. See the graphic drawn through this section.

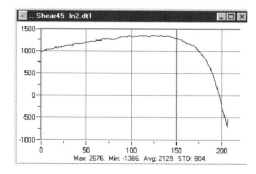

Figure 3. Stress profile along the arrow shown in Figure 1.

this enhanced sensitivity, up to a maximum of ± 0.002 fringe, much thinner coatings can be used to produce sufficient birefringence for successful photo-reflective analysis. This has resulted in simplifying the coating procedure and it is now possible to brush on epoxy resin coatings with a thickness of between 0.1–0.4 mm, taking only a few minutes to prepare and just a few hours to fully cure.

When using tinted coatings the GFP devices can measure directly the coating thickness, which is proportional to the measurement sensitivity and the measured light retardation. The practical advantage is that if there is a need to apply a coating, this can be done using mostly epoxy based fast cure resins, see Figure 1. Using color extraction routines the intensity variation measured by a sensitive color CCD camera indicates the real thickness, with a resolution of a few 1/100 mm.

3 EXAMPLES OF APPLICATION WITH A GFP

3.1 Glass inspection

Today all automobile car and truck windscreens have a black obscuration band around the edge to prevent ultra-violet light from degrading the adhesive. This thin ceramic band is bonded into the glass. It is also very important that all windshields have a significant compressive edge stress to prevent crack generation and propagation.

To measure with conventional polariscopes the edge stresses, this band would have to be removed, ordinarily by abrasion and sometimes scraping the windscreen. This is a destructive process making the screen useless.

Most measurements were made manually using hand held polariscopes and just gave point or small area measurements. The very high sensitivity of the GFP

principle can use the low level of light reflected off the black surround to measure edge stresses, Figure 2 and 3, with a need to destroy the obscuration coating.

Now, and for the first time, an instrument can verify non-destructively and on-line in production the edge stresses of the glass sheets and compare with the tolerance band. Should a mounted glass crack or even break, this problem can basically have three different sources: glass problems that for instance can arise from discontinuities within the manufacturing process, stresses generated during the assembly or, later, operational stresses generated by thermal mismatch, vibrations or mechanical loading, see Figure 4.

Assembly stresses can be measured with an additional birefringent coating on the glass surface, off the production line. Using the same coating technique, the raise and growth of operational stresses can be analysed quantitatively, with the real-time version even on a driving car.

3.2 Development of mechanical components

The same analysis tool can be applied on various ductile or brittle materials, as well as on plastic components,

Figure 4. Assembly stresses around ball joint button for tailgate backlight struts. Over torquing creates high stress concentration. Assembled backlight was coated. The torque was then released so assembly stresses would be revealed.

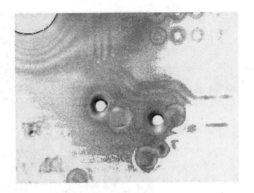

Figure 5. Section (12 mm × 8 mm) from a printed circuit board illuminated from the back. Beside the stress pattern around a large assembly hole (top left), this stress image shows two residual stress pattern around soldered contact points (center). Since they have no stresses, there is no reason for all other contact points to really show up.

to study the influence of different kinds of loading. For metallic or in general non-transparent materials a reflective coating is used, while transparent media with birefringent properties can be analysed directly in transmission, without extra coating. The applied loading generates fringes that can be count by the viewer or using special software, generating strain or stress values. This application and procedure is widely known since decades.

The main difference and progress that happened over the last few years is that the high increase of sensitivity allow thinner coatings to be applied and also measure in the subfringe range, using grey level evaluation software. At the same time, spread-on, tinted coatings were introduced, tolerating inhomogeneous thicknesses and thus a faster preparation work. This allows the engineer to get much faster the results he expects from this measurement technique.

3.3 Electronical components

The high sensitivity allows not only to visualize residual stresses on mechanical components. It is also interesting to observe residual, assembly, thermal or mechanical stresses on electronic components, like around drilled holes or soldering points on printed circuit boards (PCB), see Figure 5.

This information about local strains or stresses help to predict fatigue problems, as they for instance often occur from thermal mismatch. But mounting a PCB into a box or frame can change stress pattern. This can be seen instantaneously. Still, like in any photoelastic measurement setup, an illumination source using circular polarized light is necessary. A portable, polarizing device connected through a flexible lightguide with a light source can, like in the application with the PCB (Figure 5), also be used for

backside illumination. A front side application is visible in Figure 6.

4 REAL-TIME PHOTOELASTICITY

4.1 Polariscope + Kaleidoscope = Poleidoscope

The second generation of the Grey Field Polariscope (GFP) uses a single, specially designed lens containing four individually imaging systems, all with different image polarization. Through a single recording lens these four images are projected on a single CCD chip and simultaneously grabbed. Since these images have no time interval inbetween them, it is now possible to perform dynamic stress analysis or to freeze moving objects, as it has become obsolete to record a sequence of images through a rotating polarizer.

First shown in 2000, the principle of this innovative photoelastic stress analysis device basically still works as a polariscope, while the idea with the multiple lenses reminds a kaleidoscope. Therefore the device is also called a POLEIDOSCOPE. Since the optical feature is integrated into the optical path, the Poleidoscope basically looks like any other standard CCD camera. Depending on the computation power available for on-line image processing, the quantitative and calibrated stress pattern can be displayed at nearly real-time frequencies, close to 20 Hz.

4.2 Real-time measurements

The possibilities of a real-time instrument for photoelasticity allows to measure under a wide range of dynamic experimental conditions. Once calibrated, measurements on moving or rotating objects can be easily achieved, since the software provides on-line

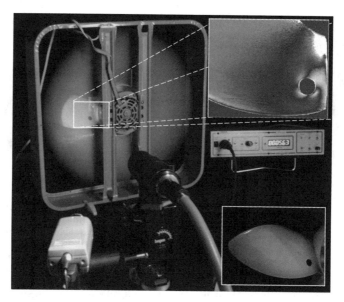

Figure 6.   Real-time stress analysis of a cooling fan at 563 rpm. An illumination head (center) projects circular polarized white light onto a coated fan blade (bottom, right). When using a stroboscopic light triggering the moving object looks steady. The Poleidoscope camera (front, left) measures the stress pattern that modulates the light retardation in the blade coating. Quantitative results can be displayed in different ways, e.g. using pseudo-colours (top, right).

evaluation and display functions, without any need for fringe counting.

When combining Poleidoscope optics with a high-speed camera and recording unit, transient loading histories can be stored first and be evaluated afterwards. Figure 6 shows a simple but state-of-the-art application on a fast rotating object, using a stroboscopic flash source.

4.3   Conclusions

Being able to record the history of strain or stress pattern is an important tool to solve engineering problems. Never before it has been so easy to work with photoelasticity, to create both qualitative and quantitative information, almost in real-time.

The latest generation of photoelastic stress analysis equipment, combined with new coating techniques, enables to work nondestructively under real industrial environment conditions.

The glass suppliers and their customers found particular use as a quality monitoring tool and they now benefit already from this universal, rapid on-line inspection technique.

Nowadays, the combination of rapid prototyping models and measurement techniques like digital photoelasticity or thermoelasticity finally offer the possibility to save development time by rapidly verifying the behaviour of the mechanical component in

operational or extreme loading situations. This information is also important to improve the predictions of finite element calculations models. Finally, as an example for using the latest generation of digital photoelasticity for monitoring quality in production environment, a glass manufacturer has reported recently that having introduced such an on-line inspection method, the rate of product failure has dropped from over 10% down to 0.2%.

REFERENCES

Lesniak, J.R. & Zickel, M.J. & Welch, C.S. & Johanson, D.F. 1997. An Innovative Polariscope for Photoelastic Stress Analysis. *Proceedings of the Spring '97 SEM Conference Seattle WA, USA.*

Zickel, M.J. & Lesniak, J.R. & Tate, D.J. & La Brecque, R. & Harkins, K. 1999. Residual Stress Measurement of Auto Windshields using the Grey Field Polariscope. *Proceedings of the Spring '99 SEM Conference Cincinnati OH, USA.*

Hobbs, J.W. & Greene, R.J. & Patterson, E.A. 2000. A novel Instrument for Dynamic Photoelasticity. *Proceedings of the SEM IX International Congress on Experimental Mechanics Orlando FL, USA, 2000.*

Calvert, G.C. & Lesniak, J.R. & Honlet, M. 2002. Applications of Modern Automated Photoelasticity to Industrial Problems. *INSIGHT Magazine,* Volume 44 (number 4, April 2002).

Automated fluorescent penetrant inspection (FPI) system is triple A

T.L. Adair & M.G. Kindrew
Pratt & Whitney, USA

D.H. Wehener
USAF – ASC/SMD, USA

ABSTRACT: This paper describes how Fluorescent Penetrant Inspection (FPI) was used in the past, how it is used now and how it will be used in the future. The United States Air Force (USAF) and Pratt & Whitney (P&W) now use a fully-automated FPI process with manual visual inspection. With direction and funding provided by the USAF/Aeronautical System Center (ASC), these fully-automated FPI processors, located at Kelly USA and PWSAIR in San Antonio, Texas were developed, manufactured, installed, qualified and put into operation by Pratt & Whitney. This paper will cover the following: the Qualification and Acceptance Tests (which include test objectives), test articles, test facility, test equipment, test results, and training.

This technology is affordable (lower processing cost resulting from increased throughput, reduced maintenance man-hours and reduced material consumption), adaptable (multiple industry applications, flexible layout accommodates simple, complex, large and small geometries), and accurate (improved reliability and detection capability through consistent processing). In addition, the FPI system is environmentally clean. P&W can design, manufacture, and install a customized FPI system for its customers. P&W is continually improving today's technologies and techniques to assure global readiness for the 21st century.

1 BACKGROUND

In the past, and in most situations even today, the USAF and P&W use an entirely manual FPI technique. Manual FPI is an eight step process consisting of surface preparation, penetrant application, pre-wash, emulsifier application, final rinse, drying developer application and inspection.

Manual FPI is a common method of non-destructive testing. However, it is hampered by liabilities such as inconsistent process flow, low throughput, susceptibility to material handling damage, and it is labor intensive.

2 INTRODUCTION

The USAF and P&W now use a fully-automated FPI process. This process is used to inspect most engine parts. Manual FPI is still used for some limited applications such as for areas that need special attention or for parts that the operator does not want penetrant to cover the entire part.

With direction and funding provided by the United States Air Force/Aeronautical Systems Center, these fully-automated FPI processors, located at Kelly USA and PWSAIR in San Antonio, Texas, were designed, manufactured, installed and put into operation by P&W. Destination, qualification and acceptance testing were performed at San Antonio-Air Logistics Center.

Fully automated FPI is essentially the same process as manual FPI except the majority of the process is automated. Automation translates into quality improvements in process consistency, reliability, productivity, and environmental cleanliness. In a word, dependability. For instance, throughput has increased by over 100 percent, as compared to the manual FPI system. In addition, there are cost savings in materials and utilities versus the manual FPI system. Except for the manual placement and removal of parts from the processing line fixtures, the actual FPI preparation is fully-automated. The test facility interface includes the following:

Fresh water:	135 GPM at 35 psi ambient temperature
Compressed air:	300 SCFM at 85 psi
Drain:	80 GPM capacity
Steam:	2000 pounds per hour at 40 psi

Electrical: 480 VAC, three-phase with
1,000 amps maximum current
draw

The following test equipment and materials were required for the qualification and testing of the processors and their associated adapter kits:

- Selected Government Furnished Equipment (GFE) engine parts
- Reliability specimen sets
- Specimen holding fixtures
- Refractometer
- Digital light meter
- Temperature switch
- Pressure gauge
- Pressure transmitter
- Temperature controller
- Long wave ultraviolet (UV) light source
- Thermometer
- Pressure controller
- Biddle Versa-cal
- Sound level meter
- FPI materials: penetrants, emulsifier & developer
- Stop watch
- Tool Aerospace Manufacturing (TAM) panels
- Probability of detection (POD) flaw locator template
- Tape measure
- Recipes for engine parts processed & POD tested
- POD Analysis Software

3 AFFORDABILITY

The best possible inspection criteria for any product that requires inspection is that it be affordable. FPI prevents premature failure, which is extremely costly. Even though the system is somewhat costly to install, the benefits include; reduced manpower, increased consistency, increased reliability, increased throughput and more accurate detection which results in fewer parts being parts with suspected defects. Return of investment can also be achieved through lower chemical consumption, water usage, disposal costs and better utilization of labor skills to operate the system.

4 ADAPTABILITY

There are three separate FPI processors. The processors were constructed using a modular concept installed with different sized station openings and layouts to enhance their adaptability to process a wide variety of part sizes. The processors are also adaptable to available floor space. The system includes a Large Parts Processor (LPP), a Small Parts Processor (SPP) and a Drum Rotor Processor (DRP). Each performs the same basic functions, except that each

is optimized to process parts of a specific size and geometry.

- The LPP, comprised of 15 stations, is designed to handle parts up to 500 pounds and a maximum envelope of 60 inches diameter by 60 inches tall. Its capacity is 12 carriers per hour, excluding inspection time. Each carrier can transport a variety of parts.
- The SPP, comprised of 15 stations is designed to handle parts up to 250 pounds, no larger than 30 inches diameter by 30 inches tall. The SPP can process up to 20 carriers per hour.
- The DRP, comprised of seven processing stations, is designed to handle drum rotors up to 500 pounds. It can process one carrier per hour.

A detailed diagram of arrangement of the Processor stations under the test conveyor loop and final configuration of the conveyor is provided in Figure 1.

5 ACCURACY

Test operators performed a Qualification and Acceptance test to provide step-by-step verification. It demonstrated that the LPP, SPP, and DRP and their associated Adapter Kits conform to functional requirements spelled out in the Configuration Item Development Specification (CIDS), and Software Product Specification (SPS). Also, the ability to operate the FPI processor computers from a software backup was demonstrated using the installation and reload instructions in the Software Users' Manual.

The successful completion of the Qualification and Acceptance Tests demonstrated that the LPP, SPP and DRP and their associated adapter kits are capable of reliably processing the family of F100-PW-229 engine parts. This is in accordance with the applicable sections of the Technical Order 33B-1-1, FPM Master and Supplement, MIL-STD-6866 and FPI Processor System Specification.

The automated FPI system is extremely accurate. The process achieved at least a 90 percent probability of crack detection (POD) at a 50 percent confidence level (CL) for a crack length of 0.040 inch, all tests. These automated FPI processors at SA-ALC are the first systems sold in the world based on POD.

6 ENVIRONMENTAL CLEANLINESS

The system uses 90 percent reclaimed water for the pre-rinse station. This equates to 14,616 gallons of water saved per year. Secondary containment is provided for the penetrate and emulsifier supply drums.

The mist collectors for penetrant spray prevent vapor from escaping to the outside air or into the facility environment. Each station that uses facility

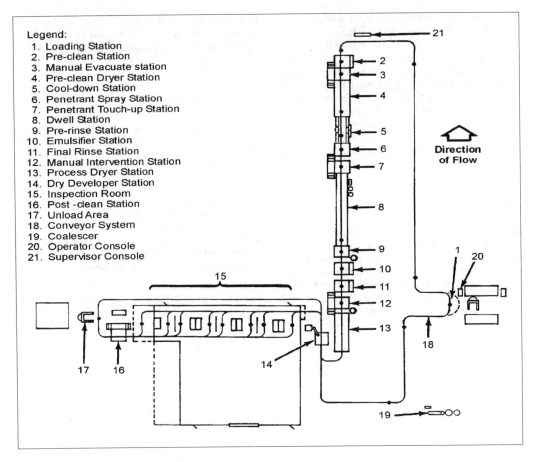

Legend:
1. Loading Station
2. Pre-clean Station
3. Manual Evacuate station
4. Pre-clean Dryer Station
5. Cool-down Station
6. Penetrant Spray Station
7. Penetrant Touch-up Station
8. Dwell Station
9. Pre-rinse Station
10. Emulsifier Station
11. Final Rinse Station
12. Manual Intervention Station
13. Process Dryer Station
14. Dry Developer Station
15. Inspection Room
16. Post -clean Station
17. Unload Area
18. Conveyor System
19. Coalescer
20. Operator Console
21. Supervisor Console

Direction of Flow

Figure 1.   Typical processor adaptable layout under the test conveyor loop.

water has a filter to remove the particulate contamination, a regulator to set the spray pressure and a backflow preventor or a three inch air gap to prevent any backflow into the facility water supply. Also, each station has a filter or regulator located in the incoming air line to regulate the air supply and to remove the moisture and particulate contamination. There are containment pans under all wet stations to contain spills. Computer control of process optimizes chemical usage to minimize generation of hazardous waste. The product can be processed and inspected without being touched by human hands, if necessary. All these features keep the work site safe, clean, and efficient.

7   TRAINING

Systems, automated or manual, require a degree of skill to operate, maintain records, and conduct systems checks that will keep the processor performing to its peak capacity, reliability and capability.

Pratt & Whitney can certify a system and provide training to familiarize operators and personnel with the maintenance requirements and characteristics of the automated FPI system, and related system software. The scope of training may include:

- Providing an overview of the maintenance procedures
- Reviewing system operation and procedures
- Identifying periodic and routine maintenance requirements
- Discussing high level troubleshooting methodology
- Providing system crash recovery procedures
- Providing a high level overview of the commercial and processor-specific software components and their interactions with the system
- Demonstrating the most common activities performed in a production environment
- Providing hands-on training

If there are a number of people requiring training in FPI, P&W can set up an American Society of Nondestructive Testing (ASNT) or International

Standard Organization (ISO) approved schools so that the lowest level of training in FPI can be accomplished. To operate the system, an operator should be at least a Level I inspector. The operator must be able to spot trouble with the FPI process before it becomes a serious problem to the inspectors performing the inspection. P&W can also provide inspector certification testing to check inspector capability. Each inspector is tested and a POD is determined for his or her capability and probability to detect flaws. Training is still one of the most important points when the investment has the potential for the best automated FPI system in the world.

8 SUMMARY

The technology is triple A – affordable, adaptable and accurate. Fully-automated FPI will provide lower processing cost with reduced inspection time, provide more precise results with improved reliability and detection capability, provide system layout flexibility to accommodate simple, complex, large and small part geometries, and improve environmental cleanliness. Just as today's weapon systems will give way to tomorrow's advancements, Pratt & Whitney is continually improving today's inspection technologies and techniques to secure global readiness for the 21st century.

9 CONCLUSION

P&W can: design, procure and install a customized automated FPI system; certify the system and personnel. With installation of an automated FPI system, the customers will enjoy several benefits.

- Lower processing cost with reduced inspection time
- Improved reliability and detection capability
- Enhanced multiple applications
- Refined layout flexibility
- Augmented accommodation of simple, complex, large and small part geometries
- Improved environmental cleanliness

NOTE: Automated FPI was one of the exhibits featured by Pratt & Whitney in its award-winning "Best Medium Commercial Booths" at AUTOTESTCON '97, AUTOTESTCON '00 and AUTOTESTCON '02.

Emerging Technologies in Non Destructive Testing, Van Hemelrijck, Anastasopoulos & Melanitis (eds)
© 2004 Swets & Zeitlinger, Lisse, ISBN 90 5809 645 9

Detection of defect applying conductivity measured by Eddy current

A. Hammouda, M. Zergoug, A. Haddad & A. Benchaala
Laboratoire d'Électronique et d'Électrotechnique, Centre de Recherche Scientifique et Technique en Soudage et Contrôle, Route de Dely Ibrahim, Algeria

ABSTRACT: Eddy current technique has got lots of advantages for resolving complexes problems in different fields. The knowledge of the electric conductivity allows characterising metallurgical states of material simply and quickly. For this purpose, a physical approach has been developed to detect different defects by measuring electric conductivity using the impedance diagram. Different experiments were realised to find correlations between conductivity measured and anomalies happened in the material's structure.

1 INTRODUCTION

Eddy currents are used in material's control and non-destructive evaluation. It allow the measurement of certain properties, dimension and detection of anomalies. Also, this process finds applications in metallurgical field of steels treatment where it can evaluate quickly damage, transformations happened during manufacturing process or by mechanical effects. Eddy current equipments are especially sensitive to three parameters (electric conductivity, magnetic permeability, and geometry of the sample). In this fact the interesting metallurgical variables have to be referred with the change of these parameters.

In this purpose, it is interesting to use electric conductivity for materials characterization and recognition of defects. These parameters can give important information about the structure's state.

It is necessary to stress that the measurement of the electric conductivity with good precision requires precision equipments, a suitable approach, and an optimisation of probe. Indeed the sensitivity to defects and the precision of detection depend on the design of this probe and the precision of the measurement. The impedance diagram informs us about variations which are produced in a material submitted to modification.

The measure of active and reactive part of eddy current sensor for an identify sample at different frequency enables us to obtain an impedance diagram and the electric conductivity thereafter, by measuring the impedance of a probe on presence of an unknown sample.

2 CONSTRUCTION OF PROBE

Each control requires a suitable probe. For that, the construction of probes for eddy current control require some conditions in order to increase their sensitivities and their resolution.

The study of the fields created in the vicinity of a probe has shown that:

– For small probes diameter, the range of field variation is small with an important strength and the lift-off affects the field by a strong attenuation.
– For probe with large diameter, the range field is large with approximately a constant intensity of the strength.

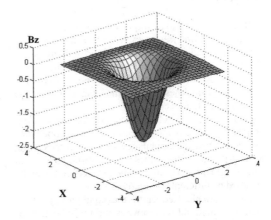

Figure 1. The induction B according to the plan of the probe.

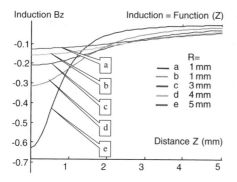

Figure 2. Induction Bz at a distance Z for different diameter.

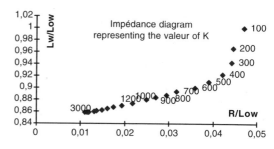

Figure 3. Impedance diagram representing the parameter K.

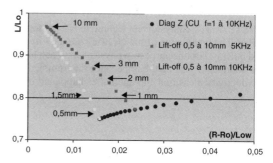

Figure 4. Impedance diagram and lift-off curve.

The study confirms that for measuring the thickness or properties of thin coatings on a conductive base, probes with small-diameter are suitable.

But for measuring conductivity, using probes with large-diameter allow reducing the lift-off effect.

3 THE MEASUREMENT OF THE ÉLECTRIC CONDUCTIVITY

The principle of the measurement is based on the impedance variation with electrical conductivity at a fixed frequency with a constant lift-off (or fill-factor).

The value of the electric conductivity of a material depends on several factors. When using eddy current to measure conductivity it is important, to ensure that these factors are kept under control, along with other factors such as the geometry, the magnetic permeability and the probe lift-off.

Active and reactive impendence parts are function of the parameter K, which is very important to measure this electric conductivity.

$$K = \sqrt{\mu_0 . \mu_r . \omega . \sigma}$$

It is evaluated into an impedance diagram obtained by variation of the frequency from 100 Hz to 100 kHz for an identified sample (paramagnetic or diamagnetic to neglect the influence of the relative magnetic permeability), which has got a known constant conductivity and a thickness three times superior than the standard constant deep.

Since conductivity variation or frequency variation has got the same trajectory in the impedance diagram. Then the impedance diagram representing the distribution of the values of K according to the real and imaginary parts of the impedance coil can be obtained by different conductivities.

At a variable frequency, the conductivity of an identified material will move along the trajectory of the normalised impedance curve.

With simulation, the impedance diagram of different conductivity at certain frequency can be represented and then the conductivity for the sample under test can be evaluated.

3.1 *The lift-off effect*

The lift-off has got an important influence on the measure of the electric conductivity. It causes the displacement of the point position representing the impedance measured of the unknown sample in the impedance diagram. This impedance's point is displacing in a linear curve and converging to the point (0,1) as it is shown in this figure representing an impedance diagram of a cooper sample with different point measured at different lift-off for two frequencies (5 kHz and 10 kHz).

To minimise the lift-off effect, the impedance measured for an unknown sample is projected to the impedance diagram using this linear curve. The intersection of the impedance diagram and the curve representing the lift-off has got two advantages; firstly, it will give the value of K equivalent for this sample and then the value of the electrical conductivity of the sample under test.

Secondly, the value of the lift-off can be measured and then thickness of metallic coating or non-conducting coating can be measured.

The choice of the frequency depends on several conditions, like the thickness of the sample under test, the range of K obtained by the impedance diagram, the

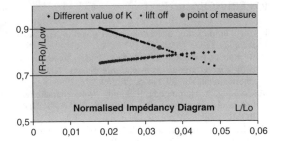

Figure 5. Representation of the measured point projected in to the diagram of impedance.

probe and the limit of our equipment. All this factors determine the precision of the measure of the electrical conductivity.

4 EXPERIMENTAL APPROACH

To see the conductivity effect, we have used a micrometric system of displacement to make a sweeping with an eddy current probe on some materials with different conductivities (Cu, Al, Br) covered with an aluminium sheet (1.8 mm thickness).

The frequency of the measure is chosen at 1 kHz to keep the constant of the standard penetration superior than the thickness of the aluminium sheet.

It is well known that defects is a variation of a conductivity, then this experience is realised to show variation of conductivities happened in deep of material.

This figure allows us to see the behavior of the conductivity measured when the conductivity of the material in deep is lower than the conductivity of the principal material (Bronze under Aluminium) or superior (copper under aluminium).

Although the thickness of the aluminium sheet is important (1.8 mm), It has been observed that:

– All materials under the aluminum sheet have been detected.
– We have to take care about the edge effect, which is very important to optimize it to perform the detection.
– Material with high conductivity is easier to detect than material with lower conductivity.
– The variation of the conductivity measured is important and follows an exponential attenuation in function of the thickness of the aluminum sheet if the difference between the two conductivities is important.

Other experiences have been realized to see the behavior of different thickness of an aluminum sample covering different materials conductivities.

EC Palpeur

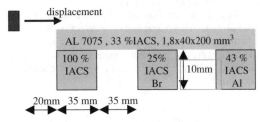

Figure 6. Detection of some samples under an aluminium sheet.

Table 1. Representation of different conductivities measured for different thickness of an aluminum sheet.

Aluminium thickness (mm)	Electric conductivity (%IACS)				
	Cu	Br	Al	Eton	Air
0	91.61	24.94	41.3	31.72	0
0.43	74.64	31.72	43.75	36.71	10.57
0.85	66.78	33.04	42.3	38.45	11.53
1.32	56.65	32.53	40.93	36.88	14.1
1.5	51.16	32.53	39.65	37.32	15.85
1.8	45.32	30.5	36.71	37.76	15.86

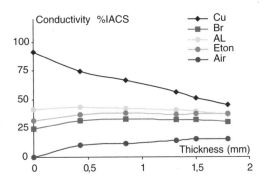

Figure 7. Representation of different curves of conductivities measured for different thickness of the aluminium sheet.

The different curves obtained allow us to deduce that the conductivity measured by our system converge from the conductivity of the material under the aluminum sheet for small thickness to the conductivity of the aluminum sheet for an important thickness.

5 CONCLUSION

Eddy current techniques are being developed to characterise materials. The knowledge of the electric conductivity allows characterising metallurgical states of materials simply and quickly.

We can deduce that this approach developed to measure electric conductivity for detecting defects or anomalies in the structure of materials has given good results and present some advantages for measuring conductivity or thickness of coating versus the standard techniques.

It has got advantages for analysing and detecting defects using electric conductivity.

It is noticed that this simulation is valid only for no ferromagnetic materials, and for that, it is necessary to saturate ferromagnetic samples in order to neglect their relative permeability.

Although, the conditions taken in consideration for building probes have given good results, the edge effect still be an important parameter to take care for increasing the sensibility of the detection.

We think that, using appropriate probes will permit to perform considerably the measure of the electric conductivity and the detection of defects.

REFERENCES

L. Hogo, *Introduction to electromagnetic nondestructive tests methods, Libby,* New York, 1977.

B.P.C. Rao, C. Babu Rao, T. Jyakumar, Baldev, *Simulation of eddy current signals from multiple defects.* NDT & E International, Vol. 29, No. 5, PP. 269–273, 1996.

M. Zergoug, A. Hammouda, F. Sellidj, Probe characterization and simulation of conductivity, 15eme, ICNDT, ROME 2000.

Non destructive testing handbook. Second edition, volume 4, Electromagnetic testing, Edition Paul Mcintire.1986.

C.V. Dodd, W.E. Deeds and J.w. Luquire, Integral solutions to some Eddy Current problems, International journal of NDT. Vol. 1, 1969–1970.

Benoit de Halleux & all, Eddy current measurement of the wall thickness and conductivity of circular non-magnetic conductive tubes, NDT&E international, Vol. 29, No.2, 1996.

M. Zergoug, Probe characterization and simulation of conductivity, 15eme, ICNDT, ROME 2000.

M. Zergoug, Conductivity simulation for material characterisation by eddy current, 3rd International conference on NDE. Seville 2001.

Emerging Technologies in Non Destructive Testing, Van Hemelrijck, Anastasopoulos & Melanitis (eds)
© 2004 Swets & Zeitlinger, Lisse, ISBN 90 5809 645 9

Risk based investigation of steel reinforcement corrosion using the AeCORR technique

M.J. Ing, S.A. Austin & R. Lyons
CICE, Loughborough University, Loughborough, Leicestershire, UK

P.T. Cole & J. Waston
Physical Acoustics Limited, Cambridge, UK

ABSTRACT: The use of a non-destructive acoustic evaluation technique as a Risk Based Inspection tool to detect the corrosion of steel reinforcement in concrete is presented in this paper. It offers the potential to save time and money for facilities owners and users. Recent research has demonstrated that AE has the ability to identify corrosion activity in concrete before conventional NDT methods, enabling faster intervention and increasing the repair options available. Monitoring a structure using the AeCORR technique, currently being researched and under development in the field, can create a digital map of part of a structure enabling an unbiased reference point for that structure for future maintenance tests as well as being able to distinguish areas of active corrosion. This paper reviews the principles and development of the new AeCORR technique for detecting and estimating the scale of corrosion induced damage and its ability as a tool to index test parts of structures.

1 INTRODUCTION

The detrimental role that corrosion of embedded steel rebar plays in the service life of reinforced concrete is well documented (Boyard et al. 1990), costing the UK an estimated £615 m per annum. Over the past 25 years, a number of methods for assessing the state of corrosion of reinforcing steel have been under development (Vassie 1978, Dawson 1983, McKenzie 1987), but to date difficulties in accuracy and reliability are still present (Dhir et al. 1993). Consequently a reliable and accurate corrosion tool that can be used to survey reinforced concrete structures is required.

The AeCORR system is a novel, non-destructive technique for detecting corrosion of embedded rebar in concrete and has been under development over the last four years. Current electrochemical techniques such as half-cell and linear polarisation are based upon the electrochemical dynamics of the corrosion reaction. In contrast the AeCORR system detects microscopic damage within the concrete created during the formation of expansive oxide at the steel/concrete interface as a consequence of the corrosion reaction.

2 PRINCIPLES OF AeCORR

Steel reinforcing bars embedded in good quality concrete are usually protected against corrosion due to the existence of a protective γ-Fe_2O_3 film on the surface of the steel formed under the highly alkaline conditions. In many cases the passive film remains intact for the life of the structure, but can be destroyed by the following mechanisms acting in isolation or together:

- The presence of aggressive species such as chlorides, that have migrated to the bar depth or were added at the time of casting.
- The reduction in alkalinity due to carbonation of the concrete cover.

Once depassivation of the steel has occurred, corrosion of the steel reinforcement can initiate if in the presence of oxygen and water. For clarity the corrosion process can be divided into two reactions, primary and secondary.

2.1 *Primary reactions*

The primary reaction involves the coupled anodic and cathodic reactions that take place on the metal surface.

At the anodic sites, the metal ions pass into solution as positively charged ferrous ions (Fe^{2+}) liberating electrons that travel through the steel to the cathodic sites.

At the cathode, oxygen (O_2) and water (H_2O) are reduced and combine with the free electrons from the anode to form hydroxyl ions (OH^-). To conserve balance of charge, the OH^- ions migrate through the pore water towards the anode and combine with the ferrous ions diffusing from the anode forming an electrically neutral ferrous hydroxide $Fe(OH)_2$.

2.2 Secondary reactions

Although the initial products of corrosion are $Fe(OH)_2$ and occasionally Fe_3O_4, these products can undergo secondary reactions by reacting with water and oxygen present in the concrete pores forming a hydrous ferric oxide $Fe_2O_3 \cdot 3H_2O$ (or haematite) – usually seen as a red-brown rust. This secondary reaction is significant to the durability of the concrete due to the change in the volume ratio of corrosion product to steel. Ferrous hydroxide has a volume of expansion of approximately 2.1:1, which can increase up to ratio of 10:1 if secondary reactions proceed to the formation of hydrous ferric oxide (Carney et al. 1990). The increase in volume at the steel/concrete interface generates large tensile stresses in the concrete, inducing microcracking in the concrete cover. Over a prolonged period of corrosion, a sufficient volume of corrosion products may accumulate resulting in the spalling or debonding of the concrete cover.

The rapid release of energy yielded by the formation of a microcrack is emitted from the source as a stress wave, detected on the surface of the concrete by a piezoelectric sensor. The magnitude and frequency of the stress waves will be related to the concrete properties (Ing et al. in prep.a) and corrosion rate (Lyons et al. in prep.). Furthermore, they will be related to and so a direct indication of the damage occurring to the concrete, enabling an early warning system against debonding and steel section loss. The detection and interpretation of these stress waves provides the basis of the AeCORR technique.

3 PARAMETERS INFLUENCING AeCORR PROTOCOLS

This paper describes the development of the AeCORR procedure by providing a review of the influential parameters and shows how they have been incorporated into the AeCORR testing procedure.

Cost-effective maintenance strategies for the repair of reinforced concrete need to be based upon reliable information about the rate of corrosion induced deterioration. The success of AeCORR to detect and monitor reinforcement corrosion will rely upon an

appreciation of the variables, which can then be incorporated into Risk Based Investigation (RBI) protocols. These factors can be broadly split into three categories: *environmental, electrochemical and material influences*. The RBI methodology incorporates the variables within these categories to enable a full assessment of the condition to be made rather than just recording instantaneous measurements.

3.1 Environmental influence

The corrosion rate of steel in concrete is highly dependent on many factors such as temperature, internal moisture content, resistivity and the availability of oxygen (Liu & Weyers 1998). These factors simultaneously influence the rate of corrosion and whilst they may be studied in isolation, in practice they are interdependent and can have significant control upon the rate of the corrosion reaction. An understanding of their influence and an ability to incorporate their effects into a testing procedure is essential if an accurate assessment of the corrosion rate is to be determined.

3.1.1 Temperature

The temperature of concrete imposes a significant effect on the corrosion rate of reinforcing steel. An increase in temperature aids solubility of the Fe^{2+} ions into the pore liquid of the concrete and consequently increases the development of the anodic reaction, increasing the corrosion rate. The change in internal temperature also induces changes in other parameters such as the concrete resistivity, oxygen diffusion in the cathode reaction and can reduce the critical Cl^- concentration required for depassivation. The overall influence of temperature is very complex but as a rule of thumb, every 10°C increase in temperature corresponds to a doubling of the corrosion rate (Bentur et al. 1997). Consequently, implications arise as to when corrosion measurements should be taken, and if normalised values are desired, adjustment factors must be used to normalise the data.

3.1.2 Resistivity

Corrosion rate is found to be strongly dependent on the electrical resistivity of concrete, which is influenced largely by temperature and the internal relative humidity (RH) of the concrete. The ohmic resistance of concrete may change significantly from more than 10^4 ohms in dry concrete to about several hundred ohms when the concrete is fully saturated (Liu & Weyers 1998). In very dry concrete the very high internal concrete resistance arrests the corrosion rate to negligible values. (Parrot 1996), suggests that the minimum RH required to support corrosion in concrete is 75% RH. An upper limit of 95–98% RH has also been suggested for very wet concrete because the corrosion rate can be dramatically reduced due to oxygen

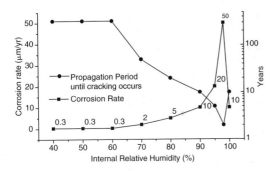

Figure 1.   Influence of internal RH on corrosion rate in carbonated concrete (Bentur et al. 1997).

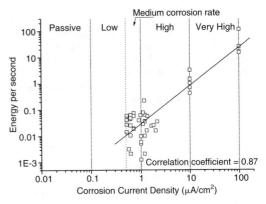

Figure 2.   Variation in energy/hour with corrosion rate.

starvation at the cathode in concrete near or at saturation. This principle has been illustrated in Figure 1.

In natural conditions, the RH and temperature are in constant flux due to diurnal and seasonal variations thus also causing the corrosion rate to be constantly changing. Small changes in either temperature or RH may significantly affect the dissolution rate, and in some instances, be significant enough to suppress the reaction. It is also clear from Figure 1 that if the internal RH is below 70%, the corrosion rate falls to negligible values in carbonated concrete, inducing a dramatic increase in the number of years required to cause cracking. As AeCORR monitors for corrosion by detecting microcracking activity, a low RH will affect the duration of the monitoring period required. Therefore protocols are required that enable an estimation of the rate of activity possible under the environmental conditions present.

This is addressed within the procedure by taking resistivity measurements. The resistivity values are grouped into three categories, red, orange and green. If the measured values fall into either of the former two, testing is possible. If they fall into the latter, monitoring is not recommended, as the conditions are not favorable for corrosion activity.

3.2   Electrochemical influence

The electrochemical influences are concerned with factors such as the corrosion rate, the corrosion type and oxide formation. Whilst there is overlap between each of these factors, singularly they are of importance.

3.2.1   Corrosion rate

Typical corrosion rate values of steel in concrete can vary from 0.1–100 μA/cm^2 on real structures, with the values between 0.1–1 μA/cm^2 being the most frequent (Andrade & Alonso 2001). The maximum corrosion rate a section of a particular structure is able to support will be strongly influenced by the water/cement ratio

of the concrete, which will determine the resistivity and the oxygen permeability values of the concrete.

The rate of corrosion has a significant influence on the time to failure of the concrete. Figure 1 illustrates how different corrosion rates dramatically reduce the time taken until cracking of the concrete cover occurs. Due to this phenomena it has been shown that AeCORR is able to estimate the rate of corrosion through measurement of the rate of damage (Energy/sec) as shown in Figure 2 (Lyons et al. in prep). The Energy/sec values are related to the rate of oxide production and microcracking within the concrete. Not only does this give a reasonable estimate of corrosion rate, but it also gives an indication of the rate of damage occurring within the concrete cover.

3.2.2   Corrosion type

Corrosion of steel reinforcement can be induced via two main processes: the chloride ion or carbonation of the cover concrete. Whilst both methods result in a loss of steel section and the production of oxides, there are crucial differences between the two for which any protocols for a measurement technique must cater.

Carbonation induced corrosion causes a general loss of section of the rebar over a relatively large area producing solid oxides on the surface of the bar. Conversely, chloride induced corrosion is usually associated with localised pitting corrosion.

The acidic condition inside the pit prevents the formation of solid oxides on the bar. Due to its localised nature and the oxide type, pitting corrosion is sometimes not detected by existing methods until extensive damage has occurred.

The AeCORR method incorporates the different corrosion types within its protocols. Research has found that the technique is able to clearly detect chloride-induced corrosion (Austin et al. in prep). Furthermore, the RBI procedures help to identify which parts on the structure are at risk from which type.

3.3 Material influences

The material properties of a structure will not only influence the resistance to corrosion and subsequent corrosion rate but will also affect the magnitude of the emissions caused by microcracking.

It is essential that differences in the emission caused by variations in for example, concrete strength, can be quantified to enable normalisation of readings between structures.

3.3.1 Strength of concrete

Reducing the water/cement ratio in concrete mix design usually results in an increase in the compressive strength, increasing the resistance to the ingress of aggressive species that may initiate corrosion.

However, good quality concrete alone is not sufficient to prevent corrosion occurring and if it occurs a greater force is required to exceed the local tensile strength of the matrix resulting in a larger stress wave being emitted during micro-fracture for a given volume of corrosion.

Previous work (Ing et al. in prep.a) has suggested that an exponential relationship exists between the compressive strength of concrete and absolute energy per gram of steel loss, as shown in Figure 3.

4 RISK BASED INVESTIGATION PROCEDURE

To ensure successful application of AeCORR, a RBI procedure is under development with the aim of determining whether active corrosion is occurring on a structure, its location and how badly it is corroding. Figure 3. Influence of Compressive Strength on absolute energy per gram of mass loss.

To achieve this aim, the procedure has four objectives:

1. To enable an accurate and efficient application of the AeCORR technology to undertake reliable and repeatable corrosion measurements on reinforced concrete structures.
2. To assess the likelihood of corrosion occurring at a selected point on a structure.
3. Using AeCORR adjustment factors, grade the extent of corrosion induced damage of the structure into one of five grades, ranging from A–E.
4. Using the grading band, and information about the type of structure and history, be able to undertake a risk assessment of the structure for the purpose of prioritising for maintenance.

These four objectives can be achieved by incorporating the influential parameters (environmental, electrochemical and material) discussed in earlier sections and combining them with structure specific information.

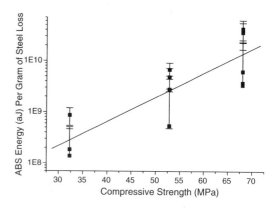

Figure 3. Influence of compressive strength on absolute energy per gram of mass loss.

The basic outline of such a procedure is presented in Figure 4.

The procedure in Figure 4 combines the environmental, electrochemical and material parameters together to form a simple test procedure that if employed correctly, will provide a reliable and accurate test technique.

The start for any structural investigation must be a comprehensive desk study. This process assembles all the information about the structure (age, exposure, orientation, concrete details etc.) which is combined with a preliminary visual survey to assess the risk posed to the structure. This study is very similar to routine maintenance or principal inspections undertaken on most reinforced structures. If the structure (or a specific element) fails a number of predetermined criteria, the investigation can proceed.

It is essential that on the day of the monitoring, conditions in the concrete are known to estimate the ability of the concrete to support corrosion. Therefore, conformance checks such, as resistivity measurements are required prior to testing. The control protocol awards each element one of three activity gradings based on the results of the conformance checks ranging from *very suitable* to *not suitable for testing*.

Obviously if the structure is *suitable* or above, testing can proceed. Each type of structure has specific test protocols developed from field experience to avoid collection of rogue or 'bad' data.

After successful data collection, post analysis of the results normalises the data for the factors discussed in Section 3. For instance, the strength of the concrete can strongly influence the rate of emission for a given mass of steel loss (Fig. 3). Consequently, the strength of the concrete needs to be normalised to enable comparisons and realistic assessments to be undertaken. The weightings and adjustment factors have been developed from extensive laboratory testing and field experience.

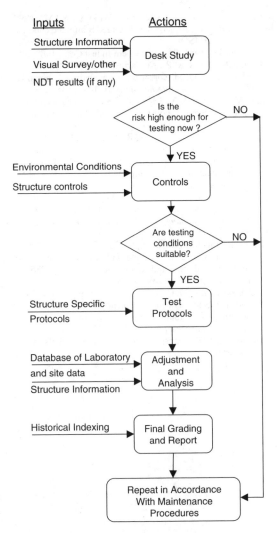

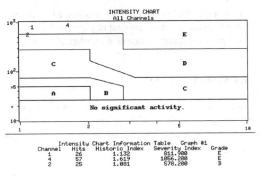

Figure 5. Activity grading chart.

	Intensity Chart Information Table	Graph #1		
Channel	Hits	Historic Index	Severity Index	Grade
1	26	1.132	811.900	E
4	57	1.619	1056.200	E
2	25	1.081	578.200	D

elements, it can provide a means of index testing, aiding the formation of maintenance schedules (Ing et al. in prep.b). An example of the grading chart is shown in Figure 5, which shows the results from an area of a reinforced concrete swimming pool wall.

The numbers 1, 2 and 4 represent three AeCORR transducers mounted on the concrete surface. Transducer 3 was the designated control and received insufficient emission to be graded due to the area being of sound nature. Large areas of the pool wall were being broken out due to significant deterioration from reinforcement corrosion. The warm, moist conditions, coupled with the chlorides in the pool water had enabled corrosion of the reinforcement to proceed unhindered leading to large areas of delamination and spalling.

The area monitored had been highlighted by the visual survey as an area of increased corrosion risk. Environmental conditions at the time of testing enabled monitoring to proceed. The results show that the area monitored indicated significant corrosion. Whilst this area had not yet delaminated the results indicate that delamination would be imminent.

Figure 4. Risk Based Investigation Procedure.

Finally the structure or structural element can be graded which will enable prioritisation for repair and provide a benchmark for future testing.

5 GRADING OF STRUCTURES

The output of the AeCORR method grades the scale of activity on a range from A–E, where A is Minor – insignificant, and E is Major – immediate intervention recommended. This gives the engineers a clear and immediate indication of the condition of the structure. Furthermore, in the situation where engineers are responsible for hundreds of structures or structural

6 CONCLUSIONS

This paper has discussed a number of important parameters that were considered during the development of a new corrosion detection method for reinforced concrete. The focus of this paper has been the RBI procedure that supports the new technique. The procedure is a result of a significant amount of research that has recently been undertaken (Ing et al. 2002a, Austin et al. 2002, Lyons et al. 2002).

1. The AeCORR method offers the potential to accurately detect active corrosion in reinforced concrete structures completely non destructively.
2. The approach of the RBI method aims to incorporate the important parameters in a well researched, structured and logical format to provide a reliable detection procedure.

3. Using this RBI procedure, the grading of structures can be made more easily and objectively, especially for planned maintenance of a number of structures.

ACKNOWLEDGEMENTS

The authors wish to acknowledge the financial assistance given by Balvac Whitley Moran Ltd and Atkins, in addition to the technical assistance and equipment loan by Physical Acoustics Limited.

REFERENCES

Andrade, C., Alonso, C. 2001. On-site measurements of corrosion rate of reinforcements. *Construction and Building Materials* 15: 141–145.

Austin, S.A., Ing, M.J., Lyons, R. The electrochemical behavior of steel reinforcement undergoing accelerated corrosion testing. Submitted to *Corrosion*, November 2002.

Bentur, A., Diamond, S., Berke, N.S. 1997. *Steel corrosion in concrete*, London: Chapman & Hall.

Boyard, B., Warren, C., Somayaji, S., Heidersbach, R. 1990. Corrosion rates of steel in concrete, ASTM STP 1065, N.S. Berke, V. Chaker, D. Whittington, (Eds) Philadelphia, P.A. 174.

Carney, R.F.A., Lawrence, P.F., Wilkins, N.J.M. 1990. Detection of corrosion in submerged reinforced concrete, London, HMSO.

Dawson, J.L. 1983. Proc Soc Chem Indust Conf, 1983, 59.

Dhir, R.K., Jones, M.R., McCarthy, M.J. 1993. Quantifying chloride induced corrosion from half-cell potential. *Cement and Concrete Research* 23: 1443–1454.

Ing, M.J., Austin, S.A., Lyons, R. 2002. Cover zone properties influencing acoustic emission due to corrosion. Submitted to *Cement and Concrete Research*.

Ing, M.J., Austin, S.A, Lyons, R. 2003. Condition monitoring of reinforced concrete structures at risk from reinforcement corrosion. Proc. 2nd Int. Conf. Innovation in architecture engineering and construction (AEC), Loughborough, UK.

Liu, T., Weyers, R.E. 1998. Modelling the dynamic corrosion process in chloride contaminated concrete structures. *Cement and Concrete Research* 28(3): 365–379.

Lyons, R., Ing, M.J., Austin, S.A. 2003. Correlation of corrosion rate of steel in concrete with acoustic emission in response to diurnal and seasonal temperature variations, In preparation for *Corrosion Science*.

Mckenzie, S.G. 1987. Corrosion Prevention and Control 34: 5.

Parrott, L.J. Some effects of cement and curing upon carbonation and reinforcement corrosion in concrete. *Materials and Structures* 29: 164–173.

Vassie, P.R. 1978. Transport and Road Research Laboratory, Report 953.

Use of wavelet packet transform for signal analysis in adhesive bond testing

S. Tavrou
School of Engineering and Science, Swinburne University, Melbourne, Australia

C. Jones
School of Mathematical Sciences, Swinburne University, Melbourne, Australia

ABSTRACT: Adhesives are widely used and often preferred as a means of fastening parts together. Aircraft applications in bonding large areas or bonding parts of vehicle chassis, has brought about the need for highly reliable techniques in non destructive testing of adhesive bonds. Among a wide range of different methods, a variety of ultrasonic techniques are used for non destructive testing of adhesive bonds including pulse echo or through signal. In all cases the ultrasonic signal response is analysed and conclusions are drawn with regard to the quality of the adhesive bond. Features of the ultrasonic signal such as amplitude and frequency are used for analysis. Even though there have been great improvements in testing adhesive bonds, it is not always easy to extract conclusive information from the signals for clear discrimination between different levels of bond quality. The method introduced in this paper, enhances the capability of discriminating between varying degrees of bond quality compared with existing methods in ultrasonic testing. Signal Processing with Wavelet Packets was used for this research and the results show significant improvements in differentiating between "good" and "bad" bonds.

1   INTRODUCTION

Non destructive testing of adhesive bonds has been the subject of research for many years (Thompson & Thompson 1991, Adams & Drinkwater 1999). A number of different ultrasonic techniques have also been developed including Leaky interface waves (Scala & Doyle 1995), Lamb waves (Heller, Jacobs & Ou 2000), pulse echo or through signal (Rose & Ditri 1992). Although any of the methods could have been used in this research, pulse echo was chosen as it requires only one transducer and access from one side of the test piece as well as the fact that this is the most common method used in testing adhesive bonds in auto-motive applications. It is also of particular interest to the authors, as this research is aimed at applications within the automotive industry. Characteristics such as the amplitude of the returning signal, its frequency, or even the signal decay (Gogio & Rossetto 1999) have been used as a measure of the quality at the bond interface. In the case of Maximum Amplitude method, the magnitude of the returning signal is taken as a measure of the level of contact between the adhesive and adherent. Higher amplitudes indicate reduced contact as less signal is transmitted through to the

adhesive. Even though this is the most widely used method, the difference in output signal can be very small and in some cases is unable to discriminate between "good" and "bad" areas. Measurement units are in terms of mVolts, that is often lost in signal noise. This paper suggests an alternative method that enhances ultrasonic signal interpretation for testing adhesive bond integrity.

2   WAVELET PACKET ANALYSIS

Fourier analysis examines how periodic signals can be reconstructed from the sum of sines and cosines. Wavelets are analogues to sine and cosine functions of traditional Fourier analysis however they offer features, which overcome some of the limitations of traditional Fourier analysis, in characterizing signal structures with sharp changes. For the analysis of irregular 1-d signals such as those from ultrasonic A-scans, it is difficult to adequately express the signal characteristics using the sum of sines and cosines since the sum may be infinite. The Wavelet transform on the other hand is a mathematical technique that is capable of numerical analysis and manipulation of both one or two-dimensional data. The transform operates like a microscope for detail examination by partitioning the signal into different frequency components, mapped onto coefficients that have different energies (Daubechies 1992).

Two orthogonal wavelet filter functions are shown in Figure 1, which differ in their degree of oscillation and regularity. The wavelet packet functions $W_b(i)$ are generated by scaling and translating of these or other available filters. Each function has a *frequency b*, which describes the number of oscillations or zero crossings the wavelet packet makes. For the "Haar" father basis (Wickerhauser 1993), the wavelet does not oscillate through zero ($b = 0$) so $\phi(i) \equiv W_0(i)$ but the mother wavelet has one zero crossing ($b = 1$) so $\psi(i) \equiv W_1(i)$. Wavelet Packet Analysis operates by approximating a signal with scaled and translated wavelet packet functions $W_{m,b,n}$ that are generated from W_b following:

$$W_{m,b,n}(i) = 2^{-m/2} W_b(2^{-m} i - n)$$

where m is a measure of the resolution level while n represents signal translation. A signal $f(i)$ can then be represented by the sum of orthogonal wavelet packet functions $W_{m,b,n}(i)$ following:

$$f(i) \approx \sum\sum\sum w_{m,b,n} W_{m,b,n}(i)$$

and the wavelet packet coefficients $w_{m,b,n}$ (in the final form identified with the notation C_n^m) are produced from the integral:

$$C_n^m \equiv w_{m,b,n} \approx \int W_{m,b,n}(i) f(i) di$$

These coefficients contain information about the energy magnitude contribution for each feature in terms of scale, frequency and position. The method suggested in this paper, uses the Total Energy, calculated by the sum of the squares of all the coefficients, as a measure of adhesive bond quality.

The coefficient and total energy calculations were done using Wavelet Packet Analysis software (Wickerhauser & Coifman 1993). The signals from ultrasonic A-scans, taken from the adhesive bond interface, were used as input to the software. The output from the software gives the decomposition of the signal and the list of its coefficients. The number of coefficients to be evaluated is user defined and depends on the length of the signal. In this case, 256 coefficients were used.

3 EXPERIMENTAL METHOD

The experimental procedures were designed to confirm the validity of Wavelet Packet test hypothesis in enhancing ultrasonic signal interpretation. The specimens were prepared from a 0.8 mm mild steel sheet and structural adhesive (Orbseal), which included defective areas at the interface of known dimensions and position. A test rig was designed and constructed, Figure 2, for performing ultrasonic c-scans. The Pulser Receiver System and Winspect software (UTEX) was used in conjunction with an XYZ scanning machine. The c-scan experiments were done in a water tank (300 × 200 × 200 mm), therefore no couplant other than water was used. An ultrasonic signal was directed onto the specimen through the steel side with the

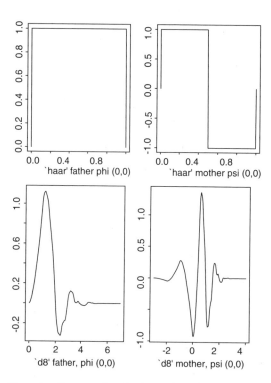

Figure 1. Two examples of orthogonal wavelet bases. Left: father wavelets. Right: mother wavelets.

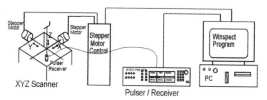

Figure 2. Experimental Instrumentation layout for ultrasonic testing of adhesive bonds.

intention of investigating the interface bond quality between the steel and adhesive. Due to high attenuation of the signal in the adhesive, the effect of returning signal from the back face of the adhesive was minimal hence it was ignored (Goglio & Rossetto 1999). The steel plate dimensions were such as to achieve as little interference between reflections of the signal within the plate (Dewen, Pialucha & Cawly 1991).

4 RESULTS

Typical results of the A-scan are shown in Figure 3 while Figure 4 shows two superimposed A-scan signals from a defective, "bad", bond and non defective, "good", bond. Figure 5 shows a c-scan of the area under test, with the lighter areas indicating defective bonding. These defective areas were confirmed by measurements of defect locations placed in the bondline during specimen preparation. The maximum amplitude was gated for the c-scan.

For the purpose of this research, results from six locations were chosen for analysis. Figure 6 shows ultrasonic signals from these six locations of the c-scan, superimposed on the same time axis. Three of these come from locations of varying degrees of defect severity while the other three come from defect free locations but again with varying degrees of bond quality. The amplitude of the returning signal was used to judge the severity of defective bonds. Defective areas return more of the signal as less ultrasound

energy is transmitted through the bond. Existing methods compare the maximum amplitude between the returning signals, to discriminate between good and bad bonds. Eleven wave cycles, "wavesets", are shown in Figure 6, where each one is a repetition of itself as the signal reverberates in the steel. Each time the signal bounces back towards the receiver, it returns with less of the ultrasound energy as some of it passes through the bondline, hence the decaying signal.

A table with values and histogram of maximum amplitude from each of the eleven wavesets is shown in Figure 7. Series 1, 2 and 3 come from defective bond areas and Series 4, 5 and 6 come from good quality bond areas. A number of observations can be made from this dataset. Firstly, the differentiation between defective and non-defective areas is clearer as the returning signal response moves further away from the initial wave. This may be explained by the fact that on well bonded areas, some of the energy is transmitted through the bondline each time it reverberates through the steel plate. The signals from poorly bonded areas have a slower decay of maximum amplitude, as less or no energy is transmitted through the bondline, leaving only lateral energy dissipation during each reverberation.

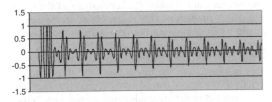

Figure 3. Ultrasonic signal from an A-scan on the metal adhesive interface.

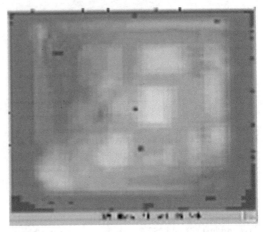

Figure 5. C-scan from a specimen with pre-prepared bonding defect locations.

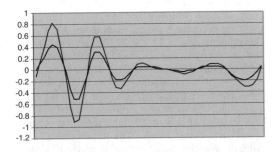

Figure 4. Superimposed signals from an A-scan test on locations with and without bonding defects.

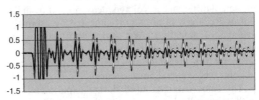

Figure 6. A series of A-scans from varying degrees of bonding quality superimposed of comparison purposes.

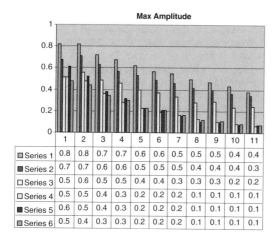

Figure 7. Table of output values and histogram of maximum amplitude of signals from six locations with varying degree of bonding quality.

	1	2	3	4	5	6	7	8	9	10	11
Series 1	0.8	0.8	0.7	0.7	0.6	0.6	0.5	0.5	0.5	0.4	0.4
Series 2	0.7	0.7	0.6	0.6	0.5	0.5	0.5	0.4	0.4	0.4	0.3
Series 3	0.5	0.6	0.5	0.5	0.4	0.4	0.3	0.3	0.3	0.2	0.2
Series 4	0.5	0.5	0.4	0.3	0.2	0.2	0.2	0.1	0.1	0.1	0.1
Series 5	0.6	0.5	0.4	0.3	0.2	0.2	0.2	0.1	0.1	0.1	0.1
Series 6	0.5	0.4	0.3	0.3	0.2	0.2	0.2	0.1	0.1	0.1	0.1

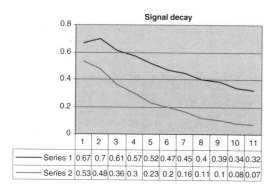

	1	2	3	4	5	6	7	8	9	10	11
Series 1	0.67	0.7	0.61	0.57	0.52	0.47	0.45	0.4	0.39	0.34	0.32
Series 2	0.53	0.48	0.36	0.3	0.23	0.2	0.16	0.11	0.1	0.08	0.07

Figure 8. Decay of ultrasonic signals from "bad" (Series 1) and "good" (Series 2) bonds.

Figure 8 confirms this difference in signal decay by using the average value from the three "bad" and three "good" bonds. The percent reduction between the first and last signal is 52.2% in the case of poor bonds (Series 1), while good bonds show an 86.8% between the first and last waveset (Series 2). Although the signal decay is evidently better at the tail end of the signal, the actual difference in voltage measurement from the Pulser Receiver, is not significant, 0.35 V from the average of defective areas and 0.48 V from good bonds. The difference is basically 0.15 V which in real terms such an advantage may easily be lost, especially in cases where signal noise occurs.

The same data was used during experimentation with Wavelet Packets to examine the capability of the technique to enhance the discrimination between "good" and "bad" bonds. A set of data from signals 3,

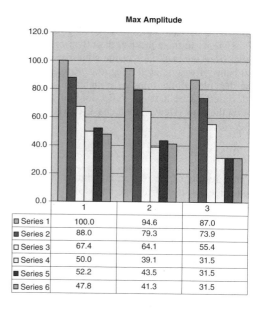

	1	2	3
Series 1	100.0	94.6	87.0
Series 2	88.0	79.3	73.9
Series 3	67.4	64.1	55.4
Series 4	50.0	39.1	31.5
Series 5	52.2	43.5	31.5
Series 6	47.8	41.3	31.5

Figure 9. Table of values and histogram from ultrasonic signals with varying degree of bond quality using the Maximum Amplitude method.

7 and 11 was analysed using the wavelet packet D8, where measurements of Total Energy from the reconstruction of Best Basis B0 were recorded. For comparison purposes, Figure 9 shows the percent maximum amplitude of signals 3, 7 and 11 while Figure 10 shows the output from Wavelet Packet analysis of the same signals. A number of significant observations can be made when comparing the results from the Wavelet Packet method to those from Maximum Amplitude method. Firstly, the average difference between signals that resulted from the three "bad" and three "good" bonds has increased considerably. It was also noted that the range between the signals coming from three defective areas has widened, while signals from areas with "good" bonds remained relatively unchanged. When the Maximum Amplitude method was used, signals at 3, 7 and 11 were 58.7%, 52.1% and 43.7% respectively, of those coming from "bad" bonds. The corresponding percentage values when using the Wavelet Packet method were 37.3%, 27.5% and 5.4%. In the best case scenario, that of signal 11, the Wavelet Packet method resulted in a reading from a "good" bond that is nearly 20 times smaller than that from a "bad" bond. The corresponding result when using the existing method of Maximum Amplitude resulted in a best case scenario of only of 2.3 times reduction between "good" and "bad" bonds. The improvement when using the Wavelet Packet method is nearly 10 fold.

A further advantage of Wavelet Packets in analysing ultrasonic signals, not shown in this paper, is its

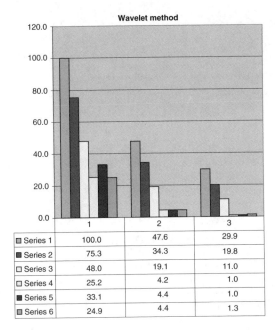

	1	2	3
◫ Series 1	100.0	47.6	29.9
■ Series 2	75.3	34.3	19.8
☐ Series 3	48.0	19.1	11.0
☐ Series 4	25.2	4.2	1.0
■ Series 5	33.1	4.4	1.0
◫ Series 6	24.9	4.4	1.3

Figure 10. Table of values and histogram from ultrasonic signals with varying degree of bond quality using the Wavelet Packet method.

adaptability to varying frequencies, amplitude and signal shape. Different types of Wavelet Packets may be chosen to suit particular signals. This is extremely important in the case of ultrasonic tests as more often than not, the signal is highly distorted by noise.

5 CONCLUSION

Signal analysis in ultrasonic testing is extremely important and researchers have given it much attention. In the case of adhesive bond testing, the interpretation of ultrasonic signals is just as crucial since it is an area that has not yet been fully resolved. Despite the many years of research there isn't yet a reliable method that can fully predict the performance of an adhesive bond under operating conditions. The method developed here is a sensitive tool providing high resolution discrimination when applied to reading of ultrasonic signals from adhesive bonds.

REFERENCES

Adams R.D. and Drinkwater B.W., Non-destructive testing of adhesively-bonded joints, Int. J. of Materials and Product Technology, Vol. 14, No. 5/6, 1999.

Daubechies, I. (1992). Ten Lectures on Wavelets, CBMS-NSF Regional Conference Series. 61, SIAM, Philadelphia.

Dewen P.N., Pialucha T.P. and Cawley P., Improving the resolution of ultrasonic echoes from thin bondlines using cepstral processing, J. Adhesion Sci. Techno. Vol. 5, No. 8, pp. 667–689, 1991.

Goglio L. and Rossetto M., Ultrasonic testing of adhesive bonds of thin metal sheets, NDT&E International, Vol. 32, pp. 323–331, 1999.

Heller K., Jacobs L.J. and Qu J., Characterisation of adhesive bond properties using Lamb waves, NDT&E International, Vol. 33, pp. 555–563, 2000.

Orbseal Australia, Structural Adhesive Orbseal 2000.

Rose J.L. and Ditri J.J., Pulse-Echo and Through Transmission Lamb wave Techniques for Adhesive Bond Inspection, British Journal of NDT, Vol. 34, No. 12, 1992.

Scala C.M. and Doyle P.A., Ultrasonic Leaky Interface Waves for Composite-Metal Adhesive Bond Characterization, Journal of Nondestructive evaluation, Vol. 14, No. 2. 1995.

Thompson R.B. and Thompson D.O., Past experiences in the development of tests for adhesive bond strength, J. Adhesion Sci. Technol. Vol. 5, No. 8, pp. 583–599, 1991.

UTEX Scientific Instruments Inc., 2319 Dunwin Drive, Unit 8, Mississauga, Ontario, L5L 1A3, Canada.

Wickerhauser, M.V. (1993). Best-adapted wavelet packet bases, Proceedings of Symposia in Applied Mathematics. 47: 155–171.

Wickerhauser M.V. and Coifman R.R. (1993). Wavelet Packet Laboratory for Windows Version:1, Digital Diagnostic Corporation and Yale University.

Emerging Technologies in Non Destructive Testing, Van Hemelrijck, Anastasopoulos & Melanitis (eds)
© 2004 Swets & Zeitlinger, Lisse, ISBN 90 5809 645 9

Surveillance of gas assisted injection moulding process by means of microwaves

C. Sklarczyk

Fraunhofer-Institute for Nondestructive Testing (IZFP), Saarbrücken, Germany

ABSTRACT: An innovative method for online surveillance and control of the gas assisted injection mould-ing process (GIT technique) has been developed. It is based on radar sensors which emit microwaves into the cavity and receive the scattered microwaves with the same antenna. Depending on sensor type the frequency is in the range between about 24 GHz and 94 GHz. Especially the correct position of the gas bubble can be moni-tored. By using a two axis scanning equipment it is possible to generate two dimensional images and to detect defects, e.g. bubbles, inside injection moulding parts.

1 INTRODUCTION

In modern civilization a big amount of plastics parts is produced by gas assisted injection moulding process (GIT technique). This technique helps to maintain the form of the part and to save plastics material. Industry is seeking for a method to online monitor and control the process with regard to the correct position and shape of the gas bubble. Presently online monitoring is performed by measuring temperature or pressure in the cavity which is filled up by the liquid plastics. This method is quite indirect and therefore supplies only insufficient information on the process. The Fraunhofer-Institute for Chemical Technology (ICT, Pfinztal) and the Fraunhofer-Institute for Non-destructive Testing (IZFP) have further developed a method described in (Diener, Busse, 1996) based on microwave technique which is able to online monitor the injection moulding process in more detail espe-cially concerning the correct position of the gas bubble. The microwave method can be an alternative for the likewise new developed ultrasound method (Volkmann, Schulz, 2000) which may exhibit problems with long term constancy of ultrasound transducer coupling.

2 RADAR SENSOR

The microwave sensor is based on a frequency modulated continuous waves (FMCW) radar module with a frequency of about 24 GHz (K-band sensor),

77 or 94 GHz (W-band sensor). The microwaves are generated by a voltage controlled oscillator (VCO) whose frequency can be tuned over a bandwidth of up to 175 MHz for the K-band and 5 GHz for the W-band sensor. The sensor works in monostatic mode, i.e. with only one antenna for both emission and reception (Tessmann et al., 2002). The microwaves are fed by a rectangular or circular waveguide into the cavity whereas a plastics window with high melting point and transparent to microwaves separates the cavity from the waveguide. The window can be regarded as a dielectric antenna. Fig. 1 gives the photograph of the

Figure 1.   Radar module with flanged horn antenna (photo: Fraunhofer-IAF).

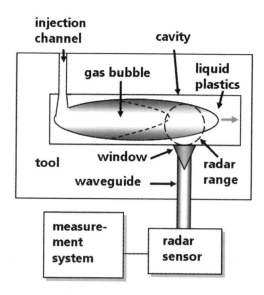

Figure 2. Scheme of microwave measurement at injection moulding.

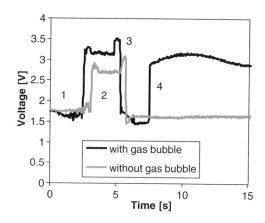

Figure 3. Behaviour of IF signal amplitude during injection moulding with polycarbonate (PC) with and without gas injection (1: cavity is open; 2: cavity is closed; 3: passage of liquid plastics over window position; 4: passage of gas bubble over window position).

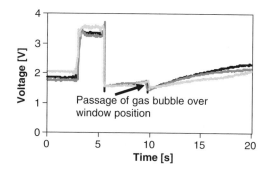

Figure 4. Behaviour of IF signal amplitude during injection moulding with perspex (PMMA) with gas injection (three injection moulding tests).

radar module and Fig. 2 the scheme of the measurement arrangement at the injection moulding machine. The sensor works according to homodyne principle in order to generate a sinus-like intermediate frequency (IF) signal which is the mixing product of the oscillator signal and of the microwave signal scattered by the object and received by the antenna. Both amplitude and phase of the IF signal depend on the position of the bubble with regard to the microwave window. The frequency of the IF signal is in the order of magnitude of a few 1000 Hz and can therefore be captured by conventional A/D converter PC cards. The detection range of the microwave sensor within the cavity is in the order of magnitude of 1 cm.

The W-band sensor was supplied by Fraunhofer-Institute for Applied Solid State Physics (IAF), Freiburg. The heart of the sensor was a microchip (MMIC, **m**onolithic **m**icrowave **i**ntegrated **c**ircuit) with a size of about $3 \times 2\,mm^2$. It may be the only radar chip in the world which integrates all high frequency components on a single chip. The chip is bonded in a metallic case which contains the rectangular wave guide aperture and the connectors for power supply, modulation signal and IF signal. The emitted power is in the range of a few mW, i.e. the material under test is not changed.

3 EXPERIMENTAL INVESTIGATIONS

3.1 *In process radar experiments*

Experiments have been performed on several plastics materials during injection moulding process.

The different process phases like closing the cavity, passage of the liquid plastics over the position of the microwave window and passage of the gas bubble over the window position could be clearly identified. At passage of the bubble the signal voltage jumped abruptly to higher amplitude values (for polycarbonate, Fig. 3) or to lower values (for polymethyl methacrylate (PMMA, perspex), Fig. 4). When there was no gas injection or when the gas bubble was too short, so that it did not pass the window position, no jump occurred. (Fig. 3). When the gas pressure is suddenly released (simulated process disturbance) the gas bubble gets foamy. This results in an additional change of voltage different from the jumps at normal bubble passage since reflection behaviour of the microwaves at the interface plastics-bubble changes. Fig. 5 shows these signal changes that occurred at about 14 s after start of the test.

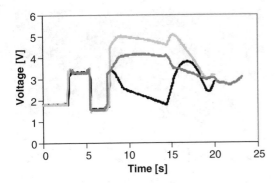

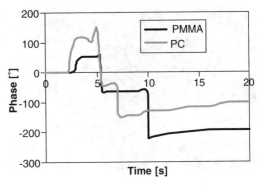

Figure 5. Behaviour of IF signal amplitude during injection moulding with perspex (PMMA) with gas injection (three injection moulding tests) with sudden release of gas pressure after ca. 14 s resulting in a foamy gas bubble.

Figure 6. Behaviour of IF signal phase during injection moulding with perspex (PMMA) and polycarbonate (PC) with gas injection; passage of the gas bubble at ca. 7 s (grey line) and 10 s (black line).

The comparison between Fig. 4 and Fig. 5 shows that the jump at passage of the bubble over the sensor position can result in both positive and negative jumps for the same plastics material under test. The difference between both tests was a much higher gas pressure at the test in Fig. 5 (30 MPa vs. 5.5 MPa) which possibly yielded a bigger diameter of the bubble. The changed bubble geometry obviously produced a change in polarity of the jump. This effect is presently investigated in more detail in the frame of some basic research.

Figures 3 to 5 show an amplitude evaluation. However, the phase of the IF signal can also be used since the interface between plastics and gas bubble produces a phase shift of the reflected microwave signal. Fig. 6 shows that the phase change at passage of the gas bubble can be considerable and can be better detectable than amplitude jump. In Fig. 4 the amplitude jumps at passage of the bubble are relatively small whereas the phase jumps are quite big (Fig. 6).

In an earlier work a similar experimental setup has already been used to detect the gas bubble during gas assisted injection moulding process (Diener, Busse, 1996). The microwave sensor had a fixed frequency and allowed only amplitude evaluation. The FMCW radar sensor used in the present study makes possible both amplitude and phase evaluation and therefore improves the detectability of gas bubbles as shown in Fig. 6. Furthermore the center frequency of the radar sensor is higher and allows smaller waveguides and thus smaller marks of the microwave window on the surface of the solidified plastics part.

A comparison of the K-band sensor with the W-band radar sensor shows that the results from the W-band sensor give better results when the amplitude of the IF signal is evaluated. This is due to the bigger frequency bandwidth. However, with K-band sensor, too, some clear differences in IF signal phase could be

observed when comparing tests with and without gas injection. One drawback of the K-band sensor is the bigger mark of the microwave window on the solidified plastics part which is due to the bigger wave guide dimensions compared to W-band.

The experimental investigations described above are very promising with regard to future application in industrial area.

3.2 Area scans on plastic parts

By using a two axis scanning device it is possible to generate two dimensional microwave images. These can be used to detect defects like macroscopic bubbles or pores in offline way. Such images have been created on test objects generated by injection moulding process without gas injection. The used test object made of polypropylene (PP) was the cover of a control device and had a thickness of ca. 3 mm. The microwave test system was based on a vector network analyzer HP8510C which was adapted for free space measurements. Frequency was swept between 75 and 100 GHz (in W-band). The measurements were performed both in reflection and transmission mode. The test object was moved in steps of about 0.5 mm while the antenna was standing still. The sending antenna was an open waveguide which had a distance of 1–2 mm from the test object. In that way nearfield effects could be used to improve the lateral resolution which is normally limited to a few mm due to diffraction effects of the microwaves (wave length 3 mm at 100 GHz). In transmission mode the receiving antenna was a normal horn positioned a few cm apart from the test object.

Fig. 7a shows the transmission optical image of an area of about 53 × 33 mm². Some bright structures (especially in the right part of the image) can be seen which are due to macroscopic bubbles in the middle

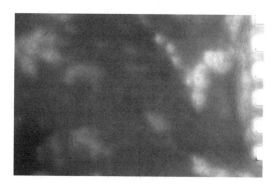

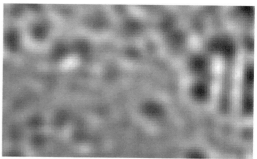

Figure 7c. Microwave image of the area shown in Fig. 7a, transmission mode.

Figure 7a. Optical transmission image of an injection moulding part containing bubbles (bright spots on the right image side). On the right border the ends of some cooling slots can be discerned. Area about $53 \times 33 \, mm^2$.

drawback is a higher expense of the measuring system. It is expected that in future the well-priced high integrated FMCW radar sensors as described in Ch. 2 can replace at least partly the high-priced network analyzer devices.

4 CONCLUSIONS

These first results show the great potential of microwave sensors and especially of high integrated FMCW radar sensors to characterize plastics materials non-destructively online and offline. Thus the microwave and radar technique may be useful both for in process control of injection moulding processes and offline materials characterization.

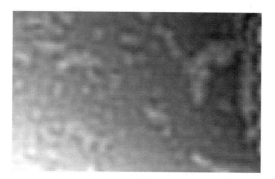

Figure 7b. Microwave image of the area shown in Fig. 7a, reflection mode (bright: high amplitude; dark: low amplitude).

layer of the specimen. The "good" material contained microscopic pores which scatter the light in all directions. Therefore the optical transmission is lower there. In the microwave image the bubbles can be discerned by brighter spots in reflection mode (Fig. 7b) and darker spots in transmission mode (Fig 7c, bright colour: high amplitude; dark colour: low amplitude).

Similar experimental arrangements like the one described above have been reported in literature (e.g. Diener, Busse, Qaddoumi et al., 1996). These measurements were done in reflection mode with fixed frequency sensors. Compared to these publications the experimental setup of IZFP offered the advantage to allow both refection and transmission measurement and both amplitude and phase evaluation. Thus the information content was increased. However, the

REFERENCES

Volkmann, K., Schulz, S. PC controlled automatic systems for thickness gauging with ultrasound, 15th World Conf. on Non-Destructive Testing, 15–21 Oct. 2000, Rome

Tessmann, A., Kudszus, S., Feltgen, T., Riessle, M., Sklarczyk, C., Haydl, W.H. 2002. Compact single-chip W-band FMCW radar modules for commercial high-resolution sensor applications, IEEE Transactions on Microwave Theory and Techniques, Vol. 50, No. 12, 2995–3001

Diener, L., Busse G. 1996. Nondestructive quality and process control in injection moulding polymer manufacture with microwaves, Materials Science Forum, Vols. 210–213, 665–670

Qaddoumi, N., Abiri, H., Ganchev, S., Zoughi, R. Near-Field analysis of rectangular waveguide probes used for imaging, Rev. of Progress in Quantitative Nondestr. Eval., Vol. 15, Plenum Press, 1996, 727–732

Emerging Technologies in Non Destructive Testing, Van Hemelrijck, Anastasopoulos & Melanitis (eds)
© 2004 Swets & Zeitlinger, Lisse, ISBN 90 5809 645 9

Applying pulsed eddy current in detection and characterisation of defect

M. Zergoug, A. Haddad, A. Hammouda & S. Lebaili
Laboratoire d'Électronique et d'Électrotechnique, Centre de Recherche Scientifique et Technique en Soudage et Contrôle, Algeria

ABSTRACT: Pulsed eddy current in the non destructive testing is very used such the defects detection, the conductivity measurement, and the estimation of the coatings thickness. The application in the industrial domain is very weak compared to conventional eddy current. Pulsed Eddy Currents excites the probe with an impulse current. In this article; we show the importance of probe construction in eddy current pulsed. The reception coil transports the induced tension of the materials magnetization and gives the possibilities to analyze pulsed eddy currents signal. The results obtained explain the behaviour of the Pulsed Eddy Currents and their influences in various electromagnetic parameters on the inspection. The technique showed their importance in the evaluation of the multi-layer structures (stuck mechanically or adhesive). Also we have studied the sensitivity of the defects and other parameters in the inspection by the pulsed method and we have showed the detection of the defects into the second and third layers.

1 INTRODUCTION

The development of non-destructive testing has a considerable importance. Pulsed eddy current measurements is very used in the non destructive testing such the detection of defects, the measurement of conductivity, and the estimation of the thickness of coatings. The basic advantages of transient system are: firstly, the circuitry, it is relatively simple compared with that needed for broad band alternating current testing.

Secondly the single transient response contains much information. In order to extract the information, and thus, realize the full potential of pulsed eddy current testing, the signals must first be analysed.

In the pulsed eddy current, a tension of a square form excites a sensor which induces an impulse with wide strip, which made an electromagnetic wave in a conductor.

Theoretically, a transient field or an induced voltage is related from the corresponding time harmonic complex amplitude through the Laplace transform. Consequently, results from an analysis of alternating fields can be used to determine the variation of the probe signal or the electromagnetic field with time. This paper present the system design, and the results of pulsed eddy current experiments performed on laboratory on multi-layered specimens. The various ways of presenting PEC data will be illustrated, such as the time based A-scan C-scan image formats and measurement thickness of coatings.

2 PHYSICAL APPROACH AND SIMULATION

The pulse unit employed to drive the transmitting probe coil. The capacitor is charged quickly from a high square voltage source, the discharges are very rapid. The voltage across the probe coil rises very fast and decreases toward zero. The resulting current in the probe produces a pulsed electromagnetic field that penetrates into the steel sheet. The important characteristics of the output pulse are the height of the peak voltage and the time delay from the pulse beginning to the peak of it.

The determination of the material's parameters by pulsed eddy current is based on the measurement of the variation response of impulse excitation.

Different parameters, which influenced the response, are the permeability, thickness or other anomalies. In physical structure. The form of induction is given by (Kim, 2001).

$$\frac{B(t)}{B_{00}} = \frac{1}{(w_0\tau^{1/2})} \frac{(\cosh^2 2k - \cos^2 2k)^{1/2}}{\cosh 2k + \cos 2k} \sin(w_0 - \vartheta)$$

$$+ \sum_{m=1}^{\infty} \frac{2}{\pi^2 (m-1/2)^2} \frac{e^{-smw0t}}{s_m + 1/s_m} , (0 \le t \le t_0)$$

$$= \sum_{m=1}^{\infty} \frac{2}{\pi^2 (m-1/2)^2} \frac{1+e^{sm\pi}}{s_m + 1/s_m} e^{-smw0t}, \quad (t \ge t_0)$$

with parameters $k = (w_0\tau/2)^{1/2}$, $s_m = \pi^2(m - 1/2)^2/ w_0\tau$, and $\tan\theta = (\sinh 2k - \sin 2k)/(\sinh 2k + \sin 2k)$.

On the other hand, if we assume that the average field follows the driving current pulse:

$B(t) = B_{00}sin(w_0t), (0 < t < t_0)$

3 SPECIMENS

Calibration test specimens were constructed from 2 mm thick Aluminium three 50 mm × 210 mm with crack, and one with different depth striated, this sheet was placed another. By inverting this combination, the specimen simulates thinning on the top of the second layer and third.

A similar approach was taken to simulate varying lift-off distances from the probe to the surface.

4 EXPERIMENTAL PROCESS

The reception sensor detects any variation or change due to the influence of the material characteristics. This response is a measurement of the total magnetic field on the surface of material, which includes transmission and reception of fields.

The penetration of magnetic field in material undergoes a widening or a delay according to the depth. Since the incident field remains constant, any variation of the signal at the edge of reception probe is due to the eddy currents. Consequently, a balanced signal is obtained by the difference of the measured signal and the reference signal.

The resulting signal represents primarily the disturbances of the reflected field due to the anomalies. The pulsed eddy currents sensors are in differential mode (excitation-reception).

The position of the reception probes are adjacent or in the center of excitation probe, (figure (1)).

Pulsed eddy currents methods consists of the potential difference measure at the boundaries of sensor. The realization of the sensors is very significant for the eddy currents testing.

The reception probe transports the induced tension of the materials magnetization and gives the possibilities to kept and analyze pulsed eddy currents signal.

4.1 Construction and characterization of probe

The most important in eddy current testing is the way in which eddy currents are induced and detected in the material under test. This depends on the design of the probe, which can contain either one or more coils. Usually the winding has more than one layer to increase the value of the inductance for a given length of coil. The main purpose of the former is to provide a sufficient amount of rigidity in the coil to prevent distortion. The region inside the former is called the core,

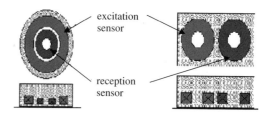

Figure 1. Diagram of the probes used one controls by PEC.

which can consist of either a solid material or just air. Small diameter coils are usually wound directly on to a solid core, the greater the sensitivity of pulsed eddy current testing.

The construction of the probe for control by pulsed eddy current requires certain conditions in order to increase their sensitivity and their resolution. Each control requires a suitable probe and condition between excitation probe and reception probe.

The study of the fields created in the vicinity of a probe permits to see the field activity and to optimise the adaptable control. (Zergoug, 2002) allow to note that:

– for the small diameter, the range of field variation is small and the strength is very height.
– for the large diameter the range variation field is large and the intensity is approximately constant.

The lift off with the diameter variation probe show that the attenuation of the field is highly for the small diameter.

We have realised an experimental device which will allow, to create and measure the pulsed eddy currents and to record signals in order to treat them numerically.

The measuring equipment was designed so as to be able to identify the response of the sensor of defects and change of structures.

4.2 Excitation system

Excitation system is composed by generator providing square amplitude which, supplies the excitation sensor, via a capacity of discharge.The choice of the square frequency must be sufficient for the load and the total discharge of the capacity.

4.3 Amplification

After the crossing of the excitation signal through the transmitting probe, the signal on the level of the reception sensor is amplified.

4.4 System of displacement

This probe is put in a mechanical system that allows using step by step three motors. The movement

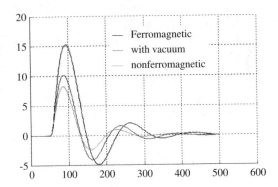

Figure 2.   Response of various materials.

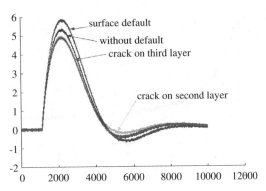

Figure 3.   A-scans representing flaw transients.

following the three axes: X, Y and Z. A software allows the acquisition of measure parameters of the signal amplitude by the means of an IEEE 488 interface and the mechanical system control by the means of the RS 232 interface.

The testing data and conditions are directly provided to the software. The results appear under the from of a color chart representing the variation with two dimensions for each one of chosen parameters.

This visualization is realized in comparison with the display parameter in respect to that of reference previously taken for each one of the sweeping points. Besides the color chart is affected by a fixed sensibility that corresponds to a color set.

This one is function of the minimal required variation.

5   MEASUREMENT AND ANALYSIS

The penetration limit of the Pulsed Eddy Currents (PEC) is significant, it is necessary in a first approach to determine the influence of the various parameters.

The signal obtained by the receiving probe is visualized in figure (2).

The signals obtained on the ferromagnetic and nonferromagnetic material testing allow noting the behavior of the PEC in the presence of a driver.

The frequencies and amplitude analysis allow the extraction of the information of the measured signal.

6   ANALYSIS OF AMPLITUDE

A significant variation of amplitude of the signal is noted on the figure (2).This variation is due to the creation of the eddy currents on the surface of the conductor. Through study is necessary to determine the various parameters influencing the variation of amplitude. We carry out modifications of certain

parameters and to see their behaviours on pulsed eddy currents thereafter.

6.1   *A-scans*

Figure (3) shows the transient signals measured by emission – reception of coil for various typical tests parameters and particular conditions of material.

The signal is a voltage measured across a reception coil, the curve can be considered as relative measurements of the tension induced into the reception coil due to the changing magnetic fields at the surface of the material. These signals, therefore, contain the total information on both the transmission and reception fields. It is immediately apparent that the signals undergo only very subtle shape and amplitude changes, even for extremely different test conditions.

The results found (A. Le Blanc,1997) have been reproduced in these experiments to show the distinctive features of the A-scans and to illustrate their potential for discriminating the different locations of the thickness.

As expected, thickness in the first, is equal to the second and the third layer.

While scanning different pieces, one will eventually encounter several structural variances with crack or other flaws, making the inspection more complex. PEC signals are extremely sensitive to the smallest amounts of plate separation. Another common noise source is the constant variation in the lift-off distance between the probe face and the surface. Substructure. Changes such as metal thickness increase due to the encounter of stringers also affect the signals differently.

6.2   *Behaviour of the lift-off*

The lift-off indicates universally the effect of separation of the sensor compared to the surface of the part, above of which it evolves. The lift-off represents a disadvantage for measurement, because the interpretation of

339

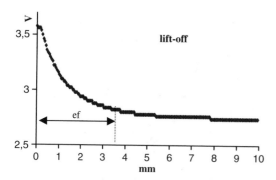

Figure 4. Variation of amplitude with displacement of sensor below the non-ferromagnetic sample.

the result becomes very difficult (fast decrease of the electromagnetic fields in the air), but it can be an advantage to detect the uniformity or measuring by the insulating coating.

One taking account of the variation of distance between the probe and the object figure (4).

We notice an exponential fall. This variation is due to the weakening of the eddy current.

We notice an exponential fall. This variation is due to the weakening of the eddy current.

We can exploit this curve for measuring the isolating coating; the thickness of the coating can be measured until the distance **ef** and to reduce the effect of the lift-off. We work at certain distance from the sample.

6.3 *Influence of defect*

For defects located at various depths, the result obtained by PEC indicates a variation of the amplitude, more the defect is deep, more the signal is significant, figure (5a, 5b) the creation of the PEC is weak, that is due to the lack of metal on the surface. We can make a difference between the surface defects and the deep defects by the behavior of the signal (amplification or attenuation of the signal compared with the state without defect).

The significant remark in this case is the depth of detection by this method. If the defect is located into the second or the third layer, we notice that the amplitude increases according to thickness added in figure (6a).

Significant parameters to be considered are the of influence of thickness on the signal and the position of defect figure (6a, 6b).

6.4 *Influence of defect size*

By Pulsed eddy currents it possible to detect defects located at various depths into the case of multi-layer. In this case we study the influence of the defects size. Same defects were tested where the length and width are constant but different size.

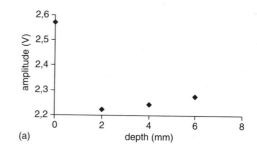

(a)

Figure 5a. Variation of the amplitude in function defect depth.

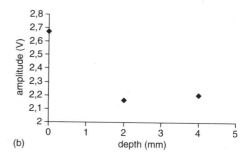

(b)

Figure 5b. Variation of amplitude according to the defect depth.

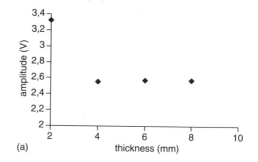

(a)

Figure 6a. Variation of the surface amplitude defect according to thickness.

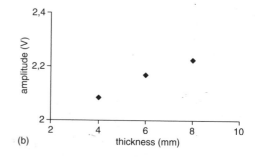

(b)

Figure 6b. Amplitude variation of defect on 2nd layer according to thickness.

340

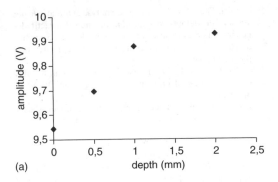

(a)

Figure 7a.   Variation de profondeur du défaut for 31A008.

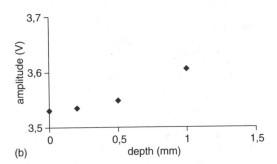

(b)

Figure 7b.   Variation de profondeur pour le 29A029.

The result obtained shows that:

More the defect is large more the amplitude is significant figure (7a, 7b).

In more one notices that the amplitude is more significant for the ferromagnetic, due to the permeability influence figure (2).

6.5   *Representation in two dimension of the defects*

Generally, in the non destructive testing, the A-Scan does not allow the diagnostic of the sample defect. In order to highlight behaviour in the detection of the defects (crack) at various depths, the image obtained by C-Scan is important

Time gates can be used to select certain portions of the A-scan data for analysis and can be generated for a portion of the A-scan which corresponds to a range of depth in the specimen.

The tension measured at the edge of reception sensor of the pulsed eddy currents is done by measuring the resulting field (incidental and reflected), the amplification or the attenuation depends on the electric and the magnetic parameters of material.

The presence of an anomaly in the structure of the material influences these parameters. The resulting field is disturbed and then the variation of amplitude.

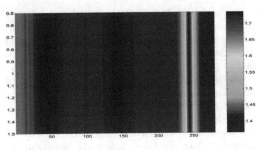

Figure 8a.   Representation of defects locates at different depth one 2D.

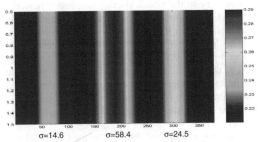

Figure 8b.   Representation of inclusion defect simulated by three type of conductivity.

In our case, we carried out the testing on fissured samples with the various depths compared to inspection surface.

The detection of defect inclusion by three types of materials, with different conductivity covered by a layer of aluminium.

The results obtained confirm that more the defect is deep more the answer is weak as it is translated by the image figure (8a, 8b).

The dimension of defect represented by the image obtained by C scan is superior to the real size of defect due to the edge effect.

The excitation field is propagated $< -2R$ until $< 2R$.

The solution suggested is the focusing the magnetic field to minimise the edge effect.

7   CONCLUSION

The technique of pulsed eddy currents presents great potentialities in the development of applications of the non destructive testing of materials in particular in aeronautic domain and the aluminium inspection.

The experimental study of the non destructive testing by Pulsed Eddy Currents into multi-layer has contributed to the determination of the reliability and the performance of this inspection.

The sensitivity to the defects and other parameters of testing can be modified by the design of the probe. The reception coil transports the induced tension of the materials magnetization and gives the possibilities to analyze pulsed eddy currents signal. The results obtained explain the behaviour of the Pulsed Eddy Currents and their influences in various electromagnetic parameters on the inspection. The technique showed their importance in the evaluation of the multi-layer structures (stuck mechanically or adhesive). Also we have studied the sensitivity of the defects and other parameters in the inspection by the pulsed method and we have showed the detection of the defects into the second and third layers.

Contrary to the classical eddy currents where the penetration is limited, the penetration of pulsed eddy currents is higher than 10 mm.

The detection of the position of the defect is conditioned by the thickness of the sample.

The decrease of the magnetic field in the presence of the various mediums is confirmed for the pulsed eddy currents, this decrease is exploited for the measurement of isolated coating and detection from the defects.

The image of defects obtained by the C scan allows to visualize the defects and to characterize them thereafter.

REFERENCES

A. Le Blanc, M. Piriou, Detection and characterisation of defect by pulsed eddy current, 1997 COFREND Congress on NDT-France

B. Benoist, Thèse Système expert d'analyse automatique des signaux de défauts lors du contrôle non destructif des générateurs de vapeur des centrales nucléaires, Soutenue le 16 octobre 1990 université de technologie de compiegne France

C.V. Dodd and W.E. Deeds, Analytical solutions to eddy current probe-coil problems, J. appl phys vol 39 No. 6 pp 2829–2838

D.L. Waidelich, Pulsed eddy current testing of stell sheets, Eds American society for testing and materials, 1981, pp 367–373

F. Thollon, F. Clauzon-CEGELY, france, Flaw characterization in metallic assemblies with pulsed eddy current NDT, 7th ECNDT 1998 PROCEEDINGS vol 1 p. 311

G. Birnabaum, G. Free, Eddy current characterization of materials and structures. American society for testing and materials, ASTM special technical publication 722

J. Bowler et M. Johnson, Pulsed eddy current response to conducting half-space IEEE transactions on magnetic published may 1997 vol 33 No. 3 p. 2258

M. Zergoug A. Haddad, A. Hammouda, Characterization of defect in multi layers by pulsed eddy current, 11th Symposium Berlin, June 2002

M. Zergoug, A. Hammouda, A. Haddad, Conductivity simulation for material characterisation by eddy current, Conférence Barcelone juin 2002

S.H. KIM Calculation of pulsed Kicker magnetic field attenuation inside beam chambers JANUARY 8, 2001

Th. Meier, G. Tober, CS-Pulsed Eddy Current Inspection for Cracks in Multi-Layered Joint Al-Alloy Aircraft Structures, Daimler-Benz Aerospace Airbus GmbH, Bremen, Germany, 7th ECNDT 1998 PROCEEDINGS vol 3 No. 9

Special applications

Emerging Technologies in Non Destructive Testing, Van Hemelrijck, Anastasopoulos & Melanitis (eds)
© 2004 Swets & Zeitlinger, Lisse, ISBN 90 5809 645 9

Feasibility of using earth-bounded NDT techniques for the space environment

V. Nikou, P.F. Mendez, K. Masubuchi & T.W. Eagar
Massachusetts Institute of Technology, Cambridge, USA

ABSTRACT: In this paper we review non-destructive testing (NDT) techniques with potential use for welding inspection in the space environment. In our comparison of NDT techniques we include fields that have not considered previously: materials to be welded and type of welding to be used.

1   INTRODUCTION

Space structures are very complicated structures that were put in space by man. One of the requirements of these structures is to have a sufficient life to provide economical exposure time in the space environment. Unfortunately space structures are not only complicated but in most cases they cannot be shut down for repair; their maintenance and repair has to be done while working. That is one of the main reasons that preventive maintenance is crucial.

For a weld to have the required reliability throughout its life, it must have a sufficient level of quality or fitness for purpose. On Earth, weld quality is often governed by codes, specifications, or regulations based on rational assessment of both economics and safety. Most of the above is experiential. In the space environment, the quality requirements are especially strict; therefore, welds that might be considered adequate on Earth might not be adequate for space. Previous work done on non-destructive testing (NDT) techniques [1–6] for the space environment involves mostly an evaluation of NDT as a monitoring process of the space structures rather than the welds. The reason is simple: welding has not yet been used as a fabrication process in space. For example, in the International Space Station (ISS), only mechanical joints will be used for the erection. Welding is not considered for the assembly, even though it could save considerable weight.

The advantages of welding as a joining process over mechanical joints are well known. It is only a matter of time before welding will be used for fabrication in space as mankind evolves to its destiny in the habitation of extra-terrestrial places. Russians have already experimented with welding techniques in space, and there is already an on going project in NASA that is dealing with using electron beam welding (EBW) as the technique to bring space fabrication to the next level.

This paper deals with the analysis of Earth-bounded NDT techniques and how they can be applied in the space environment. An analysis of the space environment and its unique characteristics is given, and an original qualitative analysis of the current NDT processes is done in order to examine the possibility of use in space. Later a similar analysis is performed dealing with the possible welding techniques to be used for welded maintenance and repair in space.

2   THE EFFECT OF SPACE ENVIRONMENT ON NDT METHODS

Nondestructive testing as well as any other activity in space faces all the peculiarities of the environment in space. The most important features of the space environment that make it unique are: (a) zero gravity, (b) vacuum, (c) radiation, and (d) composition of the residual atmosphere. All the above factors have to be accounted in order to select an Earth-bounded NDT process for use in the space environment.

2.1   Zero gravity

Strictly speaking, in space there are never zero gravity conditions (the condition under which the forces acting on an object are zero). It is better to use the term low gravity condition characterizing the condition under which the sum of the forces acting on a body is considerably smaller than on the Earth's surface. This state is usually evaluated by the value of the ratio of acceleration given to the body by the acting force (g) in relation to the force of gravitational attraction on the Earth's surface (g_0). For objects freely flying in space, the value of this ratio is of the order of 10^{-5} to 10^{-7} [7]. Under low gravity conditions, a lot of the physical processes in liquid and gaseous media related to density, convection, and surface tension change greatly.

2.2 *Space vacuum*

The mean pressure in the height range of 250–500 km is 5×10^{-4} Pa ($\sim 5 \times 10^{-9}$ Atm) [7]. The unusual features of space vacuum are the composition of the residual atmosphere and the extremely high pumping rate (diffusion rate) of gases generated in it. The pressure of the residual atmosphere surrounding space structures in low orbits is easily achieved on Earth by the use of vacuum chambers. The thing that significantly differs though is the composition of the two "vacuums". The rarefied atmosphere generated in vacuum chambers on Earth differs from the space atmosphere by the absence of atomic oxygen and the low mobility of molecules. In space, the content of atomic and ionized oxygen is very high and may exert a strong effect on joining of materials both during welding and in further service of joints.

Since space is an open infinite volume, the gas molecules generated at the surface of a space system rapidly move into space. Thus, the thickness of the natural residual atmosphere of space systems is very small. In addition local pressure gradients are almost instantaneously equalized. Therefore substances with high vapor pressure rapidly evaporate in space. That is why it is very difficult to use welding or NDT methods requiring gases or liquids with vapor pressure.

2.3 *Space Radiation*

Space radiation refers to the vacuum ultraviolet radiation (VUV) of the Sun, which greatly intensifies the oxidation processes on the irradiated surfaces, as well as the radiation coming from the radiation belts of the Earth.

The VUV as well as the absence of the atmosphere outside the space system are the reasons of another interesting phenomenon in space: the wide temperature variations of the space structures. The illuminated sections of a space structure, when in the Sun section, may be heated to a temperature of 420 K (150°C). On the other hand, when the structure is in the shadow section, it may have a temperature of 160 K (−110°C). If any part of the structure is oriented for a long period in the same direction and is in the shadow for instance, it may have even lower temperature. Any NDT machine has to account for this temperature.

2.4 *Composition of the space environment*

The space environment in Low Earth Orbit (LEO) (any earth orbit up to approximately 1500 Km) consists of neutral atmosphere, atomic oxygen, atomic hydrogen, meteoroids, and space debris. The content of atomic oxygen in space is responsible for the increased corrosion strength of the space environment. Atomic oxygen results from the interaction of solar radiation with oxygen. In space, the content of atomic and ionized oxygen is very high and may exert a strong effect on the service life of the equipment. Atomic oxygen is considered one of the most damaging aspects of the space environment experienced thus far [2].

3 NDT METHODS

3.1 *Visual*

Visual inspection is the easiest and fastest way to inspect a weld and is the most commonly used method on earth. These techniques apply to welds with discontinuities on their surface. Gross surface effects, such as severe undercut or incompletely filled grooves, can lead to immediate rejection of a weld, before any more detailed testing is undertaken. Considering these limitations as well as the necessity for quality welds in the space environment, visual methods should be used, if at all, only as a preliminary examination before the use of a more elaborate method, especially for critical joints.

3.2 *Radiographic*

In this NDT technique, X-rays, or gamma rays are used to in a manner similar to a medical X-ray to detect subsurface flaws. The changes in density are indicated on a film, or stored digitally. Even though radiographic methods produce radiation hazards, the radiation protection of the space suits and spacecraft walls might provide sufficient protection. This is true at least for the X-ray inspection of thinner sections. This method is very sensitive and portable, and devices are readily available for earth-bound applications.

3.3 *Ultrasonic*

The Ultrasonic method refers to techniques that use high frequency sound waves, which are transmitted through or reflected from objects and interfaces such as bond lines or cracks between objects. Nearly all methods of ultrasonic flaw detection use the pulse technique in which a short ultrasonic pulse is propagated from a transmitter probe, through a coupling medium, into the material under test.

Another variation of the method is the one in which the transducer is not in contact with the material to be inspected (this is called airborne or non-contact technique). Airborne techniques are not suitable for EVA (Extra Vehicular Activity) space applications because of the lack of atmosphere to transmit the vibrations to the workpiece.

Although vacuum and large temperature do not permit the use of common liquid couplants, in a NASA study [1] it was found that some space-graded compounds commonly used as lubricants are suitable for long-term use in the space environment as couplants for the ultrasonic methods. Those kinds of couplants are mainly silicone greases, or fluorinated oils.

3.4 *Magnetic*

Magnetic methods reveal structural defects through the orientation of magnetic particles on the surface of the work piece. There are two types of Earth-bound magnetic methods: wet and dry. Wet methods are difficult to be implemented in space. Perhaps the use of low vapor pressure oils might enable its use. Because of this difficulty, these methods were ruled out of consideration in a NASA study as a possible NDT technique for the space environment. Dry methods should work in space as well as on Earth.

3.5 *Penetrant*

There is general agreement in the literature that liquid penetrant methods are unsuitable for NDT applications in space. These kinds of methods can only operate down to 10^{-2} torrs ($\sim 10^{-5}$ Atm) [1], far higher pressure than existing in the residual atmosphere of space. Thus, they will not be further considered in this study.

3.6 *Electrical (Eddy current)*

Eddy current methods are the electromagnetic method where a magnetic coil is used to induce an electric current in a material. Eddy current methods are useful in surface analysis and shallow crack detection, but not as sensitive as ultrasonic or radiation methods for deep cracks.

3.7 *Acoustic Emission*

Acoustic emission (AE) is a different concept from the pre-mentioned methods in the sense that it is an entirely passive test (no external signal is put into the material to be tested). AE is elastic radiation generated by the rapid release of energy from sources within a material, which increases as the material approaches fracture.

The AE has to be detected in real time. The amount of emission produced may be affected by the background noise. Another important phenomenon, which must be considered in the AE detection, is the Kaiser effect. According to this effect, no new AE occurs when the specimen is re-stressed until the previous maximum stress has been exceeded. Even though AE was one of the top candidates for in space use in a past study [1], taking the above into consideration, and in particular the fact that AE has to be detected as it occurs, it will not be further considered. This technique may apply in health monitoring of a space structure rather than in weld flaw detection.

4 NDT METHODS EVALUATION

In 1985, the NASA and the TRW space technology group published a final report of a nondestructive equipment study [1]. Part of the report was a table summarizing and comparing various NDT methods, mainly qualitatively. Inspired by this approach a similar comparison table (Table 2) was constructed in this paper attempting to compare the considered NDT techniques more quantitatively than qualitatively. Furthermore, some different areas of comparison, not mentioned in the previous study are considered (materials, and welding techniques used in space). The visual, penetrant, and acoustic emission techniques were ruled out based on the considerations mentioned in the previous section.

In Table 2, various performance factors are considered for the four types of NDT methods selected. These factors deal with the ability that each NDT method has to detect a flaw, the materials welded in space, the geometry of the welds, the ease of operation and the user's safety. A brief analysis of each performance factor gives a more detailed understanding of the ability of each NDT process to perform an inspection in the space environment.

4.1 *Flaw detection*

The ability of each NDT process is examined by presenting actual values of the minimum size of a flaw that can be detected by each process as well as the maximum depth where the process is effective [8, 9]. The depth for the radiographic method is selected so as to have a reasonable big, portable device, in the size of 500 kV. In the ultrasonic method the maximum depth (150 mm), in order to scan for the one side of the weld, was used. The magnetic method is most suitable for cracks open to surface, while some large voids slightly in the sub-surface area might be detected. Finally the values for the eddy current method refer to the standard (37%) depth of penetration [8].

4.2 *Materials*

The materials considered are those used for welding in space or have the potential to be used in the future (Aluminum, Titanium, Metal matrix composites, Austenitic stainless steel, and Martensitic stainless steel). The radiographic method can be used in all metals. Ultrasonic inspection of austenitic welds is difficult because of the microstructure. In the cooling process, after welding, large grains with a high degree of orientation may develop in the form of large elongated columnar crystal with a fibre texture. Thus, the ultrasonic velocity is different in different directions and the material exhibits much greater scattering effects. If ultrasonic flaw detection to austenitic steel welds is performed the results will be [9]:

(a) much higher attenuation, leading to loss of echo pulses;
(b) much greater noise due to scattering;
(c) large spurious signals due to either grain boundary reflections or to beam bending;
(d) changes in the beam shape and beam path direction.

Only ferromagnetic materials could be inspected using the magnetic method, consequently, only two of the materials are not ruled out. As far as the eddy current method goes, the only limitation is that it can only inspect conductive materials.

4.3 Geometry of welds

An analysis of the weld geometry of the possible welding techniques for welded repair and maintenance in space is performed. The assumptions made were: (a) the weld geometry of electron beam and laser beam welding methods in keyhole mode is similar. The weld geometry of defocused electron beam is similar to that of arc welding processes. (b) various arc welding processes (gas tungsten arc welding (GTAW), metal arc welding (GMAW), plasma arc welding (PAW), and gas hollow tungsten welding (GHTAW) have also similar weld geometries and (c) the geometry of resistance welding (RW) joints made in space is similar to that on earth. The values of the weld width and depth for each process were found using the minimum and maximum values of several welds performed in simulated space environment and cautiously relaxing the limits [10–13]. The actual geometry values for each welding process are summarized in Table 1. LBW stands for Laser Beam Welding and AW for Arc Welding.

4.4 Ease of operation

Currently the NDT technology is advanced enough to make equipment portable, light and easy to get familiar with and use. Possible problems might the process of the data after the NDT process is used. Even in cases like that, the new techniques allow to transfer the results of the tests (i.e. digital X-rays or ultrasonic spectrums) to the earth based station for further interpretation.

Table 1. Summary of geometry welds values for the various processes.

	Weld pool diameter or width of weld (mm)	Penetration (mm)
B	<100	0.001–0.1
EBW	1–3	0.5–50
LBW	1–5	0.5–20
AW	5–15	0.5–10
RW	5–10	0.1–1

Table 2. Comparative analysis of the NDT techniques considered for use in space.

Methods	Radiographic	Ultrasonic	Magnetic	Eddy current
Flaw detection				
Size	2% of thickness	>1–5 mm depending on frequency	>0.5 mm	>0.1 mm
Depth	25 mm SS 80 mm Al	<500 mm	Surface of near surface cracks	<13 mm SS <3.5 mm Al
Materials				
Al	Y	Y	N	Y
Ti	Y	Y	N	Y
MMC	Y	Y	N	N
SS$_1$	Y	N	N	Y
SS$_2$	Y	Y	Y	Y
Geometry of welds				
B	Y	N	N	N
EBW	Y	Y	N	Y
LW	Y	Y	N	Y
AW	Y	Y	N	Y
RW	N	N	N	Y
Ease of operation	Good	Good	Good	Good
User's safety	Radiation	None	None	None

4.5 Safety

Of the NDT processes considered, the only one that raises safety considerations is the radiographic method. The inspection of thick sections, especially of steels and titanium-base alloys, might require doses that exceed the protective capabilities of standard space suits. Since X-rays propagate in "line of sight" trajectories, simple protections, such as lead-lined walls might enable safe operation.

The following abbreviations are used in Table 2:

MMC	(Metal matrix composites),
SS_1	(Austenitic stainless steel),
SS_2	(Martensitic stainless steel),
B	(Brazing),
EBW	(Electron beam welding, keyhole mode),
LBW	(Laser Beam welding),
AW	(Defocused electron beam welding and Arc welding, including gas arc welding, metal arc welding, plasma arc welding),
RW	(Resistance welding),
Y	(Yes),
N	(No).

5 SUMMARY

Even though welding in space is in its primary stages there will not be long before the need of repairs arises. Moreover the only way to erect big structures in space that will support habitats for a long period of time is by moving from the mechanical joints, currently used for space structure erection, to welds and together the inevitable need of non destructive inspection as a preventive maintenance.

There are earth-bounded NDT processes that could be used in space. Although there is not a single cure-all technique to be used to non-destructively inspect welds or structures in space. Like on Earth, different techniques are superior in some aspects but inferior in others. This could be also seen from the comparative analysis in Table 2. Furthermore no technique could be actually proven reliable, unless it will actually operate in the environment that it is designed for.

Finally more work should be done as far as the comparative analysis is concerned in the aspect of weld defects. Every welding method generates different defects that may not necessarily be inspected from a NDT method. Thus, another important performance factor that should be taken into account is how the defects associated with the various welding processes preclude the use of different NDT methods.

REFERENCES

1. Chung, T., Kwan, M. et al, Nondestructive Equipment Study, Final Report, NASA-CR-171865. TRW, Space and Technology Group, NASA L.B. Johnson Space Center, 1985.
2. Lynch, C.T., On-Orbit Nondestructive Evaluation of Space Platform Structures. Vitro Technical Journal, 1992. 10(1): p. 3–16.
3. Ithurralde, G., Simonet, D., EMATs for on orbit wall remaining thickness measurement after an impact-feasibilty study. NDT.net, 1999. 4(1).
4. Finlayson, R.D., Friesel, M., Carlos, M., Health Monitoring of Aerospace Structures with Acoustic Emission and Acousto-Ultrasonics. Roma, 15th World Conference on NDT, 2000.
5. Simonet, D., Ithurralde, G., Choffy, J.P., Non destructive testing in space environment. Roma, 15th World Conference on NDT, 2000.
6. Georgeson, G.E., Boeing Co. (USA), Recent advances in aerospace composites NDE. Proceedings of SPIE, Nondestructive Evaluation and Health Monitoring of Aerospace Materials and Civil Infrastructures, 2002. 4704(18): p. 104–115.
7. Paton, B.E. and Lapchinskii, V.F., Welding in space and related technologies. 1997, Cambridge, England: Cambridge International Science. 121.
8. Welding Technology, Welding Handbook. 8 ed. Vol.1. 1987, New York:American Welding Society.
9. Halmshaw, R., Introduction to the Non-Destructive Testing of welded joints. 1988:Edison Welding Inst.
10. Russell, C., International Space Welding Experiment (ISWE),Science Requirements Document. 1996, NASA-Marshall Space Flight Center.
11. Nishikawa, H., Yoshida, K.,Ohji, T., Fundamental characteristics of GHTA under low pressure. 1995.
12. Nishikawa, H., Yoshida, K., Maruyama, T. et al, Gas hollow tungsten arc characteristics under simulated space environment. Science and Technology of Welding and Joining, 2001. 6(1).
13. Nishikawa, H., Yoshida, K., Ohji, T. et al, Characteristics of hollow cathode arc as welding source: arc characteristics and melting properties. Science and Technology of Welding and Joining, 2002. 7(5).

Emerging Technologies in Non Destructive Testing, Van Hemelrijck, Anastasopoulos & Melanitis (eds)
© 2004 Swets & Zeitlinger, Lisse, ISBN 90 5809 645 9

Non-contact ultrasound imaging for foreign object detection in cheese and poultry

B. Cho & J. Irudayaraj
The Pennsylvania State University, USA

ABSTRACT: Novel non-contact air instability compensation transducers were utilized to detect foreign object and defects in cheese and poultry. Non-contact ultrasound velocity and energy attenuation images could be successfully assessed for metal and glass fragments in chicken breast and holes, cracks, and foreign objects in cheese. Results demonstrated that the non-contact ultrasound imaging technique has a good potential for nondestructive and rapid detection of foreign objects and defects in food materials and could provide to be yet another alternative to conventional X-ray methods.

1 INTRODUCTION

Ultrasound is one of the most promising sensing technologies for food quality measurement due to its rapid, nondestructive, and on-line potential. Its energy may be reflected and transmitted as they pass from one medium to another. The amount of energy reflected and transmitted through materials depends on their relative acoustic impedances described by propagation velocity and density of the media within and between material interfaces. In addition to the ultrasound energy attenuation, the ultrasound velocity could also serve as good indicators of material property or a change in material characteristics since ultrasound velocity is dependent on elastic property and density of the medium (Povey & McClements, 1988). Hence, in the context of foreign objects such as bone, glass, or metal fragments in cheese and poultry, one would expect a strong reflection and refraction at the interfaces of the host tissue and foreign object interface. Objects such as bone, metals, and glass have different physical properties compared to the medium, as a result of which changes in ultrasound time-of-flight and velocity is expected. To date, the conventional ultrasound measurement uses a coupling medium (oils or gels) or an immersion method to overcome the high attenuating power due to the large acoustic impedance mismatch between air and material. The use of coupling liquids and water immersion methods is undesirable in food quality measurement and process control since the chemical agent has the potential to contaminate or interact with the food sample due to absorption, interaction, or its mere presence.

Recently, non-contact ultrasound has become available as an exciting alternative to conventional contact ultrasound, especially for food quality measurements since it eliminates the contamination problem due to contact which makes the measurement cumbersome.

Since non-contact ultrasound utilizes air as its medium, the transmitted signal is highly dependent on the conditions of the air, such as temperature, humidity, and air flow. The instability of the air column causes errors is prone to errors in non-contact ultrasound velocity and thickness measurements. Hence, an air instability compensation technique needs to be developed. In order to compensate for air instability, the transducer performance in varying temperature conditions should be examined in real time for measurement optimization. By placing a fixed reference in front of the transducer and collecting the reflected signal with respect to the reference, it is possible to obtain information of the air column between the transducer and the reference (Cho & Irudayaraj, 2002a). In this study, air instability compensation transducers were used for improved non-contact ultrasound velocity and attenuation images of cheese and poultry samples with foreign objects and/or irregularities.

The objectives of this study were to investigate non-contact ultrasound imaging as a tool to detect a variety of foreign objects and internal defects in cheese and boneless chicken breast.

2 MATERIALS AND METHODS

2.1 Non-contact air instability compensation ultrasound system

The non-contact ultrasound system consists of two 1 MHz piezoelectric non-contact ultrasound transducers (NCT 510, SecondWave System Inc., Boalsburg, PA) with a ring shape reference placed in front of the transducers' surface. The transducers are connected to an analyzer (NCU1000-2E, SecondWave systems, Boalsburg, PA) which can generate synthesized signals and analyze the detected signals simultaneously. The analyzer can measure the time-of-flight up to an accuracy of ± 20 ns under ambient air and ± 1 ns under constant air property conditions. Each transducer operates both as a transmitter and as a receiver, and hence provides four operational modes: two reflection (one for each of the two transducers) and two transmission modes (one used as transmitter and the other as a receiver and vice-versa). Using the four operation modes, the sample thickness and velocity can be calculated directly.

$$D_m = S - [(S_1 \times t_1 / t_3) + (S_2 \times t_2 / t_4)] \tag{1}$$

$$V_m = D_m / [t_c - (t_1 + t_2) / 2] \tag{2}$$

where, D_m is the sample thickness, V_m is the respective velocity of ultrasound through the sample, S is the distance between transducer 1 and transducer 2, S_1 is the distance between the reference and the transducer 1, S_2 is the distance between the reference and the transducer 2, t_m is the time-of-flight in the test material, t_c is time-of-flight between transducer 1 and transducer 2 with sample, t_1 is the round trip time-of-flight between the transducer 1 and the sample, t_2 is the round trip time-of-flight between the sample and transducer 2, t_3 is the round trip time-of-flight between transducer 1 and the reference, and t_4 is the round trip time-of-flight between transducer 2 and the reference (Fig. 1).

In addition to thickness and velocity measurements, the ultrasound energy attenuation through the material can be estimated using the integrated response. The integrated response (IR) is a measurement of the area underneath the most significant peak above -6 dB of the cross-correlated transmitted signal in dB units (Bhardwaj, 2000). The IR provides information of the decrease in energy of the transmitted ultrasound signal through the sample in the time domain. The relative attenuation is calculated by subtracting the integrated response of the transmitted signal through the sample from that of the air column divided by sample thickness. The relative attenuation contains information of the attenuation by reflection on the upper and lower surface of the material as well as the attenuation by absorption and scattering in the material throughout the thickness. To minimize the error caused by diffraction on the sample surface, the sample surface should be kept parallel to the transducer.

2.2 Image scanning system

An X–Y positioning system was integrated with the non-contact ultrasound system to scan ultrasound image of samples. All scanning and data acquisition were controlled by a Pentium II computer via a parallel interface and a real-time operating system (QNIX). Parameters measured are velocity and relative attenuation through the sample. Usually the scan area is set as 20 mm × 20 mm, at spatial intervals of 1 mm by default. The scanned data were transferred to a PC and visualized and post-processed using the Matlab software (Version 6.0, The Mathworks Inc., Natick, MA).

2.3 Calibration of ultrasound measurement

The non-contact ultrasound system was calibrated before measurement with a 25 mm polystyrene block. After aligning the transducers parallel to each other and verifying the shape of the received signal, the polystyrene block was inserted between the two transducers.

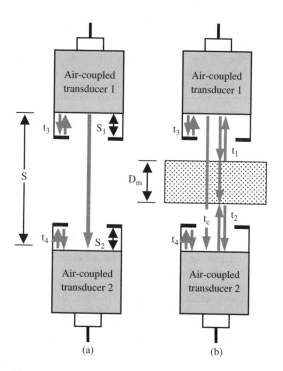

Figure 1. Schematic of the non-contact air instability compensation ultrasound measurement (a) without and (b) with a sample material.

The calculated thickness was compared with the actual thickness, and the calculated value was adjusted by changing the system parameters until the error was within 0.1%.

2.4 *Sample preparations*

Three varieties of Cheddar cheese (Sharp, extra sharp and Vermont sharp white Cheddar) were obtained from the local grocery store. The purchased cheese samples were equilibrated at room temperature ($25 \pm 1°C$) for 12 hours before measurement. From each block of cheese, a cube $45 \times 45 \times 25$ mm was cut consistently using a specially designed wire cutter to provide an even surface for minimizing the diffraction effects. In addition to cheese samples, boneless and skinless poultry breasts were purchased from a local grocery store and pieces of size $50 \times 50 \times 15$ mm were sectioned off and equilibrated at room temperature for 12 hours. Each sample was wrapped in a plastic film to minimize moisture loss. Usually, poultry breasts are often of uneven thickness and shape which causes ultrasound diffraction at the sample surface and makes the measurement difficult. To avoid unreasonable measurements, thin polystyrene panels (0.8 mm in thickness) were placed on both sample surfaces to maintain parallel thickness.

To investigate foreign objects in food materials, three types of foreign objects, such as a metal rod, metal fragments, and glass fragments were prepared. Several dimensions of each object were used to observe the sensitivity of non-contact ultrasound images for a variety of foreign object detection experiments.

3 RESULTS AND DISCUSSION

In figure 2 raw NCU images and its contours of the relative response and velocity values of a 8×10 mm glass fragment in a poultry breast. In general, the relative response provides better images than those of velocity. This is to be expected, since the ultrasound energy attenuation is more sensitive to the difference in product characteristics than velocity. In addition, the induced pressure produced when the foreign objects were inserted into the sample may cause other measurement errors. The differences between the actual and scanned shapes are due to diffraction and refraction effects as well as the 12.5 mm active diameter of the ultrasound transducer, which provides overlapped images (Gan et al., 2002).

Ultrasound attenuation includes ultrasound absorption and scattering. Absorption involves the conversion of the propagating energy permanently into heat energy. Usually, internal friction caused by viscosity is the main reason for the absorption of ultrasound by solids. When an ultrasound wave travels through a non-uniform medium, such as a foreign object embedded material, there is scattering, in which part of the wave changes its initial direction and propagates separately from the original incident wave, distorting and interfering with the initial wave. The discontinuity within a medium, plays a role in scattering of the signal, hence in non-homogeneous material, scattering affects the attenuation more than the absorption.

Figure 3 shows the non-contact relative attenuation images of a steel rod. The NCU image of relative attenuation could clearly detect a cylindrical object as

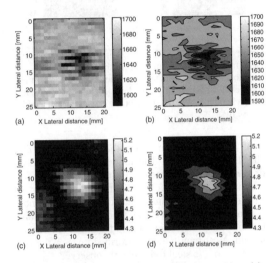

Figure 2. Comparison of raw velocity image (a) and its contour plot (b) and raw relative attenuation image (c) and its contour (d) of a 8×10 mm glass fragment in a poultry breast.

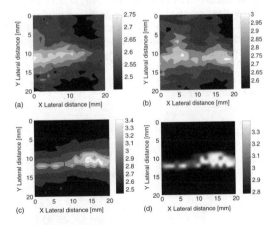

Figure 3. NCU relative attenuation images of a steel rod in diameter of (a) 1.5 mm, (b) 2.3 mm, (c) 4 mm, and (d) its modified image in the middle of sharp cheddar cheese.

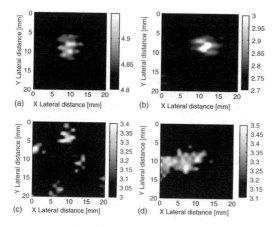

Figure 4. Modified NCU relative attenuation images of a metal fragment (5 × 3 mm²) (a), in a poultry breast and vertical hole (4 mm in diameter) at the bottom of a 2% fat milk sharp cheddar cheese (b), sporadic holes in a Vermont sharp white cheddar cheese (c), and a crack in extra sharp cheddar cheese (d).

small as 1.5 mm in diameter in the food material. The objects and holes were detected with the correct location and reasonable dimensions. In figure 4d, the original image was smoothed using interpolation in which the value of an interpolated point was a combination of the values of the sixteen closest points, and the color ranges were optimized for improved visualization. As shown in the transformed images, the linear shape of the steel rod is clearly observed.

Other foreign objects and artificial defects were introduced at the surface or embedded inside cheese and poultry samples as displayed in figure 4a, b. In addition to artificial objects and holes, some unidentified blocks of Vermont sharp and extra sharp cheddar cheese were scanned to investigate their non-uniform characteristics since air bubble and cracks were frequently observed in the cheese samples examined. The sporadic holes and a relatively long crack embedded in the cheese blocks were detected with NCU images as demonstrated in figure 4c, d (Cho & Irudayaraj, 2002b).

4 CONCLUSION

The non-contact air instability compensation ultrasound transducers, improved the velocity and thickness measurements of the calibration standard in unstable air property environment. Even though the advantages of the air instability compensation transducer did not affect the images of velocity due to low quality images, accurate images of relative attenuation could be obtained with the stable thickness calculation. Foreign fragments of a minimum size of 3 × 3 mm² in cheese and poultry and cylindrical objects of minimum size of 1.5 mm in diameter in the food materials could be clearly detected, even though the images were distorted by diffraction and refraction effects. To improve the detection accuracy and scanning speeds, a better scan system and a higher frequency ultrasound transducer integrated with advanced signal processing technique need to be developed.

The results of NCU images demonstrate its potential as a non-destructive, rapid, and economic tool to detect the presence of foreign objects and defects inside food materials. To the authors' knowledge, this is the first application of non-contact air instability compensation ultrasound imaging of foreign object detection in solid foods.

REFERENCES

Bhardwaj, M.C. 2000. High transduction piezoelectric transducers and introduction of non-contact analysis. The e-Journal of Nondestructive Testing & Ultrasonics [serial online]. 1: 1–21. Available from NDT net (http://www.ndt.net). Posted January 2000.

Cho, B. & Irudayaraj, J. 2002a. Design and application of a non-contact air instability compensation ultrasound transducer using spatial impulse response. Transaction of the ASAE (in press).

Cho, B. & Irudayaraj, J. 2002b. Foreign object and internal disorder detection in food materials using non-contact ultrasound imaging. Journal of food science (in press).

Gan, T.H., Hutchins, D.A. & Billson, D.R. 2002. Preliminary studies of a novel air-coupled ultrasonic inspection system for food containers. Journal of Food Engineering 53: 315–323.

Povey, M.J.W. & McClements, D.J. 1988. Ultrasonics in food engineering. Part 1: Introduction and experimental methods. Journal of Food Engineering 8: 217–245.

Emerging Technologies in Non Destructive Testing, Van Hemelrijck, Anastasopoulos & Melanitis (eds)
© 2004 Swets & Zeitlinger, Lisse, ISBN 90 5809 645 9

Non-destructive assessment of Manchego cheese texture

J. Benedito, A. Rey, P. García-Pascual, J.A. Cárcel & N. Perez-Muelas
Food Technology Department/IAD, Univ. Politécnica of Valencia, Spain

ABSTRACT: Low intensity ultrasounds (LIU) has been applied to different stages of the cheese manufacturing process. In this work the feasibility of carrying out non-destructive measurements on Manchego cheese in order to assess the degree of maturity was assessed. Ultrasounds and non-destructive surface measurements (needle and semispherical probe) were assayed. The ultrasonic velocity related well to the maximum in puncture and the work in puncture ($R^2 = 0,82$ and $0,79$ respectively). The probe that showed to be the best predictor of the textural characteristics of Manchego cheese was the semispherical one. The best relationship was found between the maximum forces of the puncture and the hemispherical probe compression curves ($R^2 = 0,91$). From the results obtained it can be concluded that the assayed ND techniques can be used for the assessment of Manchego cheese maturity.

1 INTRODUCTION

Public awareness with respect to food quality and safety has increased dramatically, because of health and economic implications resulting in stringent standards. The technologies used for assessing food quality are among others: Raman, mid-infrared and NIR spectroscopy, NMR, z-nose, e-nose, ultrasonics, biosensors, multiespectral and computer vision.

The interest for the application of low intensity ultrasounds (LIU) in the food industry has increased during the last decade. The ultrasonic techniques have been used to assess the structure, concentration, location and physical state of different components in food products (McClements, 1997).

The applications found in the literature include all kind of food products such as vegetables and fruits, meat and fish, drinks, oils and also dairy products (McClements, 1997; Mulet et al., 1999).

Food texture is widely recognized as one of the most important parameters for determining the quality and identity of food products. In the past, food texture has been evaluated by destructive sensory and instrumental measurements, Texture Profile Analysis (TPA), uniaxial compression and puncture test being the most frequently used instrumental techniques. At present new non-destructive techniques are emerging. Vibrating rheometers, small displacement probes and sonic and ultrasonic measurements are frequently used for this purpose (Davie et al., 1996; Mulet et al., 1999).

Ultrasonic velocity and attenuation have been used to assess the textural properties of carrots, melons, meat and cheese (Mizrach et al., 1994; Nielsen & Martens, 1997; Benedito et al., 2000a) among other products.

Manchego cheese is a typical Spanish cheese from the region "La Mancha". As happens with other types of hard cheeses, Manchego cheese matures unpacked in temperature and relative humidity controlled chambers. During this stage there is an increase in some textural properties which can be used to indicate the degreee of cheese maturity. Sometimes it is difficult to assess the state of maturity of a batch because the lack of uniformity in the maturing chambers and differences in milk and manufacturing conditions resulting in inhomogeneous batches which are refused by retailers and consumers.

The objective of this work was to assess the feasibility of carrying out non-destructive measurements on Manchego cheese, in order to determine the degree of cheese maturity. Ultrasound and non-destructive surface textural measurements were assayed.

2 MATERIALS AND METHODS

2.1 Raw material

The pieces of the Certified Origin *Manchego* Cheese used in the study were manufactured by a company located in Albacete, Spain. The twenty pieces used ranged from 29 to 300 days of maturity. The pieces were cylindrical shaped of approximately 10 cm height and 20 cm diameter.

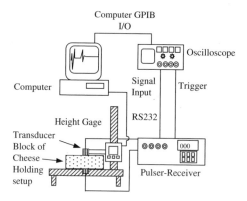

Figure 1.   Ultrasonic set-up for cheese measurements.

2.2   *Ultrasonic measurements*

Figure 1 shows the experimental set-up used for the ultrasonic measurements. It consisted of a couple of narrow-band ultrasonic transducers (1 MHz, 0.75″ crystal diameter, A314S-SU Model, Panametrics, Waltham, MA) , a pulser-receiver (Toneburst Computer Controlled, Model PR5000-HP, Matec Instruments, Northborough, MA) and a digital storage oscilloscope (Tektronix TM TDS 420, Tektronix, Inc., Wilsonville, OR) linked to a personal computer using a GPIB interface. To compute the time of flight, five acquisitions were performed and averaged. A digital height gage with an accuracy of 4×10^{-5} m (Electronic Height Gage, Model 752A, Athol, MA) was used to measure the sample thickness and the value was sent to the computer through a RS232 interface. The system delay produced in the transducers, generator–receiver and digitizer was computed, measuring the pulse transit times across a set of calibration cylinders of different thickness. Ultrasonic measurements were carried out on the whole pieces of cheese at 12°C and on cheese cubes of 2 cm side at four different temperatures (4, 8, 12 and 17°C).

2.3   *Texture*

All the textural measurements were performed in a temperature controlled chamber at 12°C.

2.3.1   *Destructive measurements*
To determine the maturity differences between the cheese pieces a TPA was performed on cheese cubes of 2 cm side. An Universal Testing Machine (TA.XT 2i, 25 kg, Stable Micro Systems, Surrey, UK) was used with a crosshead speed of 0,2 mm s^{-1}. The cubes were compressed 1 cm (50%). The deformability modulus (DM), hardness (H) and the compression work (CW) were computed (Benedito et al., 2000a). For the puncture measurements the same texture analyzer and load cell were used, a cylindrical probe of 6 mm

$\varnothing$ was fitted to the cell (crosshead speed 0,2 mm s^{-1}, 2 mm of compression). Maximum in puncture (MP), puncture work (PW) and the slope in puncture(SP) were calculated. Ten punctures for each cheese were performed on cheese slices of 2 cm height.

2.3.2   *Non-destructive measurements*
For the non-destructive measurements the same analyser and load cell were used. A small needle (1,5 mm $\varnothing$, 0,5 mm s^{-1} speed and 20 mm penetration) and a semispherical probe (6 cm $\varnothing$, 1 mm s^{-1} speed and 4 mm penetration) were fitted to the cell. For the puncture test, the maximum force (MFN), the minimum force (MiFN), the work during puncture (WNP), the work during needle extraction (WEP) and the slope (SN) were computed. For the semispherical probe tests, the maximum force (MFS), the compression work (WS) and the compression slope (SS) were determined.

3   RESULTS AND DISCUSSION

3.1   *Ultrasonic measurements*

The ultrasonic measurements carried out on the whole pieces of cheese showed that the structure of this type of cheese is different to the cheeses where ultrasound has been previously used for non-destructive testing such as Cheddar or Mahón cheese (Benedito et al., 2000 a,b). In fact, the ultrasonic signal was only detected with enough amplitude on a few pieces of cheese, making impossible for the remaining ones the determination of the ultrasonic velocity. This is due to the high number of small holes (porosity) of this cheese that scatters the ultrasonic waves to an undetectable level. Figure 2 shows the ultrasonic signal for a typical Manchego cheese (a) and that of a Cheddar cheese block (b).

More experiments should be done to determine the possibility of using the attenuation of the ultrasonic waves to assess porosity of this type of cheese, allowing the classification in homogeneous batches according to this characteristic. Experiments were carried out on cheese strips in order to determine the maximum length that could be traveled using the experimental set-up previously described. This length depended on the particular examined cheese and ranged from 2 cm to 10 cm.

Ultrasonic measurements were carried out on cheese cubes of 2 cm side. Four temperatures were considered. At 4°C ultrasonic velocity ranged from 1680 to 1640 m s^{-1}, from 1650 to 1625 m s^{-1} at 8°C, from 1643 to 1605 m s^{-1} at 12°C and from 1621 to 1580 m s^{-1} at 17°C. The average standard deviation for the ultrasonic velocity at 12°C in different samples of the same cheese was 3,4 m/s, which shows how the variability in structure/composition within the pieces can affect the ultrasonic measurements. The repeatability for the ultrasonic velocity measurements in the same

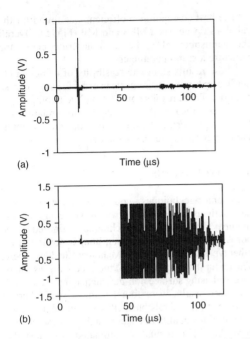

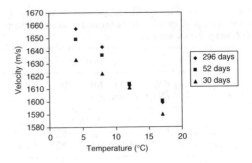

Figure 3.   Influence of temperature on ultrasonic velocity in Manchego cheese.

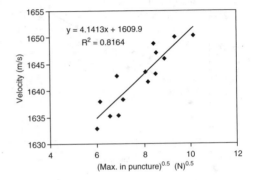

Figure 4.   Relationship between ultrasonic velocity and the maximum in puncture.

(a)   Time (μs)

(b)   Time (μs)

Figure 2.   Ultrasonic signal of Manchego (a) and Cheddar (b) cheese.

sample was estimated to be ±0,1 m/s (average value for the different types of cheese). It must be considered that the temperature control greatly affects the repeatability of the ultrasonic measurements.

Figure 3 shows the influence of temperature on ultrasonic velocity measurements in the cheese cubes. The temperature coefficient of ultrasonic velocity ranged from $-3,3\,\mathrm{m\,s^{-1}{}^{\circ}C^{-1}}$ for the less matured cheese to $-4,7\,\mathrm{m\,s^{-1}{}^{\circ}C^{-1}}$ for the most matured one. The temperature coefficient decreases for the most matured cheeses because the water content is lower and consequently the fat content higher. The negative temperature coefficient for fat involves the decrease of this coefficient for cheeses with higher fat contents. These results are similar to those found by Benedito et al. (2000a) for Mahon cheese. When fat melting occurs, several sections can be found in the velocity–temperature curve (McClements, 1997). In this study the temperature range was chosen to cover only the first section and therefore only a linear relationship was found (Fig. 3).

3.2   Textural measurements

As expected, the destructive textural measurements showed that all the parameters increased with the curing time of cheese, which makes this type of measurements very suitable for predicting cheese maturity (Benedito et al., 2000a). This behavior is the usual for

hard cheeses. The higher increase corresponded to the deformability modulus (450%) for the TPA textural measurements and to the slope in puncture (400%) for the puncture tests.

As happened with the TPA and puncture tests the non-destructive parameters also increased during maturation. The parameters that showed the higher increase were SN (10.100%) for the needle measurements and MFS (270%) for the semispherical probe.

3.3   Non-destructive assessment of texture

As long as the differences between the ultrasonic velocity for cheese pieces with different maturities decreased for higher temperatures (Fig. 3), the ultrasonic velocities at 4 and 8°C were averaged to relate velocity to the destructive textural parameters.

The ultrasonic velocity related better to the puncture test parameters than to the TPA ones. The best relationships were found for the maximum in puncture (Fig. 4) and the puncture work ($R^2 = 0,79$). The relationships between velocity and the TPA parameters were always significant ($p < 0,05$) but the percentage of explained variance was lower than 73% and consequently poorer than those found for puncture. This can be due to the fact that puncture involves lower

Table 1. Correlation coefficient (R^2) for the relationships between the hemispherical probe textural parameters and the TPA and puncture analysis.

	TPA			Puncture		
	H (N)	CW (N mm)	DM (kPa)	MP (N)	PW (N mm)	SP (N/mm)
MFS (N)	0,648	0,762	0,775	0,916	0,852	0,845
WS (N mm)	0,625	0,745	0,745	0,897	0,834	0,824
SS (N/mm)	0,610	0,725	0,711	0,887	0,824	0,805

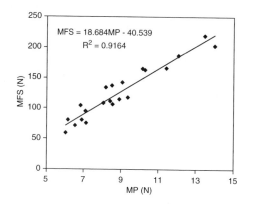

MFS = 18.684MP - 40.539
R^2 = 0.9164

Figure 5. Assessment of Manchego cheese texture using a non-destructive semispherical probe.

displacements of the probe and therefore lower displacements of the sample, mechanism which is closer to what happens when the ultrasonic waves travel across the samples.

Similar results are expected when testing the whole pieces of cheese. To apply these results to practical use in the cheese industry a higher penetration capability of the equipment should be considered. Also lower frequencies could be assayed to assess the viability of this technique in porous media.

The non-destructive textural measurements also showed to be good predictors of the cheese maturity. For the needle tests the best relationships were found between WNP and SP (74% e.v.) and the WNP and PW (72% e.v.). The cheese dough remained practically unaffected after the tests. The measurements of MFN and MiFN could be carried out with a simple manual dynamometer allowing a quick and portable assessment. The inconvenience of this method is that the needle must be disinfected for each puncture in order to avoid contaminations in the cheese. Less invasive measurements were carried out with the semispherical probe.

Table 1 shows the results for the estimation of the TPA and puncture textural parameters using the hemispherical probe.

The best correlation coefficient was found for the relationship between MFS and MP (Fig. 5). Overall the hemispherical probe tests describe better the puncture textural parameters.

These results show the feasibility of using simple surface textural measurements in order to assess cheese maturity. These results could be applied to other types of cheese regarding there is a good relationship between the surface textural properties and the internal properties of the cheese pieces.

4 CONCLUSIONS

Ultrasonic techniques could be used to non-destructively assess the degree of Manchego cheese maturity. The use of these techniques is limited by the porosity of the pieces, problem that is not found for other types of cheese such as Mahon or Cheddar cheese. The textural measurements of the cheese pieces can be assessed using surface non-destructive measurements. The best results were found using a semispherical probe. Simple and reliable methods could be developed to control the quality of cheese batches. These systems could be either manual or automated to be installed on-line.

ACKNOWLEDGEMENT

The authors would like to acknowledge the financial support from the INIA (CAL01-077-C3-1), Ministerio de Ciencia y Tecnología, Spain.

REFERENCES

Benedito, J., Carcel, J., Clemente, G. & Mulet, A. 2000a. Cheese maturity assessment using ultrasonics. J. Dairy Sci. 83: 248–254.

Benedito, J., Cárcel, J., Sanjuan, N. & Mulet, A.2000b. Use of ultrasound to assess Cheddar cheese characteristics. Ultrasonics 38 (1–8): 727–730.

Davie, I.J., Banks, N.H., Jeffery, P.B., Studman, C.J. & Kay, P. 1996. Non-destructive measurement of kiwifruit firmness. New Zealand Journal of Crop and Horticultural Science 24: 151–157.

McClements, D.J. 1997. Ultrasonic Characterization of food and drinks: Principles, methods and applications. Critical Reviews in Food Science and Nutrition 37: 1–46.

Mizrach, A., Galili, N. & Rosenhouse, G. 1994. Determining quality of fresh products by ultrasonic excitation. Food Technology December: 68–71.

Mulet, A., Benedito, J., Bon, J. & Sanjuan, N. 1999. Low intensity ultrasonics in food technology. Food Sci and Technology International 5: 285–297.

Nielsen, M. & Martens, H.J. 1997. Low frequency ultrasonics for texture measurements in cooked carrots (Daucus carota L.). J. Food Sci. 62: 1167–1175.

Emerging Technologies in Non Destructive Testing, Van Hemelrijck, Anastasopoulos & Melanitis (eds)
© 2004 Swets & Zeitlinger, Lisse, ISBN 90 5809 645 9

Acoustic testing of rectifier transformers

P.T. Cole, J. Iqbal & S.N. Gautrey
Physical Acoustics Limited, Cambridge, England

ABSTRACT: Rectifier transformers used in chlorine production are a major capital item, costing ~£1 m each, the cost of a failure is high both in direct costs, and downtime. Production losses due to lack of transformer availability can be substantial, especially as very small stocks of product are maintained for both safety and economic reasons, repair of a transformer may take months, and replacement up to a year, requiring decisions well in advance of failure. For this reason transformers are monitored using a variety of methods, the main one being analysis of dissolved gas in the oil, which indicates the presence of internal partial discharge, insulation breakdown, or arcing. Dissolved gas analysis does not indicate the location of the gas generation however, only that a problem exists. This paper discusses the use of acoustic methods to locate the source of gassing and gives a number of case histories.

1 INTRODUCTION

Rectifier transformers used in chlorine production are a major capital item, costing ~£1 m each, the cost of a failure is high both in direct costs, and downtime. Production losses due to lack of transformer availability can be substantial, especially as very small stocks of product are maintained for both safety and economic reasons, repair of a transformer may take months, and replacement up to a year, requiring decisions well in advance of failure. For this reason transformers are monitored using a variety of methods, the main one being analysis of dissolved gas in the oil, which indicates the presence of internal partial discharge, insulation breakdown, or arcing. Dissolved gas analysis does not indicate the location of the gas generation, only that a problem exists and its relative change over time. It is a powerful method when used correctly, to indicate that something has changed, or is wrong. The time lag between fault generation and measurable change in dissolved gas means that the precise conditions resulting in the fault may not be possible to identify. Sophisticated analysis of many dissolved gases can indicate probable type of damage, for example if the fault is low-temperature or high temperature such as arcing. Certain faults that generate gas may be no threat to the transformer integrity, others may result in subsequent failure. However, without knowing the location and likely cause, a decision on what action to take is difficult, and much time and money may be wasted investigating problems

without success, and the transformer ultimately may still fail, which may also occur if no action is taken.

All major transformer manufacturers use Acoustic partial discharge location (ref. [1]) during production, in order to identify and locate discharge activity during initial electrical test. Without acoustics major damage could occur, or much time be wasted attempting to locate a fault. The method uses an array of high-frequency acoustic sensors, typically responding at up to 150 kHz, attached to the outside of the tank. These sensors are sensitive to the transient acoustic signals resulting from partial discharge or arcing, whilst being insensitive to vibration and general noise. They contain electronics to amplify and buffer the signal making them immune to electrical noise. Time arrival methods are used to locate the sources, in the same way that earthquakes are located, but extended to 3-dimensions. In the past, operators looked at the signals captured on digital oscilloscopes to perform the source location, a highly skilled job that required a continuous source. Specialist signal acquisition systems such as those manufactured by Physical Acoustics (ref. [2]) are now used to locate and characterise emissions.

Acoustic analysis is also used in the field, both for troubleshooting when gas analysis indicates a problem exists, and for "fingerprinting" new, or repaired transformers. Faults may take a long time to become obvious, so early identification of unusual behaviour by acoustic monitoring helps support warranty claims. Sources in operating transformers are often discontinuous, being dependent on operating conditions such as load, tap position, and temperature. For this reason the systems used monitor and record the data required to locate any characterise sources, and the conditions under which they occur. These systems are able to acquire and analyse transient emission data simultaneously on 20+ channels at many thousands of transient signals per second, giving a detailed picture of any source.

One company started using acoustics on rectifier transformers in 1994, and since then has built up substantial experience using the method in conjunction with contractor Physical Acoustics who manufacture the test equipment and provide the field-services. The rectifier transformers are 33 kV primary, to 400 V 50,000-amp. secondary. Tank size is approximately 3 m × 3 m × 4 m containing the three phase main windings, mid-point autotransformer windings, plus other items such as tapchanger, (sometimes in a separate tank section), range-changer, and current transformers.

The usual approach to monitoring a transformer is to mount sensors at appropriate positions around the tank wall, it is important to know what's inside the transformer, as this affects propagation of the acoustic signal, and may have an influence on where sensors are best positioned. Sixteen to twenty sensors are normally mounted, then the system is calibrated, and the transformer taken from zero to maximum load, noting the tap and load. Tap changes are characteristic markers in the acoustic data. The transformer is monitored up to its normal operating temperature, as some faults may occur only over a narrow temperature range. If a fault is known to exist from the gas analysis, and it is not immediately obvious acoustically, then the transformer is run under as many conditions as possible to try and find the one under which the fault is active.

In some cases multiple source locations are found, in which case the analysis of signal characteristics, in conjunction with knowing their location, helps to provide information indicating their relative significance.

The behaviour of a source as a function of tapchanger, and time on-load, also gives clues to the significance of the source.

2 CASE HISTORIES

Some Case histories from the use of the method on one site are given below:

2.1 *July 1994 – Transformer Z1*

High gas in oil readings, acoustic measurement indicated a source in the mid-point auto-transformer after the transformer had been running for long periods at full load, the source was active only on odd taps.

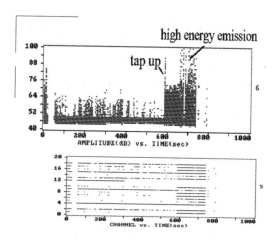

Figure 1. Acoustic source in 50 MW rectifier transformer located at mid-point autotransformer, and active on odd taps.

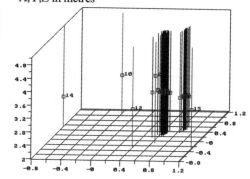

Figure 2. 3D location of a range-changer fault in a 50 MW rectifier transformer.

In June 1995 the auto-transformer was replaced, and the old winding stripped, confirming an embedded fault in the winding.

2.2 *July 1994 – Transformer Z3*

High gas in oil readings, acoustic measurement indicated an intermediate source, ~1000 hits/second 74 dB, at the range-changer location, Figure 2. The range-changer sets the output to 100% or 50% range, (500 or 250 volts), and is rarely used. This is a common fault, due to arcing contacts, and if it cannot be cleared with a few operations of the unit, the contacts are bolted or welded together. Although not a threat to the transformer, leaving the range-changer arcing generates fault gases that mask more serious faults in the transformer that could lead to failure.

360

2.3 March 1996 – Transformer Z4

Fault on DCCT, which fed measurement and protection circuits causing transformer to go to top tap resulting in one side overload and LV winding failure. Transformer replaced with old one from another unit.

2.4 October 1996 – Transformer Z4 (previously at another location)

Two acoustic sources identified, one near yellow phase HV connection, one near HV auto connection.

August 1997- Transformer failed in service, tripped on AC o/l RYB buchholtz surge. Transformer HV/LV yellow winding damaged beyond repair. The breakdown was yellow phase HV connection to earth insulation failure (at one of the locations previously identified by the acoustic test).

2.5 December 1998 – Transformer Z4

Complete new transformer and rectifier installed, new design and manufacturer with thyristor rectifier.

September 1999, acoustic fingerprint obtained, no sources present.

2.6 October 1996 – Transformer Z5

Gas levels high, acoustic analysis indicated the range changer to be the source (see example fig.2), it was repaired and gas levels returned to normal.

2.7 July 1997 – Transformer Z1

Acoustic fingerprint of a new transformer to the old design, source near yellow phase top connections.

August 1999, gas levels showed increase in low temperature fault gases 3–4x December 1997 levels.

2.8 June 1999 – Transformer Z3

Hydran on-line gas analyser showed increasing levels of fault gases, detailed sample analysis indicated high temperature thermal fault.

July 1999- Acoustic analysis indicated no sources present. Oil dropped for replacement and inspection, nothing found, inspection did show a gas path from the selector to the main tank. The selector had previously been "hunting", operating at >10x normal rate, due to a control fault which had now been fixed.

Transformer returned to service, gas levels remained low, conclusion was the selector hunting has caused gas levels to rise, and these had "leaked" into the main tank.

2.9 September 1999 – Transformer Z5

New transformer and rectifier to new design and manufacturer with thyristor control. Acoustic fingerprint obtained-no sources present.

3 CONCLUSIONS

Acoustic monitoring is a useful tool to support dissolved gas analysis in the diagnosis of faults in large rectifier transformers, and to provide "fingerprints" for these expensive capital items. It thus helps in the planning and management of the overall plant and prediction of maintenance and replacement needs. Identification of sources and the conditions under which they are active gives invaluable information to help the transformer specialist diagnose faults and decide when a transformer should be removed for service. Acoustic monitoring is not straightforward, there are many sources of noise in a transformer, and experience and great care are required to execute the test properly. It is essential, once any sources are identified, that specialist knowledge of the particular transformer is available in order to reach the right conclusions regarding further action. Acoustic transformer testing is becoming increasingly common in other areas, such as power distribution (ref. [2]), and continuous monitoring is gaining in popularity as the cost of instrumentation and communications reduces.

REFERENCES

Cole P.T., Location of Partial Discharges and Diagnostics of Power Transformers Using Acoustic Methods. IEE Colloquium on Power Transformer Diagnostics (1998).

Nunez A., Miller R.K., Detect and Locate Sources of Power Transformer Deterioration Using High Speed Acoustic Emission Waveform Acquisition With Location and Pattern Recognition. EPRI Substation Equipment Diagnostics Conference X (2002).

Emerging Technologies in Non Destructive Testing, Van Hemelrijck, Anastasopoulos & Melanitis (eds)
© 2004 Swets & Zeitlinger, Lisse, ISBN 90 5809 645 9

Education for humantarian demining

V. Krstelj & J. Stepanić
University of Zagreb, Croatia

A. Lypolt
Croatian Society of NDT, Croatia

ABSTRACT: Humanitarian demining is a necessary activity to assure human right to live in mine free world. This is an extraordinary difficult task since with current technology of mine detection we will need more than 1000 years for demining several millions of Antipersonnel mines left in more than 90 countries. The only future prospect is intrinsically linked to the new technology. Taking into consideration the possibility of implementation of many NDT methods, recognizing that NDT is a powerful resource of methods with promises to be even greater in the future, this paper describes the challenge and possibilities for NDT experts and needs for capable and competitive cadre in due cause.

1 INTRODUCTION

Mine affected areas exist in around 90 countries world-wide, containing about 80 millions of mines, causing significant impact on humans and regional economics (ICBL 2001). The negative effects range from physical and psychological to missed economic opportunity and always create additional hardship in the areas/countries that are already behind in development. Challenge for the International community is demining and overcoming of the mine problem in next 10–15 years in all mine affected countries. Regarding that, plan of deactivating all mines until year 2010 is declared by most of the obligatory documents.

The only answer to resolve the problem of such a large number of buried mines is a humanitarian demining what is roughly and concisely defined as a set of operations conducted in order to transform mine affected region into region safe for regular life of people. The present day humanitarian demining needs 1250 years and app. 30 billion dollars in order to achieve the proclaimed "Mine free world".

There are two approaches to improve the humanitarian demining, and contribute of achieving the proclaimed goal, Fig. 1. In the first approach, figure 1a, the emphasis is put on the mine detection, especially antipersonnel mine detection (Bruschini and Gross 1997, Krstelj and Stepanić 1999, Kingsbury 2001) as it has been recognized as a part of humanitarian demining responsible for the majority of its duration and accompanying risk (Krstelj and Stepanić 2000).

In the second approach, Fig. 1b, the systemic character of humanitarian demining is emphasized. In it, the humanitarian demining is interpreted as a service-providing activity, which is conducted by a number of firms, on a number of terrains, in a number of different situations, with a legislative which is usually in-complete owing to the rather immature age of the humanitarian demining.

While the first approach has a large potential, it is still questionable into what amount it can be realized in a rather short time period. R&D so far failed to provide more advanced equipment for area reduction and mine detection.

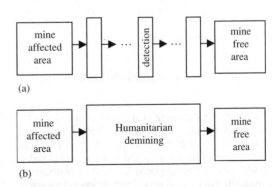

(a)

(b)

Figure 1. Approaches to humanitarian demining improvement.

The inefficiency of the corresponding time scale for the first approach implementation is not the problem for the second approach.

Let us take the following data as illustration of the present situation: in the last decade US$ 1.4 billion were spent for humanitarian mine action. 2001 was the first time since 1992 that funding level didn't increase significantly. Due to more efficient programmes the area returned to the population in 2001 has increased, but not to the level required to fulfill the 2010 goal – at the current rate of 3–4 mines/man-day it would take 20–30 million man-days to complete the task if we stop spreading of new mines. Despite that success, it should be noted that it is not uniform, because reached level of humanitarian demining organization is highly local as it depends prevalently on the experience of the local personnel. Current best practices could not be copied or disseminated in other regions due to lack of skilled professionals that could disseminate the knowledge.

Moreover, two somewhat complementary requirements are spontaneously imposed onto the improvement; the tendency to shorten the implementation time onto as small as possible value, yet not to lower the redundancy for safety checking and risk minimizing procedures. The experience gathered from the practical work in the field, which includes work in regulating and organizing of the mine detection and related technical activities of humanitarian demining, active involvement in the R&D work on humanitarian demining detection methods and techniques, design of the detection equipment and systems testing procedures, as well as the experience in the cover-field of non-destructive testing, and in the techno-economic problems, led us to conclude that the establishment of the humanitarian demining education centre is the optimal solution to the problem of humanitarian demining improvement within the second approach.

In this paper, we present the results achieved regarding the humanitarian demining education center initiative. The basic premises configuring it, as well as the work already conducted are given. In addition, the lines of future development as projected from the present point of view are given and discussed.

2 EDUCATION FOR HUMANITARIAN DEMINING

Humanitarian demining is an complex activity often requiring people with different backgrounds and education to collaborate against a common goal. The complexity of the mission and many involved stakeholders are increasing a need for an efficient demining operation requiring: (i) good organization, (ii) planning, (iii) guidance at all levels of demining operations conduction and (iv) a technical know-how. Within this approach, the goal is to improve humanitarian demining through improvement of its organizational aspects.

One important characteristic of humanitarian demining is that it lacks professional education. Therefore, personnel conducting humanitarian demining is recruited from the variety of professional fields, including social work; management; mechanical, civil and other engineerings; medical doctors, etc. While it is a fact that all these, as well as some other, professions are needed in order to contribute to efficient humanitarian demining, it is also a fact that every (or at least a majority of) person(s) within the humanitarian demining system should be capable in performing some, yet unspecified level of activities in different fields. E.g., engineers working in humanitarian demining are capable of performing tasks about the machines and other equipment maintenance, but lack professional skills in management. In practice that means that a person with a given background learns necessary knowledge and develops needed skills through the *trial-and-error* mode, what is not acceptable.

There are two realizations of humanitarian demining specific education, Fig. 2. Time units in Figs 2a and 2b are the same, and point to different duration of the two processes. In reality, the time unit corresponds to app. 3 months period. The two realizations are divided on the basis of different duration.

In the first realization, Fig. 2a, the objective is to provide the humanitarian demining system with a relatively large number of personnel trained in different

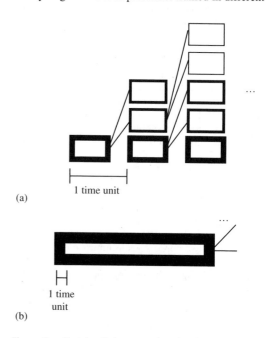

(a)

(b)

Figure 2. Sketch of the two educational processes for humanitarian demining improvement.

topics, listed before. After training, these trainers will proceed with similar trainings in the different mine-affected countries within the national or regional humanitarian demining systems. The capacity and success of trained local deminers could be increased by leveraging the most relevant and effective expertise (i.e., methods and technology) through improved communication channels, aligning various existing training activities and standardizing training requirements and level of qualification. The boxes on Figure 2a represent one education process, and their thicknesses correspond to the different generations. In other words, the thickest box belongs to the primary education center, the middle thick boxes correspond to the courses which trainers trained in the primary center conduct, etc.

In the second realization, Fig. 2b, the goal is to educate interested persons within the thorough, high-education system, in all aspects of humanitarian demining. The corresponding program, the *Mine Action Academy* (MAA), a three year college institution providing a proper education for people working in various fields of humanitarian demining, was worked upon for some time in Croatia, in collaboration of national and international authorities. MAA will in three years provide students with organizational, management and technical skills to perform all the tasks required for the demining. More specific MAA will cover tasks such as financial management, quality control, use of new methods and technologies in demining process, standardisation, demining operation organization (i.e., in demining company).

MAA focus will be on both: (1) increasing the output of trained deminers and (2) creating a core capability of the future operational excellence (train the trainer/expert/manager/leader).

The core group of skilled demining professionals i.e., humanitarian demining engineers, with abilities covering the whole Mine Action Activity Chain will be able to rapidly deliver or grow, depending on the context, the capacity and the success rate (reliability, speed and cost effectiveness) needed to reach the targets set by the international community, Fig. 3.

It is to be mentioned that the proposed education is not confronted to the current courses for pyrotechnicians and deminers, which are regularly conducted. On the contrary, the proposed education system is complementary to the regular deminer courses, as the later are usually highly concentrated on the very in-the-field process, i.e. mine detection. Current training activities, while led by highly motivated seasoned professionals, are not often well aligned and standardized and also are mostly focused on detection and clearance activities.

3 MINE ACTION ACADEMY SPECIFITIES

MAA is planned as an institution providing organizational, management, and technical skills for people working in various fields of humanitarian demining. Its proposed location is in the suburb of Zagreb, the Croatian capitol.

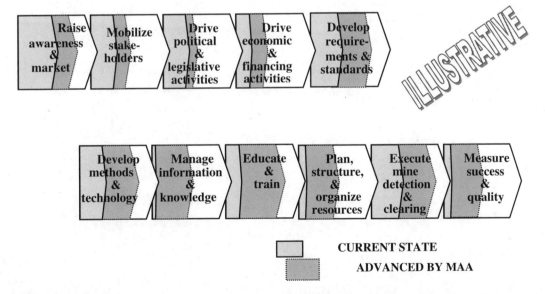

Figure 3.   Mine action activity chain and the sketch of the impact of MAA. The relative proportions of the shaded areas are of illustrative character.

The institution is backed up on:

- vast knowledge and experience from the humanitarian demining in Croatia,
- proximity to practice/test fields,
- international mine action expertise,
- cooperation with Croatian Mine Action Center,
- support by Zagreb University,
- involvement and interest of about 30 demining companies in Croatia and neighboring countries,
- central European location and near-by Zagreb Airport.

MAA will provide its candidates with:

- Three year practice intensive mine action education creating a professional cadre – Humanitarian mine action Engineer,
- Variety of training and certification courses (from few weeks to a couple of months) creating qualified mine activists – Humanitarian mine action Specialist,
- It will cover tasks such as demining operation organization (i.e., in demining company or MAC), financial management, quality control and use of existing and new methods and technologies in demining process through an extensive field training.

Presently MAA is in the process of formation. It is approved by national authorities and is planned to start with the education program in the autumn 2003.

REFERENCES

Bruschini, C.; Gros, B., 1997, A Survey of Current Sensor Technology Research for the Detection of Landmines, SusDem'97 – International Workshop on Sustainable Demining, Zagreb, Proceedings, pp. S6.18–S6.27.

International Campaign to Ban Landmines, Landmines Monitor Report 2001, Landmine Monitor Core Group, USA, 2001.

Kingsbury N., Land Mine Detection DOD's Research Program Needs a Comprehensive Evaluation Strategy, US GAO Report, GAO-01 239, 2001.

Krstelj, V., 2002, Demining – International objective, 8th ECNDT, Barcelona, Proceedings – CD edition.

Krstelj, V.; Stepanić, J., 1999, Non-Destructive Testing in Antipersonnel Landmine Detection, MATEST 99, Cavtat, Proceedings, pp. 109–115.

Krstelj, V.; Stepanić Jr. J., 2000, Humanitarian de-mining detection equipment and working group for antipersonnel landmines detection, Insight 42(3), pp. 187–190.

Zvizdić, D.; Krstelj, V.; Grgec-Bermanec, L., 2002, Modular education programme for setting up mine detection "field" laboratories, 8th ECNDT, Barcelona, Proceedings – CD edition.

Author index